This item must be returned or renewed by the last date shown below. The loan period may be shortened if it is reserved by another reader.   A fine will be due if it is not returned on time.

DATE OF RETURN

30/09/2011

Chemical Sensors

An Introduction for Scientists and Engineers

Peter Gründler

Chemical Sensors

An Introduction
for Scientists and Engineers

With 179 Figures and 25 Tables

 Springer

Peter Gründler
Hallwachsstraße 5
D-01069 Dresden
Germany
e-mail: gruendler.dresden@freenet.de

Library of Congress Control Number: 2006933730

ISBN 978-3-540-45742-8 Springer Berlin Heidelberg New York
DOI 10.1007/978-3-540-45743-5

Springer is a part of Springer Science+Business Media

springer.com

Cover design: KünkelLopka GmbH, Heidelberg
Typesetting and production: LE-TEX Jelonek, Schmidt & Vöckler GbR, Leipzig, Germany
Printed on acid-free paper   52/3100/YL - 5 4 3 2 1 0

Preface

When this book appeared in German, its main task was to bridge the gap between the traditional ways of thinking of scientists and engineers. The differences in how scientists and engineers think stem from the fact that chemical sensors may be interpreted, on the one hand, as a kind of artificial sense organ developed by engineers to equip automatic machines. On the other hand, chemical sensors are not unlike the other myriad small analytical instruments common in analytical chemistry.

The book was written with the aim of providing students of technical disciplines with a basic understanding of certain aspects of chemical science as well as providing students of chemistry with a basic knowledge of electronics and other technical aspects of their discipline.

When the book first appeared, the author was unaware of a single publication that he could recommend to his students without reservations. It seems that this situation has not changed markedly in the time since then. Thus, the author feels encouraged in presenting this English version of his textbook, which is more or less a translation of the German edition.

November 2006 *Peter Gründler*

Contents

1 Introduction

Sensors and Sensor Science

Sensors belong to the modern world like the mobile phone, the compact disc or the personal computer. The term 'sensor' is easily understood. People may imagine a sensor similar to a sensing organ or a tentacle of an ant. *Chemical sensors*, as a special variety of sensors, can be found, for example, in a cold storage place in the form of a freshness sensor which detects spoiled food. A generation ago, the word sensor was not widely used. Today, however, sensors are becoming ubiquitous in our daily lives. Our world is changing rapidly, and sensors play an important role in this process.

Chemical sensors analyse our environment, i.e. they detect which substances are present and in what quantity. Generally, this is the task of analytical chemistry, which aims to solve such problems by means of precise instruments in well-equipped laboratories. For a long time there was a trend towards increased centralization of analytical laboratories, but in certain respects we now see a reversal of this tendency away from instrumental gigantism. Such a tendency to build smaller devices instead of ever bigger ones occurred many years ago when the personal computer appeared and started to replace the large, highly centralized data processing centres. A similar development brought about a rapid expansion in the use of chemical sensors.

1.1.1
Sensors – Eyes and Ears of Machines

The term 'sensor' started to gain currency during the 1970s. This development was caused by technological developments which are part of a technical revolution that continues to this day. Rapid advances in microelectronics made available technical intelligence. Machines became more 'intelligent' and more autonomous. There arose a demand for artificial sensing organs that would enable machines to orient themselves independently in the environment. Robots, it was believed, should not execute a program blindly. They should be able to detect barriers and adapt their actions to the existing environment. In this respect, sensors first represented *technical sensing organs*, i.e. eyes, ears and tentacles, of automatic machines. With our senses we can not only see, hear and feel, but also smell and taste. The latter sensations are the results of some kind of chemical analysis of our environment, either of the surrounding air or of liquids and solids in contact with us. Consequently, chemical sensors can be considered *artificial noses* or *artificial tongues.*

If we accept that sensors are technical sensing organs, then it might be useful to compare a living organism with a machine. When we do this, it is plausible that the term sensor came into use simultaneously with the advent of the microprocessor and the mobile personal computer. Figure 1.1 illustrates the similarities between biological and technical systems.

In a living organism, the *receptor* of the sensing organ is in direct contact with the environment. Environmental stimuli are transformed into electrical signals conducted by *nerve cells (neurons)* in the form of potential pulses. Strong stimuli generate a high pulse frequency, i.e. the process is basically some kind of frequency modulation. Conduction is not the only function of neurones. Additionally, signal amplification and signal conditioning, mainly in the form of signal reduction, take place. In the brain, information is evaluated and, finally, some action is evoked.

We see many similarities between living organisms and machines when we compare how modern sensors and living organisms acquire and process signals. As in a living organism, we find a *receptor* which is part of the technical sensor system. The receptor responds to environmental parameters by changing some of its inherent properties. In the adjacent *transducer*, primary information is transformed into electrical signals. Frequently, modern sensor

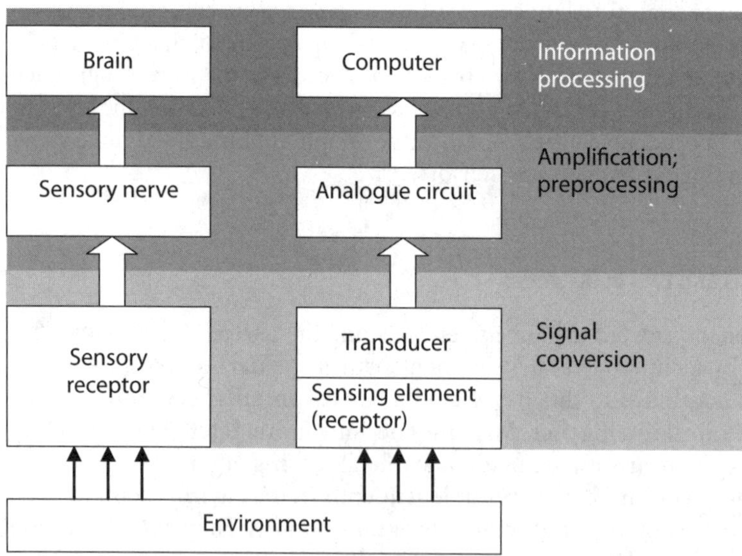

Figure 1.1. Signal processing in living organisms and in intelligent machines

systems contain additional parts for signal amplification or conditioning. At the end of the chain, we find a microcomputer working like the central nervous system in a living organism.

The above considerations, although simplified, demonstrate that signal processing by electronic amplifiers or by digital computers is indispensable for sensor function, like the indispensability of neurones and the brain for physiological processes in organisms. As a consequence, we should accept the fact that 'sensor' does not mean simply a new expression for well-known technical objects like the microphone or the ion selective electrode. Indeed, use of these objects takes on a new meaning in the emerging sensor era.

1.1.2
The Term 'Sensor'

It would not be sufficient to see sensors merely as some kind of technical sensing organs. They can be used in many other fields besides just intelligent machines. A modern definition should be comprehensive. Actually, there is still no generally accepted definition of the term. On the other hand, it seems to be rather clear what we mean when we talk about a sensor. We find, however, differences regarding whether the receptor alone is a sensor or whether the term encompasses the complete unit containing receptor plus transducer.

Regardless of such differences, there is broad agreement about attributes of sensors. Sensors should:

- Be in direct contact with the investigated subject,
- Transform non-electric information into electric signals,
- Respond quickly,
- Operate continuously or at least in repeated cycles,
- Be small,
- Be cheap.

It seems astonishing that sensors are expected to be cheap. Such an expectation can be understood as the expression of the self-evident requirement that sensors be available in large quantities, above all as a result of mass production.

1.2
Chemical Sensors

1.2.1
Characteristics of a Chemical Sensor

The term 'chemical sensor' stems not merely from the demand for artificial sensing organs. Indeed, chemical expertise was necessary to design chemical sensors. Such expertise is the subject of analytical chemistry in its modern, instrumental form. Initially, chemists hesitated to deal with sensors, but later their interest in them grew. The field of chemical sensors has been adapted and is now largely considered a significant subdiscipline of analytical chemistry. On the other hand, the field is given little space in analytical chemistry textbooks. This is true mainly in European textbooks. Sensors do not fit smoothly into traditional concepts and appear to belong to an unrelated field. Up to now, they have not been a typical constituent of analytical chemistry lectures in Europe. There is no doubt, however, that chemical sensors comprise a branch of analytical chemistry. The latter by definition aims to '... *obtain information about substantial matter, especially about the occurrence and amount of constituents including information about their spatial distribution and their temporal changes ...*' (Danzer et al. 1976).

There are two obvious sources for the formation of sensor science as an independent field. One of these sources is the above-mentioned development of microtechnologies, which stimulated a demand for sensing organs. The second source is a consequence of the evolution of analytical chemistry which brought about a growing need for mobile analyses and their instrumentation. Figure 1.2 attempts to outline the formation of sensor science as a bona fide branch of science.

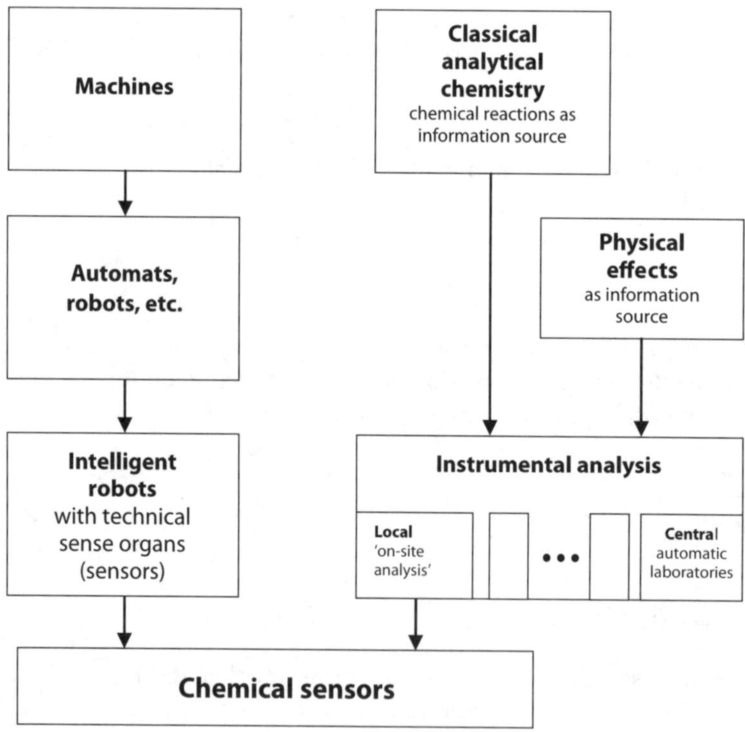

Figure 1.2. Two sources in the development of chemical sensors

As a science, chemistry from the very beginning required information about chemical composition. In other words, analytical chemistry comprises one of the earliest foundations of chemical science; it is as old as general chemistry. The high degree of importance of this field is a result of the natural interest of humans in the composition of our environment. The systematic development of analytical chemistry started with the work of Robert Boyle in the 17th century. Since that time, when we speak about 'analysing' mixtures, we do not mean simply decomposing them. In fact we often carry out a chemical reaction with the express purpose of obtaining knowledge about the composition of the chemicals or materials involved in the experiment. So, for example, since ancient times a well-known indication of the presence of the element chlorine has been the formation of a white precipitate with the addition of silver nitrate solution. Since that time, Boyle's 'wet analysis' has reached a high degree of perfection. Much later, in the middle of the 19th century, the arsenal of analytical chemistry was perfected by the addition of new tools, namely the evaluation of physical properties like light emission. Meanwhile, today instrumental techniques are a significant part of analytical chemistry. Spectroscopy and chromatography are examples of such techniques.

In the final decades of the last century, a strong tendency towards automation appeared in chemical analysis. Big central laboratory complexes were established. One reason for centralizing the resources is the high costs of modern instruments. However, at a certain point, it became obvious that not every problem could be solved smoothly in this way. In many cases, it was difficult or impossible to transport samples long distances without decomposition. This is a typical problem in environmental analysis, a branch of growing importance. Commonly, it proved to be much more simple to bring the instrument to the sample rather than the sample to the instrument. Mobile techniques of chemical analysis attracted increasing interest. Analysts started to look for small, transportable analytical probes which could be stuck smoothly into a sample. Probes of this kind are e.g. the well-known ion selective electrodes, among them the glass electrode for measuring pH. With the growing popularity of sensors in technical applications, it turned out that analytical chemistry already possessed some types of chemical sensors. Thus, during the 1970s, the term 'sensor' became increasingly popular for well-established devices as well.

Now, having discussed the problems associated with defining the sensor in general, we can seek a definition of the chemical sensor. Such a definition was given by IUPAC in 1991:

> *A chemical sensor is a device that transforms chemical information, ranging from concentration of a specific sample component to total composition analysis, into an analytically useful signal.*

This is rather general. Thus, many pragmatic descriptions exist in the literature. Consider the following definition by Wolfbeis (1990):

> *Chemical sensors are small-sized devices comprising a recognition element, a transduction element, and a signal processor capable of continuously and reversibly reporting a chemical concentration.*

The attribute of *reversibility* is considered important by many authors. It means that sensor signals should not 'freeze' but respond dynamically to changes in sample concentration in the course of measurement. The following characteristics of chemical sensors are generally accepted. Chemical sensors should:

- Transform chemical quantities into electrical signals,
- Respond rapidly,
- Maintain their activity over a long time period,
- Be small,
- Be cheap,
- Be *specific*, i.e. they should respond exclusively to one analyte, or at least be *selective* to a group of analytes.

The above list could be extended with, e.g., the postulation of a *low detection limit*, or a *high sensitivity*. This means that low concentration values should be detected.

Classification of sensors is accomplished in different ways. Prevalent is a classification following the principles of *signal transduction* (IUPAC 1991). The following sensor groups result:

- Optical sensors, following absorbance, reflectance, luminescence, fluorescence, refractive index, optothermal effect and light scattering
- Electrochemical sensors, among them voltammetric and potentiometric devices, chemically sensitized field effect transistor (CHEMFET) and potentiometric solid electrolyte gas sensors
- Electrical sensors including those with metal oxide and organic semiconductors as well as electrolytic conductivity sensors
- Mass sensitive sensors, i.e. piezoelectric devices and those based on surface acoustic waves
- Magnetic sensors (mainly for oxygen) based on paramagnetic gas properties
- Thermometric sensors based on the measurement of the heat effect of a specific chemical reaction or adsorption which involves the analyte
- Other sensors, mainly based on emission or absorption of radiation

Alternative classification schemes do not follow the principles of transduction but prefer to follow the appropriate application fields or receptor principles. In this way is the large and important group of biosensors defined.

Biosensors are often considered to be an independent group in sensor science. In this book, however, we will regard them as a special type of chemical sensor. Consequently, they will be integrated into the appropriate chapters of the book. This concept corresponds to that of the responsible IUPAC commission which has expressed in an official document (IUPAC 1999): *Biosensors are chemical sensors in which the recognition system utilizes a biochemical mechanism.*

Since there cannot be found a perfect classification scheme for chemical sensors, in what follows an attempt will be made to find a compromise between the various concepts.

1.2.2
Elements of Chemical Sensors

Section 1.1.1 showed that the functions of a chemical sensor can be considered to be tasks of different units. This is expressed typically in statements like the following (IUPAC 1999): *Chemical sensors usually contain two basic compo-*

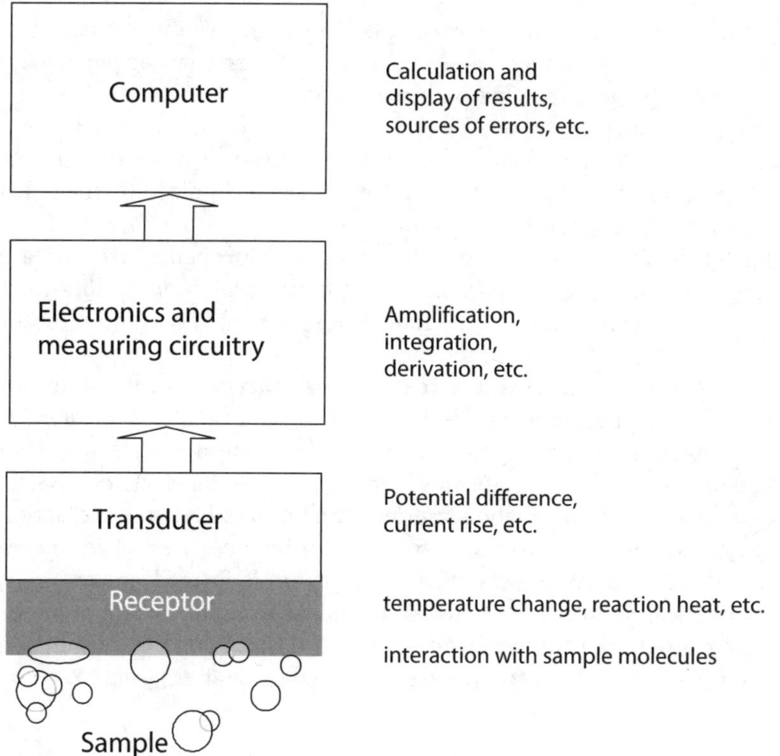

Figure 1.3. Scheme of a typical chemical sensor system

nents connected in series: a chemical (molecular) recognition system (receptor) and a physicochemical transducer. In other documents, additional elements are considered to be necessary, in particular units for signal amplification and for signal conditioning. A typical arrangement is outlined in Fig. 1.3.

In the majority of chemical sensors, the receptor interacts with analyte molecules. As a result, its physical properties are changed in such a way that the appending transducer can gain an electrical signal. In some cases, one and the same physical object acts as receptor and as transducer. This is the case e.g. in metallic oxide semiconductor gas sensors which change their electrical conductivity in contact with some gases (Chap. 5, Sect. 5.2). Conductivity change itself is a measurable electrical signal. In mass sensitive sensors, however, receptor and transducer are represented by different physical objects. A piezoelectric quartz crystal acts as transducer. The receptor is formed by a sensitive layer at the crystal surface. The latter is capable of absorbing gas molecules. The resulting mass change can be measured as a frequency change in an electrical oscillator circuit.

Receptor

The receptor function is fulfilled in many cases by a thin layer which is able to interact with analyte molecules, catalyse a reaction selectively, or participate in a chemical equilibrium together with the analyte.

Receptor layers can respond selectively to particular substances or to a group of substances. The term *molecular recognition* is used to describe this behaviour. Typical for biosensors is that molecules are recognized by their size or their dimension, i.e. by *steric recognition.*

Among the processes of interaction, most important for chemical sensors are adsorption, *ion exchange* and *liquid–liquid extraction (partition equilibrium).* Primarily these phenomena act at the interface between analyte and receptor surface, where both are in an equilibrium state.

Instead of equilibrium, a chemical reaction may also become the source of information. We find this, for example, in receptors where a catalyst accelerates the rate of an analyte reaction so much that the released heat from the reaction creates a temperature change that can be transduced into an electrical signal.

Processes at the receptor-analyte interface can be classified into interaction equilibria and chemical reaction equilibria. The differences are not significant for work with sensors. A true chemical equilibrium is formed, for example, in electrochemical sensors where receptor and analyte are partners of the same redox couple. The fundamental chemical relationships in connection with signal formation at the receptor are discussed in Chap. 2, Sect. 2.2.

Transducer

Today, signals are processed nearly exclusively by means of electrical instrumentation. Accordingly, every sensor should include a transducing function, i.e. the actual concentration value, a non-electric quantity, must be transformed into an electric quantity—voltage, current or resistance.

The pool of transducers can be classified in different ways. Following the quantity appearing at the transducer output, we encounter types like 'current transducer', 'voltage transducer' etc. In the international literature, there exists no systematic concept for classification. In what follows, an attempt is made to find a classification scheme which reflects the inner function of the transducers using only a few transducer principles. It is based on a scheme developed by electronics engineers but has not been applied to sensors till now (Malmstadt et al. 1981). Among the examples given are those that develop their sensor function only in combination with an additional receptor layer. In other types, receptor operation is an inherent function of the transducer.

Energy-Conversion Transducers The principle of energy conversion means that electrical energy is produced by the sensor. Many of these kinds of sensors are able to operate without external supply voltage. In Fig. 1.4, we see two examples together with their measuring circuitry.

The photovoltaic cell, taken as an example of an energy-conversion transducer, converts radiation energy into electrical energy. It is intended to mea-

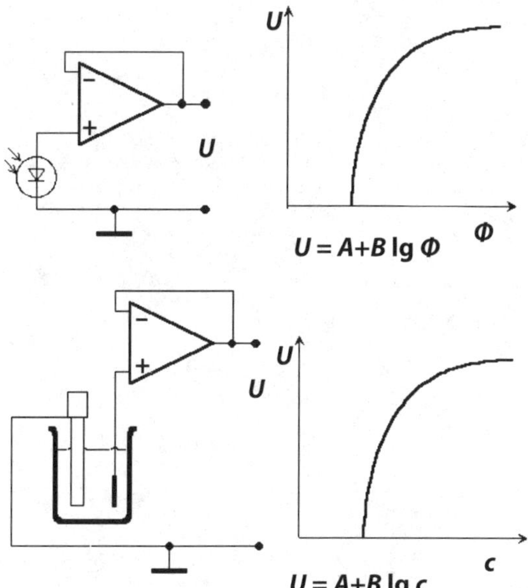

Figure 1.4. Examples of energy-conversion transducers. *Top*: Photo element, *bottom*: galvanic cell

sure the luminous flux Φ. The transducer generates an electrical voltage U as a measure of the quantity to be determined. This voltage can be measured, in many cases, without any amplifier circuit. A similar transducer is the galvanic cell (Fig. 1.4, bottom). Potentiometric sensors are simply galvanic cells. They generate an electrical voltage as a measure of the concentration of a type of ion. The amplifier circuit shown in Fig. 1.4 is not mandatory. The intention was to demonstrate that voltages should be measured without charging them with a current, i.e. voltage should be measured by a unit with high resistance input. As shown in the figure, with energy-conversion transducers, we often find a logarithmic relationship between the voltage formed and the quantity to be determined.

Further examples of energy-conversion transducers are the thermocouple, where heat energy is transformed into electrical energy, and the tachometer generator, where an AC or DC voltage is generated as a measure of the mechanical energy of a rotating body.

Limiting-Current Transducers Voltage sources can reach a limiting state if they are short circuited. Many transducers of the energy conversion type show this behaviour. In the limiting state, a maximum current flows which cannot be increased even if an additional voltage is supplied. If we short circuit the photovoltaic cell (Fig. 1.5, top), then a limiting current arises that depends on the amount of photons hitting the light-sensitive area per time unit. This means

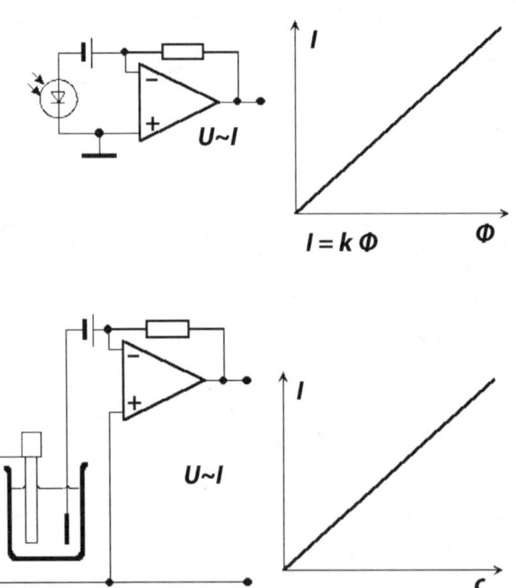

Figure 1.5. Examples of current-limiting transducers. *Top*: Photo diode, *bottom*: electrolysis cell

that the signal current becomes a measure of the *illumination*. The resulting sensor is called the *photo diode*. For the galvanic cell (Fig. 1.5, bottom), we get a similar state when short circuiting. In this case, it is more common to speak about an *electrolysis cell* instead of a *galvanic cell*. The current generated by the electrolysis cell cannot be larger than the value controlled by the amount of reducible or oxidizable charges arriving at the electrode surface. The measured signal of limiting-current transducers typically is linearly dependent on the quantity to be followed over many decades. A chemical sensor based on this operating principle is the Clark probe for determining dissolved oxygen (Chap. 7).

Further examples for limiting-current transducers are the vacuum phototube and the flame ionization detector.

Resistive Transducers In many cases, electrically conducting materials change their conductivity (or, in other words, their resistivity) when environmental properties change. Specific conductance of metals decreases with increasing temperature, whereas semiconductors tend to increase their conductance with higher temperature. In both cases, resistance change can be used to determine temperature. The well-known semiconductor thermistors react sensitively to small temperature differences. They may be converted to give chemical sensors by coating them with a catalyst layer which catalyses a heat-generating chemical reaction. The local temperature increase at the thermistor surface comprises a measure of the concentration of one of the reactants, e.g. for the partial pressure of hydrogen in air.

1.2.3
Characterisation of Chemical Sensors

The performance of chemical sensors should be expressed in the form of numbers. The criteria defined by traditional analytical chemistry were established primarily for characterizing analytical results and analytical procedures, but not for describing devices. We must distinguish whether a process (the analysis) must be evaluated or a device (the sensor) is the subject of consideration.

Some of the traditional criteria, like *sensitivity*, can be applied to procedures as well as to devices. Others, like *accuracy*, have been defined clearly for validating measurement results. A measured value may be correct or false, but a sensor itself may be neither correct nor false.

Validation of Analytical Results

The following units are commonly used for characterizing the validity of analytical results.

- *Accuracy*: an expression of the agreement between the measurement result (given as the average value of a measurement series) and the *true value*. It is also a measure of the systematic error, i.e. deviation from the true value (normally given as a percentage).

- *Precision*: an expression of the random error of a measurement series or, in other words, of the scattering of single values around the average value. The generally accepted way to express precision is with the *standard deviation* (STD). The latter is the mathematical term for the width of the Gaussian error distribution curve given in the form of σ (the distance between the centre and the inflection point of the Gaussian curve). For practical purposes, instead of σ, the estimated value s is determined from a finite population of single values. The approximate value s is given by the equation $s = \sqrt{\frac{\sum (x-\bar{x})^2}{n-1}}$, where x is every individual value, $\bar{x}$ the average value and n the number of measurements.

Parameters of Chemical Sensors

The following list contains static as well as dynamic parameters which can be used to characterize the performance of chemical sensors.

- *Sensitivity*: change in the measurement signal per concentration unit of the analyte, i.e. the slope of a calibration graph.
- *Detection limit*: the lowest concentration value which can be detected by the sensor in question, under definite conditions. Whether or not the analyte can be quantified at the detection limit is not determined. Procedures for evaluation of the detection limit depend on the kind of sensor considered.
- *Dynamic range*: the concentration range between the detection limit and the upper limiting concentration.
- *Selectivity*: an expression of whether a sensor responds *selectively* to a group of analytes or even *specifically* to a single analyte. Quantitative expressions of selectivity exist for different types of sensors. For potentiometric sensors, e.g. (Chap. 7, Sect. 7.1), it is given by the selectivity coefficient.
- *Linearity*: the relative deviation of an experimentally determined calibration graph from an ideal straight line. Usually values for linearity are specified for a definite concentration range.
- *Resolution*: the lowest concentration difference which can be distinguished when the composition is varied continuously. This parameter is important chiefly for detectors in flowing streams.
- *Response time*: the time for a sensor to respond from zero concentration to a step change in concentration. Usually specified as the time to rise to a definite ratio of the final value. Thus, e.g. the value of t_{99} represents the time necessary to reach 99 percent of the full-scale output. The time which has elapsed until 63 percent of the final value is reached is called the *time constant*.

- *Hysteresis*: the maximum difference in output when the value is approached with (a) an increasing and (b) a decreasing analyte concentration range. It is given as a percentage of full-scale output.
- *Stability*: the ability of the sensor to maintain its performance for a certain period of time. As a measure of stability, drift values are used, e.g. the signal variation for zero concentration.
- *Life cycle*: the length of time over which the sensor will operate. The maximum storage time (*shelf life*) must be distinguished from the maximum *operating life*. The latter can be specified either for continuous operation or for repeated on-off cycles.

1.3
References

Danzer K, Than E, Molch D (1976) Analytik – systematischer Überblick. Akademische Verlagsgesellschaft Geest & Portig K.-G. Leipzig
IUPAC (1991) Pure Appl Chem 63:1247–1250
IUPAC (1999) Pure Appl Chem 71:2333–2348
Malmstadt HV, Enke CG, Crouch SR (1981) Electronics and Instrumentation for Scientists. Benjamin/Cummings, Menlo Park, CA
Otto M (1995) Analytische Chemie. VCH, Weinheim New York
Schwedt G (1995) Analytische Chemie. Thieme, Stuttgart
Wolfbeis OS (1990) Fresenius J Anal Chem 337:522

2 Fundamentals

2.1
Sensor Physics

2.1.1
Solids

Many phenomena occurring at the surface of solids are important for sensors. The components of solids are typically arranged regularly. These components, i.e. atoms or molecules, are fixed by bonding forces so that they form regions with a regular lattice structure. Amorphous substances like glasses or polymers are not considered to be solids, although they have some properties of solid bodies.

The different sorts of solids are characterized preferably by the type of chemical bonding that predominates. In *metallic solids*, free electrons are delocalized in a framework of regularly arranged cations. Metallic bonding brings about a rigid but ductile structure. The crystal structure forms as a result of the attempt of spherical atoms to approximate each other as close as possible and to form a package of maximum density. In an *ionic solid*, ions of opposite charge are held together by Coulomb forces. Every ion tends to be surrounded as uniformly as possible by oppositely charged counterions, thus forming an electrically neutral structure. The variety of existing structures is a result of the fact that ions often have quite different radii and often are not shaped like a ball. In solids with an *atomic lattice*, atoms are connected by covalent bonding forces forming a regular network involving the complete crystal. The crystal structure is now determined by the tendency of orbitals to overlap rather than to follow geometric principles. The diamond structure is a typical example of an atomic lattice. In the diamond crystal, every carbon atom is interconnected with four equal neighbours in the form of a tetrahedron. Each atom is in an sp^3 hybridization state and forms four σ bonds. Solids with an atomic lattice often are very hard and chemically inert. Crystals with *molecular lattices* are formed by molecules interconnected by intermolecular forces. Such forces are much weaker than chemical bonding forces. The vast majority of organic solids are molecular crystals. Generally they are soft and have a low melting point.

Solids with mobile electrons are *electronic conductors*. They can be classified following the type of temperature dependence of their electric conductivity. The conductivity of metallic conductors decreases with increasing temperature, whereas semiconductors show the opposite behaviour. Their temperature dependence is higher than that of metals. Elements like silicon or germanium as well as compounds like gallium arsenide are typical semiconductors. Substances with a very low conductivity (e.g. diamond) are isolators. Their conductivity tends to increase with temperature, like those of semiconductors.

Energy Band Model

Following the molecular orbital (MO, see textbooks on general chemistry) theory we see that two molecular orbitals with different energy levels are formed when two atomic orbitals overlap such that each atom provides one electron in the resulting chemical bond. For three atoms, three MOs are formed, and so forth, until *bands* with N very closely arranged energy levels are formed if a very large number N of atoms is present. If there are unoccupied levels in one of the bands, only a tiny quantum of energy is necessary to lift one electron to this level. In this case, electrons are *mobile*, i.e. they will start to move when an electric field exists between two points of the solid body. Consequently, an electric current will flow. According to this explanation, partially filled bands are the reason for electric conductivity. The *Fermi Energy* E_F (*Fermi level*) refers to the relative energy level of an electron in the material. With a half-filled band, E_F lies at the centre position of the band, in the highest occupied state. The band with the higher energy state is named *conduction band*, the band with lower energy is the *valence band*.

According to the energy band model, we can distinguish metallic conductors, semiconductors and isolators by considering whether conductivity and valence bands are separated by a gap (the *band gap*) or whether they overlap (Fig. 2.1).

The band gap corresponds to the amount of energy necessary to transfer one electron from the valence band to the conduction band. This amount results from the energy difference between the lower edge of the conduction band and the upper edge of the valence band $E_g = E_C - E_V$. E_g is very large for isolators (higher than $\sim 5\,eV$). Owing to this high value, there are only a small number of electrons that stay in the conduction band of isolators, under standard conditions. There must be a reason why some electrons stay in this band at all. In conductors with inherent conductivity (e.g. *intrinsic semiconductors*) some charge carriers have an energy level sufficient to leave the crystal lattice. They leave *positive holes* in the solid. Isolators and semiconductors behave similar in this respect. They differ only gradually, according to the magnitude of their band gap. In substances with intrinsic conductivity, the number of holes p should correspond to the number of mobile electrons n. This process is shown schematically in Fig. 2.1 (left).

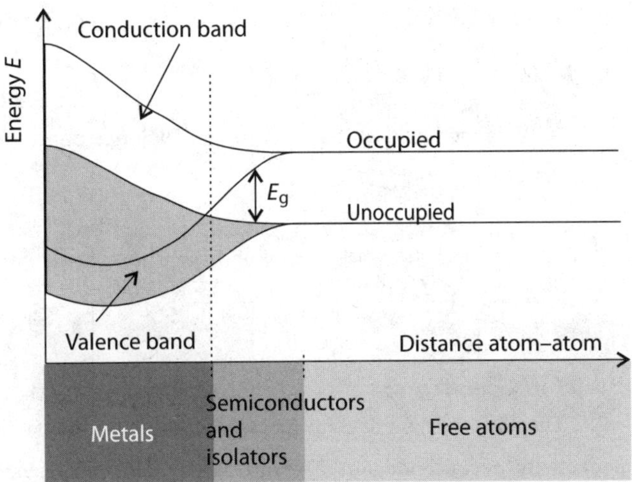

Figure 2.1. Origin of energy bands by combination of atomic orbitals

The conductivity of semiconductors can be controlled strongly by *doping*. Traces of foreign elements are inserted into the highly purified material for that purpose. The foreign element added purposely as a 'contaminant' can act either as an electron scavenger or implant electrons into the lattice. In the former case, each dopant atom catches one electron from the filled valence band so that a positive hole arises. This happens, e.g., when silicon is doped with gallium or indium. As a result, a so-called *P-type semiconductor* is formed. Their conductivity is caused by electrons that can move along the positive holes. Alternatively, it could be interpreted in such a way that the holes are mobile. If an element like arsenic with five outer electrons is added, then an extra electron is contributed. The additional electron passes over to the conduction band that was empty before. An *N*-type semiconductor is formed this way, as shown schematically in Fig. 2.2 (right), which shows that the dopants exhibit their own narrow energy bands. Distances between the interacting bands are small; thus charge carriers can migrate easily between the bands. Doping acts upon the position of the Fermi level. For *N*-type semiconductors, E_F is located close to the conduction band, whereas it lies near the valence band with *P*-type semiconductors.

There is an alternative explanation for the conductivity enhancement of semiconductors when doped. We consider an analogue taken from chemical equilibrium. The intrinsic conductivity of semiconductors can be considered to be analogous to the intrinsic conductivity of pure water given by its self dissociation. In dissociation, the ions H_3O^+ and OH^- are formed, however in a very low concentration. By application of the law of mass action it follows that the product of their concentrations is a constant in equilibrium:

$$K_W = [OH^-] \cdot [H_3O^+] \, . \tag{2.1}$$

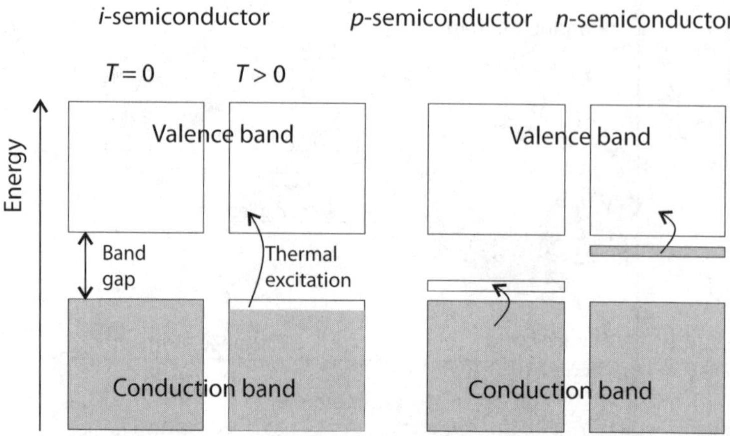

Figure 2.2. Conductivity in semiconductors. *Left*: Intrinsic conductivity (*i*-semiconductors). *Right*: *p*- and *n*-semiconductor function with doping

In chemical equilibria, the electric neutrality condition is always valid. If an acid or a base is added to an aqueous solution, either the concentration of H_3O^+ or that of OH^- is increased. As a result, ionic conductivity increases strongly. In solids, similarly, the dopant amplifies the concentration of either holes $h^\bullet$ or mobile electrons e'. Again, the product of both concentrations is a constant, as given in Eq. (2.2). Also, the electric neutrality condition is valid:

$$K_{el} = [e'] \cdot [h^\bullet] . \tag{2.2}$$

Lattice Defects, Ionic Conductance, Hopping

An alternative interpretation of electric conduction in solids is based on the fact that charge transfer is possible only in non-ideal lattices. The solid must have crystallographic defects. Foreign atoms as dopants in semiconductors can be considered as a defect itself, but not as the only one.

Defects are mandatory for additional transport processes in solids, among them *diffusion* and *ionic conductance*. The latter is an important property for sensors. It is encountered mainly in ionic crystals, e.g. in metal oxides.

For every temperature above 'absolute zero' ($T_0 = 0\,K$), in all solids there exists a finite concentration of defects. This is a consequence of a fundamental law, the third law of thermodynamics (Nernst's theorem).

Reasons for crystallographic defects can be

- Unoccupied lattice sites (vacancies)
- Interstitials (atoms or ions that occupy a site in the crystal structure at which there is usually no atom)

- Foreign ions (impurities or doping agents) incorporated at a regular atomic site in the crystal structure
- Ions with charges not corresponding to the stoichiometric composition of the crystal

If there no external electric fields act on the solid, and if there do not exist concentration gradients, then electric neutrality is encountered. Charges caused by the crystallographic defect must be compensated for by opposite charges. This does not mean that there could not be local partial changes inside the crystal.

For chemical sensors, ionic crystals are of particular significance. Their most important defect types are the following ones.

The Schottky defect (Fig. 2.3, top) is characterized by the presence of an equal number of cationic and anionic vacancies. Characteristic of the Frenkel defect (Fig. 2.3, bottom) is the presence of only one sort of charge in the vacancies, either cationic or anionic. The lack of charge is compensated for by an interstitial ion.

Non-stoichiometric compounds arise if one of the compound-forming elements is present in a deficient amount according to stoichiometric composition. The number of positive and negative lattice sites is constant, regardless of whether there exists non-stoichiometry. The lack of charge must be compensated, therefore, by oppositely charged 'electronic defects'. A typical example is ferrous oxide FeO, which always possesses a composition $Fe_{1-x}O$, where $x > 0.03$. As demonstrated in Fig. 2.4, electric neutrality is achieved since, for each Fe^{2+} missing, an Fe^{3+} ion is incorporated into the lattice.

If electrons or holes are localized in the lattice, as with $Fe_{1-x}O$, then a special semiconductor property arises, the so-called *hopping*. A common visualization is of electrons 'hopping' from hole to hole.

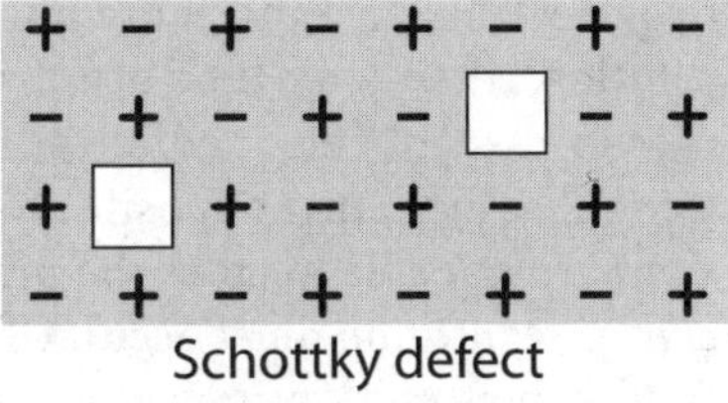

Schottky defect

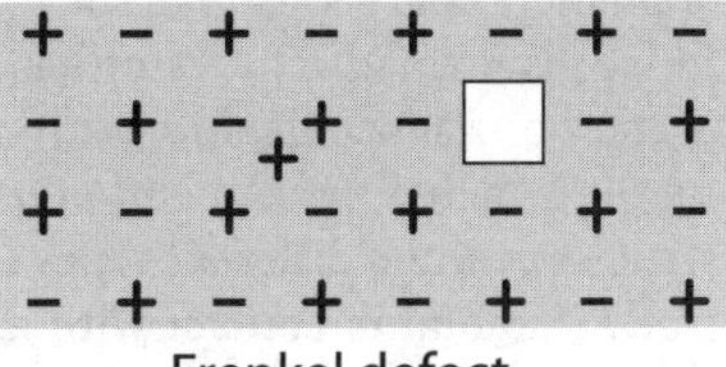

Figure 2.3. Schottky and Frenkel type defects

Frenkel defect

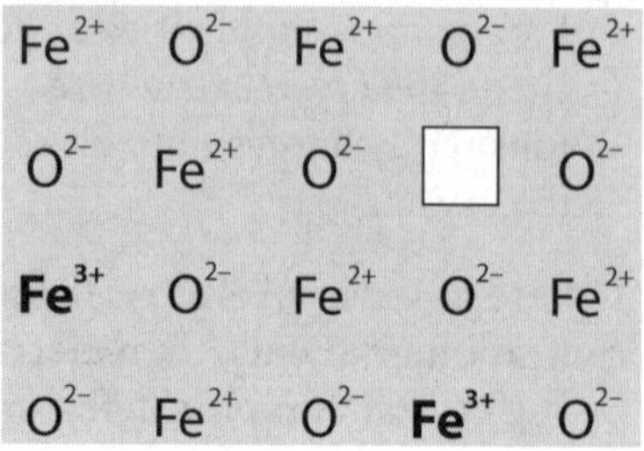

Figure 2.4. Ferrous oxide as an example of compounds with metal deficiency

Junctions and Potential Barriers

Semiconductor materials are generally not used in the form of homogeneous bodies. Their special potentialities become usable if zones of different conduction types come into contact.

The contact of zones with different of charge carrier mobility gives rise to the formation of a voltage across the interface. Such junctions often behave like diodes, i.e. an external voltage brings about a current only in one direction, the *conducting direction*. If biased in the opposite way, current flow is blocked. Characteristically, junctions of this kind act like an electric capacitor with a defined *capacity C*.

The formation of junctions and of potential differences across an interface is not a property of semiconductors alone. We find them also with metals in contact with electrolytes, metals with semiconductors, or electrolyte solutions with different electrolyte solutions. Unfortunately, the theories of such interfaces are based traditionally on quite different models. This suggests the association of completely different phenomena. There is, however, a common simple way to visualize the conditions for all the examples mentioned. It is simply necessary to imagine that electric neutrality must always persist if phases come into contact. If charges of one sign have different mobility in both phases (e.g. electrons in a metal are much more mobile than anions in an electrolyte solution), then some kind of a 'drift' tendency will arise which drives the charge carriers towards the medium where the mobility is higher. For simplification, we may assume that opposite charges (e.g. the cations in the electrolytic solution) are not able to cross the interface at all. At the phase boundary, a partial *charge separation* must occur. Of course, this will stop soon, since the growing Coulomb attraction prevents the counterions from completely drifting away. In the resulting stationary state at the interface, a *double layer* consisting of oppositely charged carriers is formed. Also, an electric voltage across the interface is formed. The latter has different traditional names, according to the branch of science dealing with the actual phenomenon. In electrochemistry, e.g., it is called the *Galvani potential difference*. Figure 2.5 sketches the layer structures when different types of conductors come into contact. With an electrode (a metal in contact with an electrolyte solution), the structure is more complex and extends over a broader range than with semiconductor interfaces.

Figure 2.5. Formation of double layers and potential differences across the interfaces between different kinds of conductors

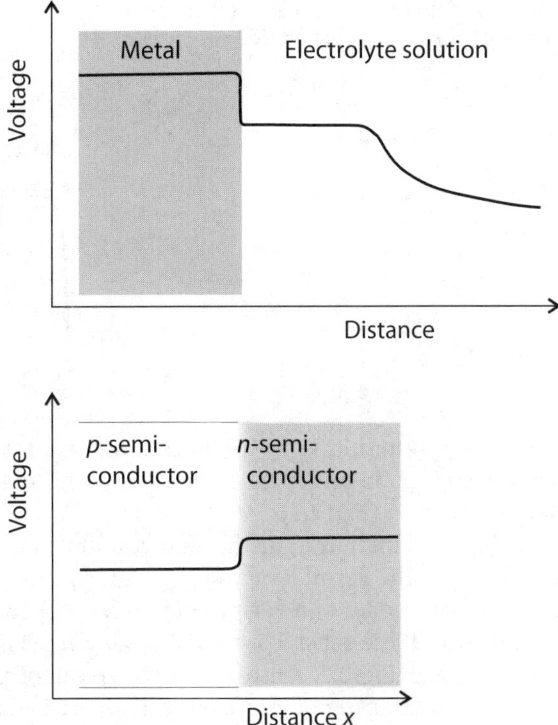

For semiconductors, the most important junction is the *p-n* junction, i.e. the interface between *N*-type and *P*-type doped materials. Such a junction can be *abrupt* (if, for example, two dishes of different materials are pressed together). More important, however, are *diffuse* junctions, which are formed, for example, when a *P*-dopant diffuses from the gas phase into a piece of *N*-type silicon forming around it a region of *P*-type material.

Application of an external voltage across the junction can bring about two different situations. If the voltage is biased in the forward direction (positive terminal at the *P*-type material), then a current can flow more or less without a barrier. With the opposite biasing (negative terminal at the *P*-type material), the junction acts as a barrier. In simpler terms, this means that the positive holes are not able to cross the interface. The lack of charge carriers leads to a *depletion* zone. In this state, only a few charge carriers cross the junction, resulting in a very low current, the *reverse current*. The *depletion width* depends on the materials' properties as well as on the dopant concentration. Altogether, the *p-n* junction behaves like an electric check valve, the *diode*. Diodes do not follow Ohm's law. The current increases exponentially with applied voltage (Fig. 2.6).

p-n junctions are sensitive to external effects. This makes them important for sensor applications. The reverse current strongly depends on temperature and on exposure to electromagnetic radiation. Such effects correspond to an

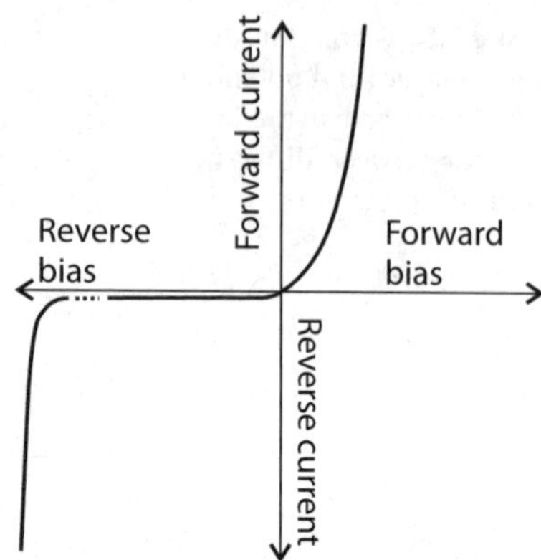

Figure 2.6. Current-voltage curve (characteristic curve) of a semiconductor diode

energy supply resulting in the formation of a free charge carrier inside the depletion layer. To understand this, it is useful to look at the energy bands at the *p-n* junction (Fig. 2.7).

If the *p-n* junction is in thermal equilibrium, and no external voltage is applied, then the Fermi level must be equal in each region. Since the distance between band edges and Fermi level merely depends on temperature, close to the junction there must arise a *band bending* (Fig. 2.7, left). With a reverse biased voltage (Fig. 2.7, centre), the Fermi levels of N-type and P-type materials are different. Transfer of electrons is hampered additionally in comparison to a non-polarized state, since an additional barrier is formed. The opposite situation arises with forward biasing (Fig. 2.7, right).

The energy band concept explains the light sensitivity of *p-n* junctions. If the junction is in thermal equilibrium (as in Fig. 2.7, left), and if it is irradiated with photons possessing energy greater than that of the appropriate band gap, then electron–hole pairs can form in the illuminated region (Fig. 2.8). Owing to *band bending*, electrons migrate into the semiconductor bulk, whereas holes form in the P-type region. The N-type region, as opposed to the P-type region, assumes

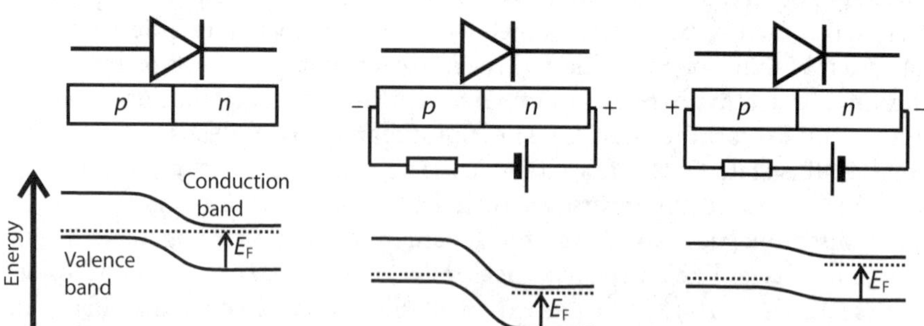

Figure 2.7. Energy bands at *p-n* junction. *Left*: without external voltage. *Centre*: Forward biasing. *Right*: Reverse biasing. E_F: Fermi energy

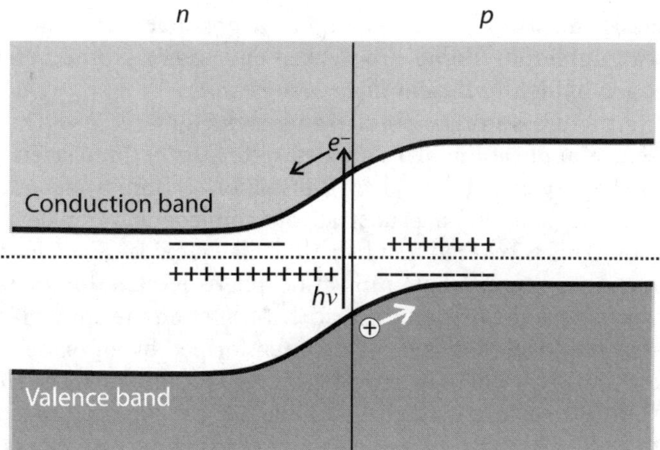

Figure 2.8. Behaviour of *p-n* junction with illumination

a positive potential. The voltage generated in this way can be measured. It depends on the logarithm of incident light intensity. If the N-type and P-type regions are interconnected via an external resistor, then a current flows with a magnitude proportional to the intensity of incident light. In this case, the transducer principle of 'energy conversion' applies. Indeed, there is really no difference between the function of a photo diode for measuring purposes and a photovoltaic cell for transforming sunlight energy into electric power.

The inverse case also exists. With a light emitting diode (LED), an external voltage across the *p-n* junction will inject additional charge carriers (electrons and holes) into the interface region. If one electron and one hole combine with each other under *annihilation*, the band gap energy E_g is emitted in the form of a photon. Red LEDs are widely used in millions of specimens. The LED colour preferably depends on the width of the band gap. Blue LEDs (on the basis of gallium nitride) could be produced only late and after considerable effort.

Structures

Semiconductors must be structured to be useful in technical applications. Zones of different conduction types must be connected to electric leads, certain regions must be covered by insulating layers, windows for illumination must be provided, and so forth. The technology of semiconductor structuring has reached a very high level of perfection. The result is that structures of amazingly complex structure can be produced by the various techniques of microelectronics. Millions of transistors find their place on chips with an area of not more than one square centimetre. The latest achievements in semiconductor processing have even brought about three-dimensional structures. Generally, such structures are built layer by layer. Typical operations are covering by masks, diffusion processes, vapour deposition of thin metallic layers etc. This way, numerous 'floors' are stacked one upon the other. The result is the production of highly efficient *integrated circuits*.

The MIS structure is an example of a semiconductor structure of high importance for sensor applications. This structure consists of three layers arranged as a stack. The sequence is metal (M), insulator (I) and semiconductor (S). Usually, the set-up starts with a substrate of a semiconductor material such as P-type silicon. This is covered first by a thin silicon oxide (SiO_2) layer formed by oxidation in an oxygen-containing atmosphere. Next, a thin metallic layer is applied by vapour deposition. Instead of MIS, often the abbreviation MOS is used, since the insulating layer (I) is often formed by an oxide (O). Thus, a field effect transistor (FET) in a MOS structure is called a 'MOSFET'.

The MOSFET structure is sketched in Fig. 2.9, a P-type silicon substrate is assumed to be the basis. By diffusion from the gas phase, two zones of N-type silicon are formed. Also, the complementary structure would be possible, i.e. N-type substrate with two P-type zones. An insulating layer I covers the substrate. This layer commonly consists of SiO_2. At the top is a metallic layer M, generally made of vapour-deposited gold.

The MOSFET has three terminals, source (S), drain (D) and gate (G). Initially, both N-type zones S and D are equal in function. The source acquires its special function when one of the zones is connected to the substrate. The gate terminal is not connected electrically to any other part of the structure, as is reflected by the graphic symbol of the FET.

If an external voltage is applied between metallic layer and semiconductor substrate, three situations may occur (Fig. 2.10).

If the negative terminal of the voltage source is connected to the metal (Fig. 2.10, left), then positive holes are accumulated at the semiconductor surface. With opposite biasing (Fig. 2.10, centre), the positive holes are displaced from the semiconductor. As a result, a depletion layer is formed. If, however, the magnitude of the positive potential becomes high enough, electrons start to enrich at the semiconductor surface. An *inversion layer* is formed. The width

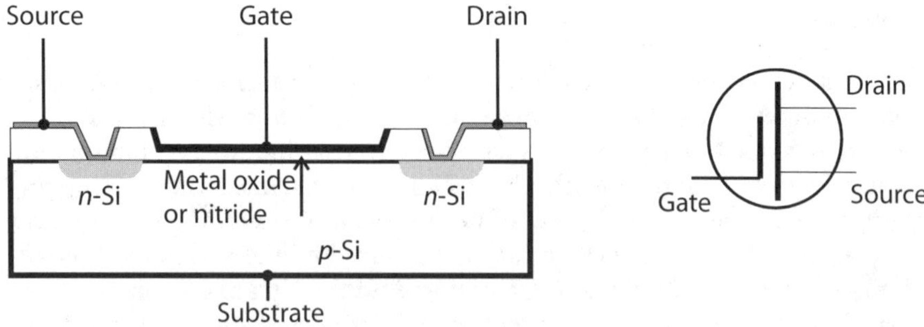

Figure 2.9. Scheme of a MISFET (or MOSFET). *Left*: semiconductor structure, *right*: graphic symbol

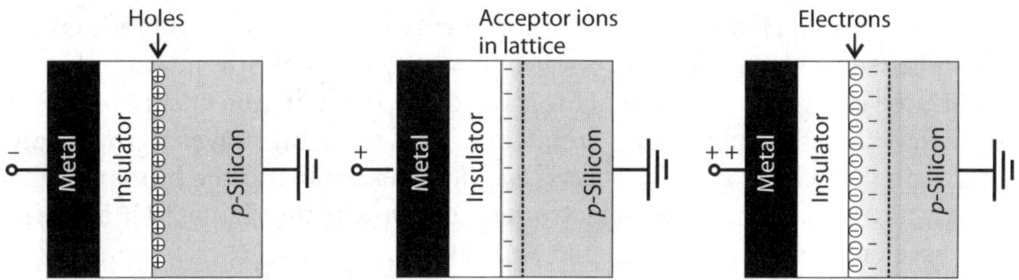

Figure 2.10. Situations when voltage is applied between metallic layer and semiconductor substrate in MOSFET structure

of the depletion layer keeps constant. Between metal and semiconductor, an electric field is built up. This is restricted closely to the thickness of the insulating layer. In the inversion state, a conducting channel forms between N-type zones due to their free electrons. This is the basis of the amplifier function of the MOSFET. Gate voltage variation causes a modulation of field strength across the insulating layer. Changing field strength brings about a changing channel resistance, and consequently an 'amplification' of the voltage applied to the gate. Indeed, MOSFETs are real voltage amplifiers since they work nearly without electrically loading the gate voltage. MOSFETs are the basis of a special chemical sensor type discussed in Chaps. 3 and 7.

The picture of *band bending* at the interface between semiconductor and insulator can be applied also to MOSFETs. Taking P-type semiconductor material as an example, the three cases sketched in Fig. 2.10 can be distinguished following the question whether the surface charge ψ at the interface between semiconductor and insulator is lower or higher than the potential ϕ in the semiconductor bulk phase. The three cases are:

$\psi < 0$ accumulation of holes
$\phi > \psi > 0$ depletion of charge carriers
$\psi > \phi$ inversion

2.1.2
Optical Phenomena and Spectroscopy

Interaction Between Radiation and Matter

Interaction between radiation and matter is a precondition for optical measurements with the aim of obtaining analytical information. For sensors, by far the most important kind of radiation is electromagnetic radiation.

Electromagnetic waves can interact with matter in two different ways, either without loss of energy by *elastic interaction* or with energy loss by inelastic interaction. Elastic interactions like *reflection* or *refraction* yield information about optical properties of the sample. In many cases, such properties depend on composition. Hence optical measurements can be utilized sometimes for chemical sensors. Much more useful, however, are inelastic interactions, as illustrated in Fig. 2.11. In the scheme given there, the radiation energy increases from left to right. Increasing frequency, or decreasing wavelength, means increasing radiation energy. The *visible light*, accessible to the human eye, is only a narrow section of the electromagnetic spectrum (Table 2.1). It covers the wavelength region between 380 and 780 nm.

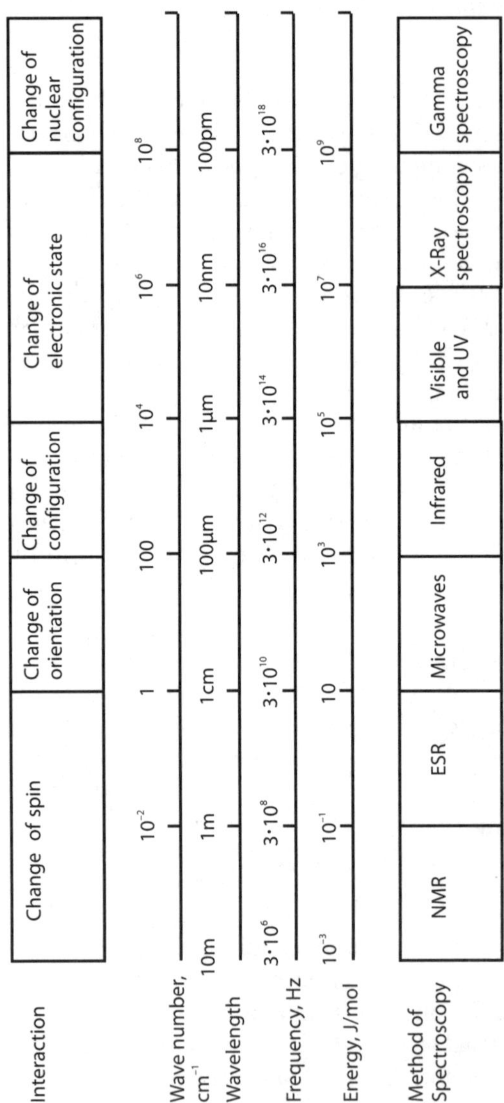

Figure 2.11. Electromagnetic spectrum and its regions useful for analytical measurements

Table 2.1. Colours of visible spectrum

Wavelength (nm)	Colour	Frequency Fresnel	Wave number (cm^{-1})
750.00	Red	400.00	13.34
620.00	Orange	484.00	16.14
600.00	Yellow	500.00	16.67
580.00	Green	517.00	17.24
500.00	Blue	600.00	20.00
440.00	Violet	682.00	22.75

Non-elastic interaction means absorption of photons by the medium studied. The amount of energy introduced into the molecule by absorption of a photon can excite various processes. Radiation with the lowest energy is only able to change the rotation of specific molecules. For excitement of oscillations, a somewhat higher energy (higher frequency of radiation) is necessary. Next, a change in energy level in the electron sheath of free atoms or of molecules and ions in solution follows. Finally, at the end of the frequency scale, we find waves of very high energy that may excite even processes in the atomic nucleus. All the interactions mentioned can be utilized by appropriate analytical instruments. For chemical sensors, the most important impacts in our experiments came from spectroscopy in the ultraviolet and visible spectral regions (UV-Vis spectroscopy) and from infrared spectroscopy (IR spectroscopy). Other spectroscopic techniques did not prove useful for application in sensors. Atomic spectroscopy, as an example, can hardly be used in sensors which require, by definition, direct contact with the medium to be investigated. Solid or liquid samples must be vaporized and atomized by application of thermal energy to perform atomic spectroscopy. This would not be a realistic scenario for a sensor.

Molecules change their energetic state when absorbing photons. Starting from a ground state, the amount of energy added generates excited states which can be symbolized as levels in an energy-level diagram. When electrons return to the ground state, the excess energy is emitted in the form of radiation, i.e. photons are emitted (Fig. 2.12).

Optical spectra are two-dimensional diagrams on which a quantity of intensity (e.g. light emission or light absorption) is plotted vs. a quantity of energy (like wavelength or frequency). The outer appearance of spectra can be quite different. Line spectra are formed by free atoms only. In such spectra only a few discrete excited energy levels exist which are separated by broad distances. Accordingly, a highly specific absorption of light occurs, i.e. only photons within a narrow wavelength region are absorbed or emitted. In a classic spectroscopic instrument, such narrow band regions appear in the form of *spectral lines*. All remaining spectra are *band spectra*. They show more or less broad absorption or emission maxima, the *spectral bands*. The width of such bands, as well as their fine structure, depends on the distance between adjacent sublevels of an

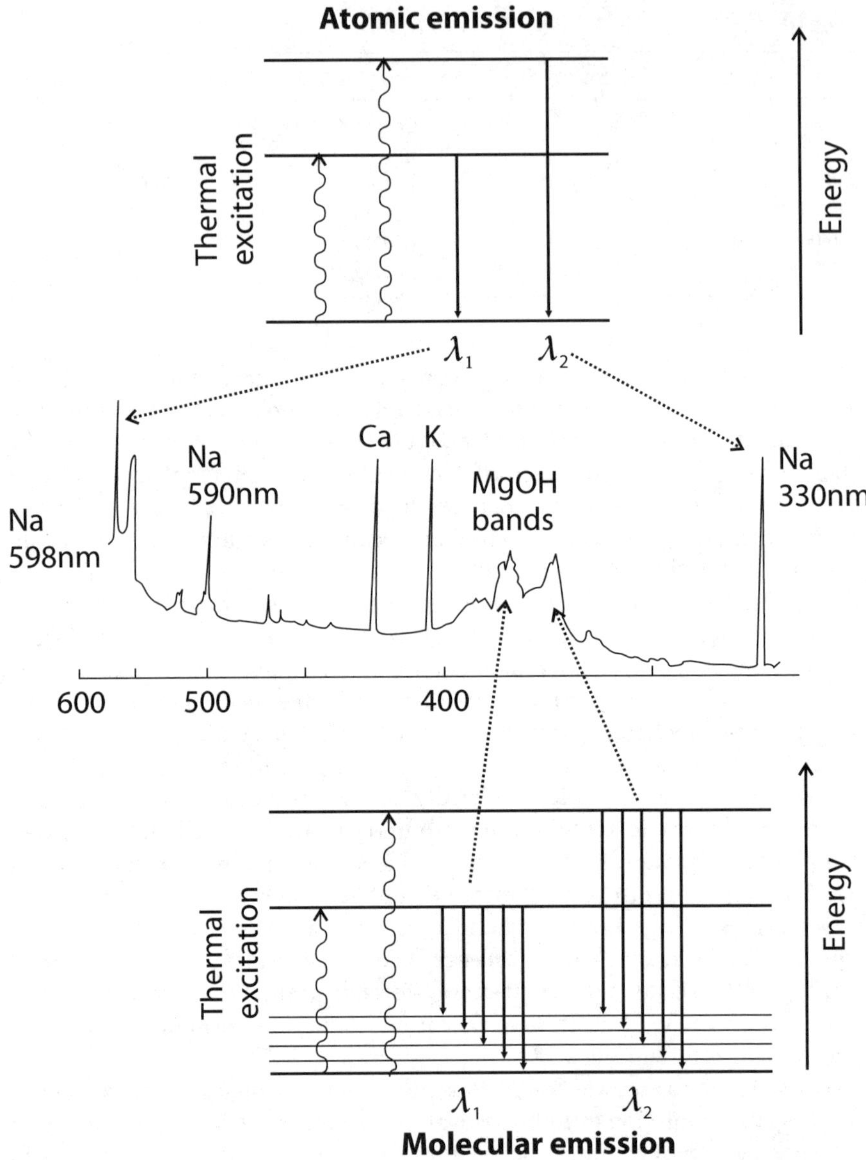

Figure 2.12. Energy level diagram of atoms and molecules with their corresponding emission spectra

energy level in the molecule. With decreasing distance, the number of possible energy states increases. A higher number of states means broader spectral bands. Consequently, the spectra of isolated molecules are more narrow band than the spectra of dissolved molecules. In solution, additional interaction with solvent molecules tends to increase the number of possible states.

Table 2.2. Spectral regions important for optical sensors. λ wavelength; $\bar{v} = \lambda^{-1}$ wave number; v frequency. The product hv corresponds to the energy of the associated photon

Range	λ/nm	$\bar{v}$/cm^{-1}	$h \cdot v$/eV
Ultraviolet (UV)	200–380	50 000–26 000	6.2–3.3
Visual (Vis)	380–780	26 000–13 000	3.3–1.6
Near Infrared (NIR)	780–3000	13 000–3100	1.6–0.4
Infrared (IR)	3000–50 000	3300–200	0.4–0.025

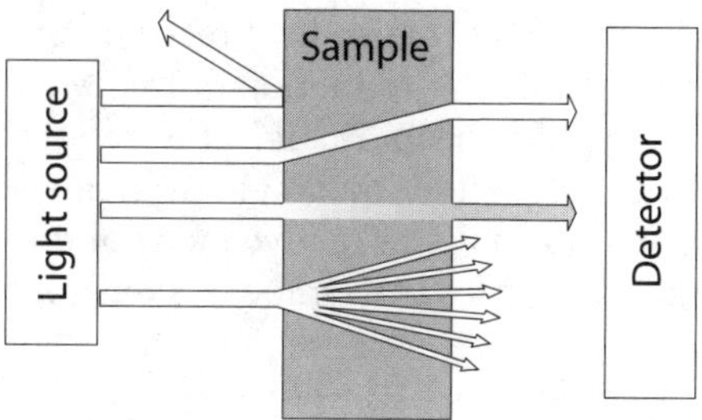

Figure 2.13. Three useful forms of interaction between radiation and sample: reflection, absorption and scattering

Not all spectroscopic measurements can be utilized in sensors, hence only part of the spectroscopic wavelength range is useful for sensor applications. The regions listed in Table 2.2 are meaningful for sensors.

From a practical point of view, we can distinguish the following ways in which radiation can interact with an analytical sample (Fig. 2.13):

- Reflection (diffuse or specular, depending on interface constitution)
- Refraction
- Absorption
- Scattering including fluorescence and phosphorescence

All these phenomena can be utilized in chemical sensors. Before discussing them, some photometric quantities must be defined.

Radiant Flux Φ, and 'Intensity'. The quantity Φ is the power emitted by a radiation source, measured in watts (W). This quantity is measured by photometric detectors. The power incident on a surface $d\Phi/dA$ (in W m^{-2}) is commonly called the *intensity* (with a non-standardized unit I), although this is not correct. The correct SI notation is *irradiance* (symbol E). It should not be confused with the *radiant intensity I*, the power per unit solid angle (in W sr^{-1}).

Transmittance T is the fraction of incident light at a specified wavelength that passes through a sample [Eq. (2.3)], where I_0 is the intensity of incident light and I the intensity of light issuing from the sample.

$$T = \frac{I}{I_0} \tag{2.3}$$

Absorbance A is defined as the common (decadic) logarithm of transparency [Eq. (2.4)]. Absorbance is a very important quality for photometric measurements in analytical chemistry.

$$A = -\log T = \log \frac{I_0}{I} \tag{2.4}$$

Reflection and Refraction

A light beam incident on an interface between two media of different optical densities will change its direction. According to the *angle of incidence* α_i, this results either in refraction or in *total internal reflection* (Fig. 2.14).

The light beam is subject to refraction according to Snell's law [Eq. (2.5)]. In this equation, the angle of incidence is related to the angle of reflection. n_1 indicates the refractive index of the optical more dense medium, n_2 that of the less dense. The symbol β stands for the *angle of refraction*.

$$n_1 \cdot \sin(\alpha_i) = n_2 \cdot \sin(\beta) \tag{2.5}$$

If the angle of incidence attains the value of the *critical value* α_c, then all the incident light is reflected in parallel to the interface between the media. For a water–air interface, α_c amounts to 43.75°. If the condition $\alpha_i > \alpha_c$ holds, then all the incident light is reflected towards the optically denser medium. This is the case of *total internal reflection*.

Reflection and refraction are meaningful for two aspects of sensors. The refractive index of liquids depends on their composition. The latter can be measured by means of microrefractometers composed of optical fibres. The second aspect is the optical fibre itself, where incident light remains 'captured' inside the fibre by multiple internal total reflections. In this way, light can be 'conducted' to any place desired. Optical fibres represent an important base for many types of chemical sensors.

Another phenomenon belonging to the interface between media of different optical densities is *evanescence*. A light beam coming from the denser medium is always bounced towards the less dense medium. The change in direction at the interface is expected to occur without any loss of energy, so that no

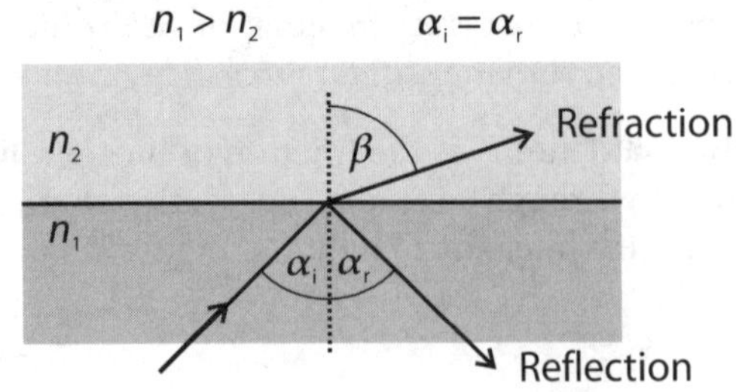

Figure 2.14. Refraction and reflection of light beam at interface of two media with different refractive indices

radiation energy is transferred to the medium with lower optical density. This is only half the truth, however. Interpreting the described optical effects in terms of wave phenomena, it is found that at the interface between the media a standing wave is formed by interference of the incident with the reflected beam. The associated electromagnetic field is called the *evanescent field*. This field intrudes to some extent into the optically less dense medium and can interact with it. The *depth of penetration* d_p can be calculated by Eq. (2.6):

$$d_p = \frac{\lambda}{2\pi\sqrt{n_1^2 \sin^2 \alpha_i - n_2^2}} \, . \tag{2.6}$$

For visible light in contact with common materials, d_p amounts to ca. 100 to 200 nm. Its quantity preferably depends on the wavelength λ of the incident light. This is important for chemical sensors where evanescent waves are utilized to obtain analytical information.

Light Absorption, Photoluminescence, Chemoluminescence

A closer look at the energy-level diagrams mentioned above reveals that numerous processes take part when electromagnetic waves interact with molecules (Fig. 2.15).

In the ground state, the electrons in the molecule are existent spin paired in the singlet state. Excited states of molecules can be singlet or triplet states, depending on the orientation of electron spin (Fig. 2.16).

To the excited state of a molecule belongs a multitude of sublevels, each of which corresponds to one vibrational state of the molecule (Fig. 2.15, left). Adjacent excited states can overlap each other, as shown e.g. for the levels S_1 and S_2. Excitation from ground state to one of these levels is generally very fast. It proceeds in about 10^{-15} s.

Excited states tend to lose energy very fast by *radiationless energy transfer* processes. *Internal conversion* (as e.g. in Fig. 2.15 at the transition from S_2 to S_1) takes place when two energy levels are so close that high-energy vibrational states of the ground level may be excited. This excess vibrational energy is lost by collision with other molecules.

Fluorescence takes place when a molecule returns from the lowest level of excited state to one of the vibrational states of the ground level. In such a process, light with a wavelength higher than that of the exciting radiation is emitted. This emission process also happens very quickly, in an interval lasting up to 10^{-8} to 10^{-6} s after excitation.

By collision of excited particles with gas molecules, *fluorescence quenching* may occur. It is caused by radiationless energy transfer (*vibrational relaxation*). Fluorescence quenching is a very important process for oxygen sensors. The light-intensity decrease caused by quenching depends on the oxygen concentration.

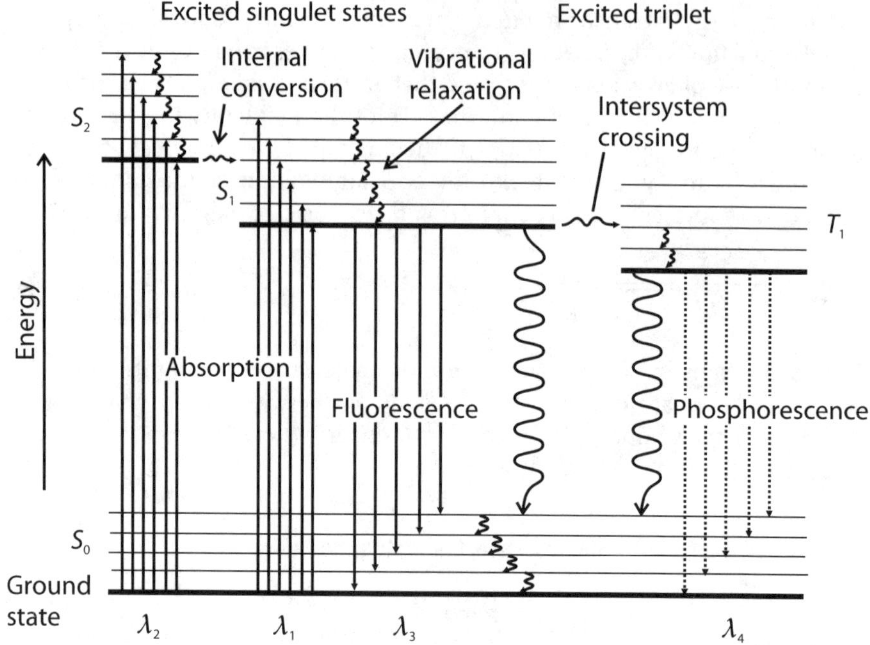

Figure 2.15. Processes participating in interaction between electromagnetic waves and molecules

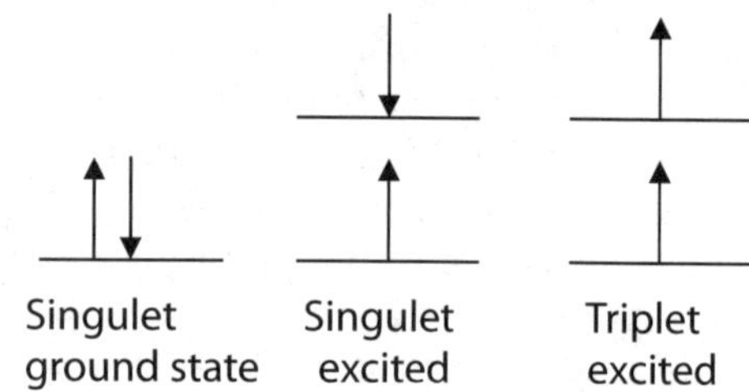

Figure 2.16. Singlet and triplet states in molecules

Spin pairing of the molecules does not change by the processes discussed above. In some cases, however, the triplet state can be occupied starting from a singlet state, as with the transition from S_1 to T_1 in Fig. 2.15. Molecules in the triplet state commonly lose their energy radiationless or by *phosphorescence*. The latter is a slow process lasting up to 10 s. Consequently, phosphorescing samples continue glow when the exciting light source is already extinct. The corresponding delay can be utilized to distinguish between fluorescence and phosphorescence. The generic term for both processes is *photoluminescence*.

Absorption. Very important for chemical sensors is *absorption* of electromagnetic radiation, preferably in the visible spectral region. Due to the multitude of vibrational levels in close proximity, molecules tend to absorb light in a broadband manner. Examples of typical absorption spectra are given in Fig. 2.17. For comparison, the line spectrum of an atomic vapour is also included.

Absorption spectra of molecules in solution have much broader bands than gas molecules due to their much larger number of interaction facilities in solution state. Even broader are the absorption bands of molecules with mobile π-electrons. This is illustrated when benzene is compared with biphenyl in Fig. 2.17 (bottom curve).

The relationship between the magnitude of light absorption and the concentration of dissolved dyes is well known. Lambert described first that the intensity of monochromatic light decreases when it crosses a light-absorbing body. The decrease is a logarithmic function of increasing length of the light path. Beer stated that the transparency of a coloured, light-absorbing solution is an exponential function of its solute concentration. Both relationships can be combined to give an Eq. (2.7) which is well known under the name *Beer's law* (also *Beer–Lambert law* or *Beer–Lambert–Bouguer law*):

$$A = \alpha l \cdot c . \tag{2.7}$$

In this equation A is *absorbance*, given by $A = -\log I/I_0$. The term α is the *absorption coefficient* or *molar absorptivity*, whereas l means the *path length*

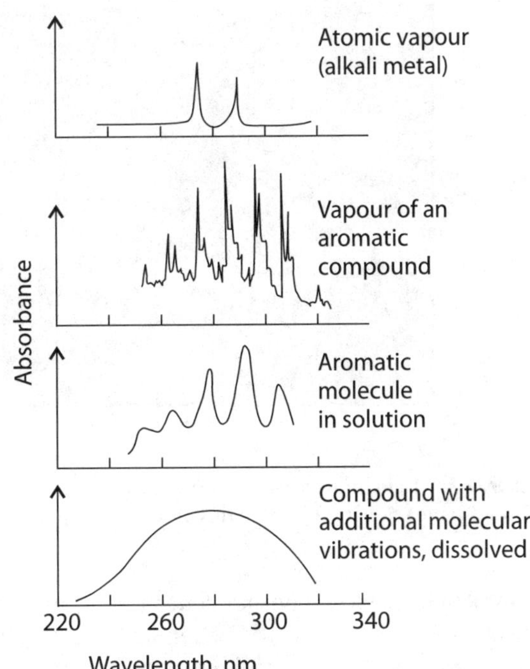

Figure 2.17. Absorption spectra of atoms, gas molecules and molecules in solution

of light through the material. The concentration of the dissolved absorbing species is given by c. Beer's law is the basis of many photometric methods in analytical chemistry. The law is valid independently of the physical condition of the analyte, i.e. it describes the behaviour of dissolved molecules as well as of free atoms in a gas plasma. The only precondition is that electromagnetic waves should be absorbed. The value of the absorption coefficient can be very high. Specially designed ligands form deeply coloured complexes. Their molar absorptivity may reach values up to $60\,000\,cm^2\,mol^{-1}$. The result is that extreme trace concentrations can be determined by photometric measurements.

Deviations from Beer's law result in non-linear calibration graphs if absorptivity A is plotted vs. concentration. Such deviations can have chemical reasons, if e.g. a chemical equilibrium is shifted with the overall concentration change of the absorbing substance. Also, if polychromatic light instead of monochromatic is used, non-linearities come about. This can be a problem with chemical sensors. If sensors are utilized on site, in the environment, often the same instrumental effort cannot be employed like in specialized photometric laboratories. One must get by with small semiconductor light sources rather than large and expensive diffraction grating monochromators. An explanation of non-linearities in photometry is given in Fig. 2.18. If the spectral resolution $\Delta\lambda$ is broader than the width of the absorption peak at half peak height, then only a smaller part of light intensity I is absorbed. This effect is less meaningful with very broad absorption peaks.

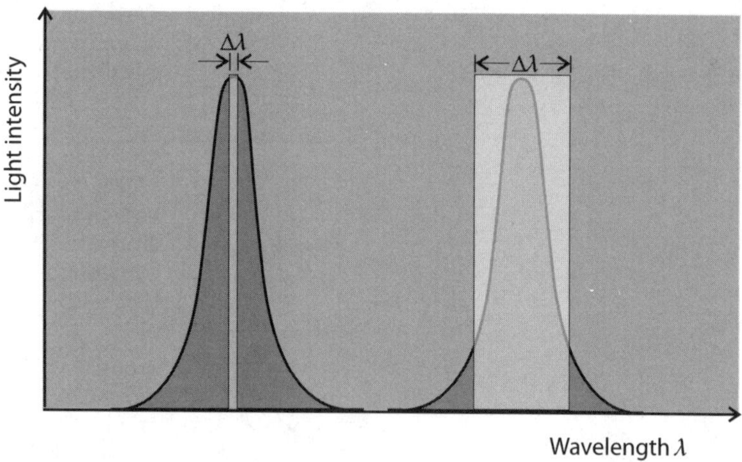

Figure 2.18. Deviation from Beer's law caused by non-monochromatic light. Polychromatic (broad-band) radiation is extinguised much less than monochromatic light

Photoluminescence. Luminescence phenomena introduced previously are the basis of extremely sensitive chemical sensors. They measure the intensity of light emitted when molecules return from excited to ground state. In luminescence studies, light intensity can be measured at an angle of 90° with respect to the exciting light beam. This means that the instrument 'looks' towards a dark background. In this way, generally signal-to-noise ratio is better than with absorption measurements. Values of the detection limit in the ppb range can be expected. As a rule, fluorescence signals are concentration proportional over a broad concentration range. The intensity of fluorescence depends linearly on the concentration of the fluorescing agent, as given in Eq. (2.8):

$$I_f = 2.303 \cdot \phi_f \cdot I_0 \cdot \alpha \cdot l \cdot c , \tag{2.8}$$

where ϕ_f is the fraction of photons causing luminescence, I_0 is the intensity of the exciting radiation, α is the absorptivity coefficient, and l is the width of the light path through the medium.

Fluorescence is encountered preferably with aromatic compounds which have $\pi–\pi^*$ transitions (*conjugated chromophores*). The wavelength shifts towards lower values as the condensation degree of the aromats increases, e.g. when going from benzene via naphthalene to anthracene in the corresponding homologous series. Typical fluorescence spectra are given in Fig. 2.19. The figure shows that excitation/absorption spectra commonly look like the mirror image of emission spectra. A three-dimensional plot with emission wavelength as x-axis and excitation wavelength as y-axis (Fig. 2.20) is useful to identify definite compounds which appear as well-recognized patterns in the figure.

Phosphorescence can be found with pesticides, enzymes and aromatic hydrocarbons.

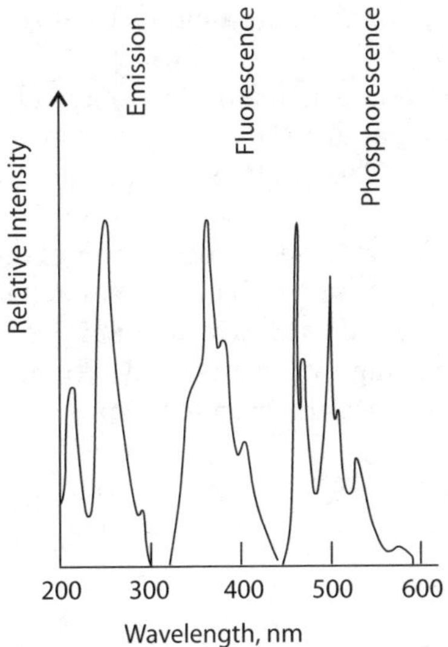

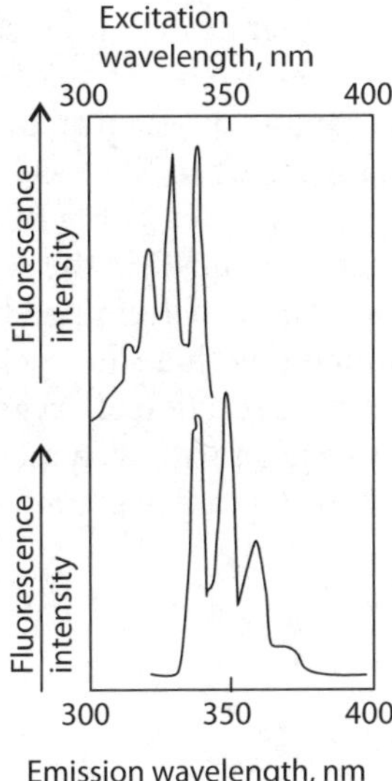

Figure 2.19. *Left*: molecular spectra of phenanthrene. *Right*: fluorescence spectrum 1 ppm anthracene in ethanol. *Right, top*: excitation spectrum; *right, bottom*: emission spectrum

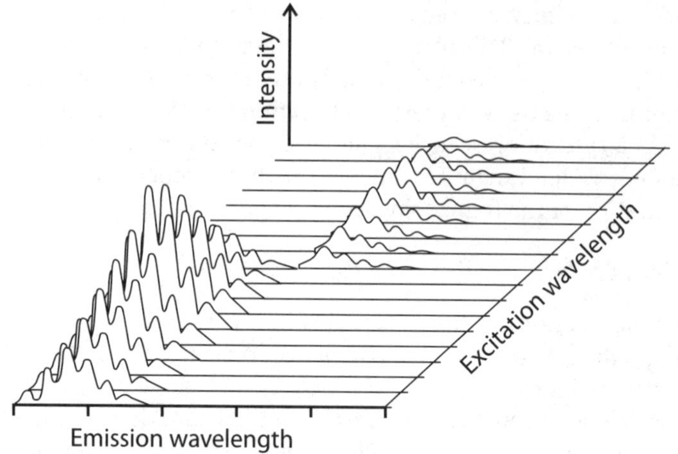

Figure 2.20. Excitation/fluorescence spectra of two compounds

Measurement must be performed at a very low temperature so as to suppress radiationless relaxation. Thus, some phosphorescence effects visible at room temperature are more interesting for use with chemical sensors. They are found with substances adsorbed at a solid surface. Adsorptive bond may stabilize the triplet state.

Chemoluminescence is not a purely physical phenomenon, in contrast to the effects discussed so far. It happens when chemical reaction energy is emitted in the form of radiation. The reaction vessel itself plays the role of light source. The only other devices necessary for measurement are a monochromator and a light detector.

An example of practical importance is the reaction of nitrogen oxide with ozone, according to $NO + O_3 \rightarrow NO_2^* + O_2$ and $NO_2^* \rightarrow NO_2 + h\nu$. Molecules marked by an asterisk are in excited state. The reaction is important for the determination of NO in the atmosphere.

Ozone can be determined by its reaction with an adsorbed layer of the dye rhodamine B with a silica gel. Another reaction frequently used in chemical sensors is the reaction of oxygen or hydrogen peroxide with luminol (Fig. 2.21). In combination with enzymatic reactions, highly sensitive and extremely selective sensors can be manufactured. Examples are given in Chap. 8.

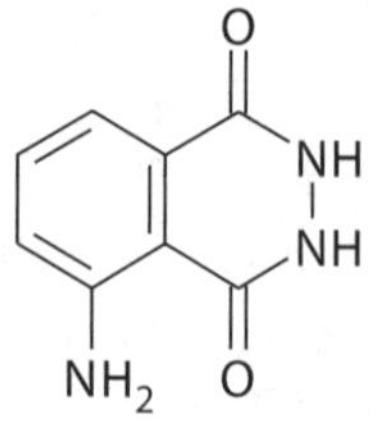

Figure 2.21. Luminol

In biosensors, the chemoluminescence of reactions catalysed by oxidases can be utilized, e.g. oxidation of glucose using glucose oxidase (GOx):

$$\beta\text{-D-glucose} + O_2 \xrightarrow{\text{GOx}} \beta\text{-gluconic acid} + H_2O_2$$

$$2H_2O_2 + \text{luminol} \xrightarrow{\text{peroxidase}} 3\text{-aminophthalate} + N_2 + 3H_2O + h\nu$$

2.1.3
Piezoelectricity and Pyroelectricity

The *piezoelectric effect* was discovered by the brothers Curie already in the 19th century. If certain crystals such as e.g. α-quartz become subject to pressure, then between opposite surfaces a decaying voltage is generated. When the pressure ceases, again a voltage appears, however with the opposite charge sign. Conversely, an external voltage will deform the crystal. If crystals of this kind are included in an electric circuit with positive feedback, then permanent mechanical oscillation of the crystal can be stimulated. The frequency of these oscillations is extremely stable. It depends nearly exclusively on the crystal's mass and is not very sensitive to temperature variations. Quartz crystals find widespread application as frequency standards, e.g. in quartz watches. In chemical sensors, frequency measurement is used to determine tiny mass changes of the quartz oscillator, also in the form of thin layers of foreign materials at the crystal surface. This device is called a *quartz microbalance*, which is the basis of mass-sensitive chemical sensors. Details are presented in Chap. 4.

For quartz crystals cut in the most common direction (the so-called AT cut) and processed in the most advantageous oscillation direction, the Sauerbrey Eq. (2.9) is valid. It describes the relationship between the frequency f and the mass m of a thin film at the crystal surface:

$$\Delta f = f_0^2 \frac{-\Delta m}{A\sqrt{\varrho_q \cdot \mu_q}}, \tag{2.9}$$

where f_0 is the resonant frequency of the crystal, A is the active area of the crystal (between electrodes), ϱ_q is the density of quartz and μ_q is the shear modulus of quartz.

Equation (2.6) demonstrates that a relatively low mass change Δm can bring about very high values of mass change Δf. For Δf given in Hz, f_0 in MHz, Δm in grams, and A in cm^2, the Sauerbrey equation acquires the following form:

$$\Delta f = -2.3 \cdot 10^6 f_0^2 \frac{\Delta m}{A}. \tag{2.10}$$

A mass change of $10\,\text{ng/cm}^2$ would bring about a frequency change of $2.3\,\text{Hz}$ with a crystal oscillating at the base frequency $f_0 = 10\,\text{MHz}$. Such frequency deviations can be measured precisely without too much effort.

The *pyroelectric effect* is similar to the piezoelectric effect. Mechanical deformations generate electric voltages. Such deformations are caused by temperature changes with pyroelectric materials.

Pyroelectric materials are a certain class of ferroelectric substances (those with permanent dipole momentum). Bodies made of such materials have two sides with somewhat different partial electric charges, a more positive and a more negative one. The corresponding surface charge generally is not measurable. Temperature variation influences the lattice distances, and consequently the dipole momentum. The resulting excess charge can be measured in the form of electric current. Under certain conditions, this current obeys the following equation:

$$I = p \cdot A \cdot \frac{\mathrm{d}T}{\mathrm{d}t} , \tag{2.11}$$

where A is the light-sensitive sensor area and p the temperature coefficient of the internal dipole, the so-called *pyroelectric coefficient*. In some materials, this coefficient itself is a temperature function, so that the measuring results depend on the ambient temperature during measurement. When the ceramic material lithium tantalate is used, the temperature dependence of the coefficient is negligible. For this material, p values are in the range of $6 \cdot 10^{-9}\,\mathrm{A\,s\,cm^{-2}\,K^{-1}}$.

2.2
Sensor Chemistry

2.2.1
Chemical Equilibrium

The most important basis of analytical chemistry is the theory of chemical equilibrium. For ca. 350 years chemists have been performing chemical operations with the intention obtaining information about chemical composition. This means, first of all, utilizing the laws and the relationships describing chemical equilibria.

Chemical equilibrium is a *dynamic equilibrium*. In a system which is in equilibrium, reactions do not stop, even if no movement is visible for an external observer. When molecules react, they form new products, and the products are decomposed again, and whereas a certain amount of products is formed, simultaneously an equal amount of the original reactants is generated as a result o the consumption of the products.

For a theoretical description of chemical equilibrium and to derive its inherent laws, there exist two fundamentally different models, namely the *thermodynamic* approach and the *kinetic* approach. Both approaches result in the same mathematical relationships.

For quantitation of mixtures of substances, the following quantities are important:

n the amount of substance, measured in the SI unit *mol* (mole). One mole means a very large number of particles, namely $6.022 \cdot 10^{23}$ pieces. This corresponds to the number of atoms in 12 g carbon or in 197 g gold, or to the number of molecules in 2 g hydrogen gas. It is advantageous to give a substance amount not by its mass but by the number of moles. One mole always indicates the same number of particles, independent of the kind of substance.

c the concentration of solutions, preferably denoted in terms of molarity (the number of moles per volume of solution in litres $c = n/v$). For example, $c = 1\,\text{mol/L} = 1\,\text{M}$.

χ the mole fraction (molar fraction). This denotes the number of moles of dissolved substance n_B as a proportion of the total number of moles in a solution ($n_A + n_B$, where the index B denotes the solvent). $\chi = \frac{n_B}{n_A + n_B}$

a the activity $a = f \cdot c$ is a concentration where the *activity coefficient f* is some kind of 'correcting factor'. Activity is measured using the same units as concentration c. For low values of c, approximately $a \approx c$ and $f \approx 1$.

Activity values in electrolyte solutions strongly depend on the concentrations of all the species present in solution and of their charge numbers. In some cases, a can be calculated using the *ionic strength I*.

The chemical reaction rate $r = \mathrm{d}n/\mathrm{d}t$, measured in $\mathrm{mol\,s^{-1}}$, depends on the concentrations of reactants as well as on the concentrations of products. For a simple reaction, if all the partners are in a gaseous state, e.g. the reaction of iodine vapour with hydrogen gas,

$$\mathrm{H_2 + I_2 \longrightarrow 2HI} \qquad \text{we can write} \quad \vec{r} = \vec{k} \cdot c(\mathrm{H_2}) \cdot c(\mathrm{I_2})$$

whereas for the opposite direction

$$\mathrm{2HI \longrightarrow H_2 + I_2} \qquad \text{the rate is} \qquad \overleftarrow{r} = \overleftarrow{k} \cdot c^2(\mathrm{HI})$$

The terms $\vec{k}$ and $\overleftarrow{k}$ are the *rate constants* of the forward and backward reactions, respectively. They depend only on temperature.

When we write a chemical equation for a reaction in equilibrium, it makes sense to write the double arrow '$\rightleftarrows$' or '$\rightleftharpoons$' instead of a '$=$', since the dynamic character of chemical equilibria is symbolized in this way: $\mathrm{H_2 + I_2 \rightleftarrows 2HI}$.

Under conditions of equilibrium, both reaction rates are equal:

$$\vec{r} = \overleftarrow{r} \quad \text{or} \quad \vec{k} \cdot c(\mathrm{H_2}) \cdot c(\mathrm{I_2}) = \overleftarrow{k} \cdot c^2(\mathrm{HI}) \, .$$

Consequently we can write $\dfrac{\vec{k}}{\overleftarrow{k}} = K$, and finally

$$K = \frac{c^2(\mathrm{HI})}{c(\mathrm{I_2}) \cdot c(\mathrm{H_2})} \, . \tag{2.12}$$

Equation (2.12) is well known under the historical name *law of mass action*.

The derivation of this law given above is not really a strict one. The result, however, is of strict validity regardless of the actual shape of reaction-rate equations. Of course, for other reactions with different stoichiometric factors (the numbers which appear in chemical equations left from the chemical formula symbols), the law may look different from the example given in Eq. (2.12).

Starting with a given 'stock' of reactant molecules, the rate of forward reaction initially must be high but decrease more and more in the course of reaction (Fig. 2.22). Just the opposite behaviour can be expected for the backward reaction.

Alternatively, the *law of mass action* can be derived on the basis of the assumption that for a definite initial state (a given set of reactants each with given concentration), a *driving force* should exist. As in electric engineering, where the *current* is the result of the 'driving force' *voltage*, in chemistry the *reaction rate* can be considered the result of a '*chemical driving force*'. Obviously, this chemical driving force depends on reactant concentrations. Since these concentrations decrease in the course of the reaction, the driving force must also decrease, and finally approach zero.

The chemical driving force has a name. It is the so-called *change of free Gibbs energy*, denoted by the symbol $\Delta_R G$. This very important quantity is a function of all reacting species appearing in a chemical reaction scheme:

$$\Delta_R G = \Delta_R G^{\theta} + R \cdot T \ln \prod f \cdot c_i^{\nu_i} \, . \tag{2.13}$$

In this equation, $\Delta_R G^{\theta}$ and R are constants. The stoichiometric coefficients ν_i appear with a positive sign for products generated and with a negative sign for reactants consumed in the course of reaction. The operator $\prod$ means 'multiply all the following numbers'. Translated into the language of our aforementioned example, the following expression results:

$$\Delta_R G_{\text{iodine-hydrogen-r.}} = \Delta_R G^{\theta}_{\text{iodine-hydrogen-r.}} + R \cdot T \ln \frac{f_{\text{HI}}^2 \cdot c^2(\text{HI})}{f_{\text{I}_2} \cdot f_{\text{H}_2} \cdot c(\text{I}_2) \cdot c(\text{H}_2)} \, . \tag{2.14}$$

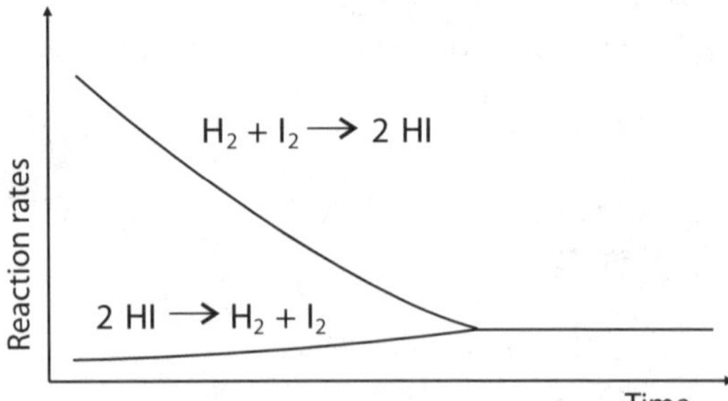

Figure 2.22. Variation of reaction rates for one set of reactants in course of chemical reaction

In equilibrium state, the driving force $\Delta_R G$ of the reaction can be set to zero. For constant temperature T and $a = c$, the logarithmic term of Eq. (2.14) must be a constant. In this way, we again obtain the equation for the law of mass action, namely the same expression as given in Eq. (2.12).

2.2.2
Kinetics and Catalysis

In the previous subchapter, the term *reaction rate* $r = \mathrm{d}n/\mathrm{d}t$ was introduced. We can easily find a way to obtain a popular visualization of the fact that the reaction rate should depend on the concentrations of reacting species. Of course the probability that molecules will collide should increase with increasing concentration. *Chemical kinetics*, as a branch of *physical chemistry*, deals with the regularities describing the processes of chemical reactions.

The *rate law* relates concentrations of reactants to the reaction rate. For a reaction (with the symbolic reactants A and B) of the form $A + B \rightarrow AB$, we could have found experimentally, e.g.

$$r_1 = \frac{\mathrm{d}n(A)}{\mathrm{d}t} = k \cdot c(A) \tag{2.15}$$

or perhaps

$$r_2 = \frac{\mathrm{d}n(A)}{\mathrm{d}t} = k' \cdot c^2(A) \cdot c(B) \, . \tag{2.16}$$

In the case of Eq. (2.15), we stated that our reaction is of *first order*, in the case of Eq. (2.16) we would deal with a reaction of *first order with respect to reactant B* and of *second order in respect to A*.

The reaction rate is not a result of the magnitude of the driving force alone. Of the same importance are kinetic hindrances. *Catalysts* are substances which reduce such hindrances. Even tiny amounts of a catalyst may significantly accelerate a reaction. Kinetic hindrances can be seen as some kind of a barrier which must be overcome. To do this, we must expend the *activation energy*. As soon as the barrier has been overcome, the reaction starts to proceed spontaneously. The function of a catalyst is to decrease the value of activation energy.

For application in chemical sensors, two aspects of chemical kinetics are important. The first is: If the reaction rate depends on concentration, then it should be possible to determine concentration values by measuring reaction rates. The second aspect is: If the amount of a catalyst affects the reaction rate, then in some cases it should be possible to determine catalyst concentrations by kinetic measurements.

Sensors that are based on the determination of reaction rates are found primarily among *biosensors*. In this field, *enzymes*, which are *biocatalysts*, play an important role. Some pecularities of enzyme reaction kinetics are considered in Sect. 2.2.8.

2.2.3
Electrolytic Solutions

Electrolytes are electric conductors, but, in contrast to ordinary conductors, they are decomposed by electric current. The charge transport in electrolytes is carried by ions which start to move if an electric field acts on the electrolyte. Many samples to be studied by sensors are electrolyte solutions. In such solutions, the ionic concentration may be high enough to cause mutual hindrance by *ionic interaction*. Such an interaction explains why the activity coefficient assumes values so different from 1.

In a certain range, f can be calculated using concentration values of all the ions present in solution. The effect of all ions is summarized in the quantity *ionic strength I* [Eq. (2.17)]. The symbol c_i means the concentration and z_i the charge number of an individual ion:

$$I = \frac{1}{2} \sum c_i \cdot z_i^2 \, . \tag{2.17}$$

The activity coefficient f (or f_i for an individual ion) can be calculated using an equation [Eq. (2.18)] first found empirically by Lewis and later theoretically verified by Debye and Hückel. For highly diluted solutions, the so-called *ideally diluted solutions*, f approaches 1, so that we may set $a = c$. A is a constant with a value of ca. 0.5 for room temperature:

$$\log f_i = -A \cdot z_i^2 \cdot \sqrt{I} \, . \tag{2.18}$$

The electric conduction of electrolytes is carried by ions. The ions migrate, each kind with its own individual velocity v_+ or v_- as soon as an electric field with a *field strength E* is applied. The ratio of the velocity of an ion and the corresponding field strength is called the ion mobility u_i. Cations as well as anions contribute to the overall conductivity of an electrolyte, each contribution being based on the individual mobility of the specified ion type. The conductivity of an electrolyte solution can be measured easily. Commonly, the *specific conductance κ* (sometimes called SC) is determined. The latter is derived from the resistance of an electric conductor R and its dimensions length l and cross-sectional area A, as given by $\kappa = l \cdot R^{-1} \cdot A^{-1}$. The quantity κ is related to the ionic mobilities of ions by the following relationship:

$$\kappa = \sum c_i \cdot F \cdot (u_+ + u_-) \, . \tag{2.19}$$

The symbol c_i (in mol $\cdot$ cm^{-3}) denotes all the concentrations of ions contained in the solution considered. F is the *molar charge* (in A s $\cdot$ mol^{-1}), also known as the *Faraday constant*.

The relationship given in Eq. (2.19) allows one to determine ionic concentrations by measuring electric conductance. However, the individual constants u_+ and u_- are not known a priori; the conductance measurement only allows for an estimation rather than a real determination of concentration values.

2.2.4
Acids and Bases, Deposition Processes and Complex Compounds

Acids and Bases

The meaning of the terms *acid* and *base* has changed in the course of the development of chemical science. Even now, they are not uniformly standardized. For interpretation of phenomena in aqueous solution, the acid-base concept of Brønsted and Lowry has proved very useful. It is the basis of the following treatment. Following this concept, acids are characterized by their function in releasing protons, whereas bases are able to accept protons. This means that, as a precondition for an acid-base reaction, an acid as well as a base must be present. Only protons can be subject to transfer from one partner to another. Acid-base reactions always follow a scheme like this:

$$A_1 + B_2 \rightleftarrows B_1 + A_2$$

as an example

$$HCl + NH_3 \rightleftarrows Cl^- + NH_4^+ \, .$$

As a result of the reaction, an acid is transformed into a base, and vice versa. This defines the existence of *conjugate acid-base pairs* A_1/B_1, A_2/B_2 like NH_4^+/NH_3.

As a solvent, water can act either as an acid or a base; it is called an *amphoteric compound*:

$$HCl + H_2O \rightleftarrows Cl^- + H_3O^+ \text{ or}$$

$$NH_3 + H_2O \rightleftarrows NH_4^+ + OH^- \, .$$

Consequently, an amphoteric compound like water can react 'with itself' in a reaction of the following type (a so-called *auto-ionization*):

$$H_2O + H_2O \rightleftarrows H_3O^+ + OH^- \, .$$

For equilibria like this, the law of mass action assumes a special form. If we apply the common conventions for notation of chemical equilibria to the auto-ionization equilibrium as given above, then we will consider water formally as a pure substance, since its 'concentration' is extremely high compared with concentration values of reacting species. Following the conventions for notation of chemical equilibria, the concentration of water as a solvent is given by the dimensionless *molar fraction* χ, and since it is nearly a pure solvent, we set $\chi = 1$. The *law of mass action* assumes the following form when applied to the *auto-ionization* of water:

$$K_w = c(H_3O^+) \cdot c(OH^-) = 10^{-14} \, \text{mol}^2 \cdot L^{-2} \, . \tag{2.20}$$

The *auto-ionization constant* of water K_w is very important for the chemistry of aqueous solutions. We need to quantify only one of the concentrations, either $c(H_3O^+)$ or $c(OH^-)$, if we intend to obtain the acidic (or basic) character of the solution. Most useful for presenting the corresponding concentration is the pH value. The pH is simply the negative of the power of 10 of the molar concentration of H^+ ions:

$$pH = -\log c(H_3O^+) \,. \tag{2.21}$$

The amphoteric character of water offers a way to standardize the *strength* of acids or bases. For that purpose, we consider how an acidic or basic substance reacts with water as a *standard reaction partner* following the general scheme $A_1 + H_2O \rightleftharpoons B_1 + H_3O^+$. This equilibrium is characterized by an equilibrium constant K_A, the *acid dissociation constant*. Often the common logarithm of K_A, pK_A, is used. The corresponding formulation of the law of mass action is given Eq. (2.22).

$$K_A = \frac{c(B_1) \cdot c(H_3O^+)}{c(A_1)} \,. \tag{2.22}$$

K_A allows classifying acids in relation to their strength. As a result, we get tables like Table 2.3, which is given as an example.

Alternatively, a classification series resulting from the reaction of bases with their 'standard partner' water following the general scheme $B_1 + H_2O \rightleftharpoons A_1 + OH^-$ can also be given. The corresponding base dissociation constant K_B can be calculated using Eq. (2.23):

$$K_B = \frac{c(A_1) \cdot c(OH^-)}{c(B_1)} \,. \tag{2.23}$$

Table 2.3. Strength of acids in aqueous solution

pK_A	Acid A_1		Base B_1
−1.74	Hydronium	H_3O^+	H_2O
−1.32	Nitric acid	HNO_3	NO_3^-
1.96	Phosphoric acid	H_3PO_4	PO_4^-
3.7	Formic acid	$HCOOH$	$HCOO^-$
4.75	Acetic acid	CH_3COOH	CH_3COO^-
6.52	Carbonic acid	$CO_2 \cdot H_2O$	HCO_3^-
7.12	Dihydrogen phosphate	$H_2PO_4^-$	HPO_4^{2-}
9.25	Ammonium	NH_4^+	NH_3
10.4	Hydrogen carbonate	HCO_3^-	CO_3^{2-}
12.32	Hydrogen phosphate	HPO_4^{2-}	PO_4^{3-}
15.74	Water	H_2O	OH^-

K_A and K_B of a conjugate pair are correlated, since their product is equal to the auto-ionization constant of water [Eq. (2.24)]:

$$K_A \cdot K_B = K_w = 10^{-14} \, .$$

(2.24)

pH is highly important for many chemical and biological processes. Often it is necessary to stabilize its value regardless of external distortions caused by a reaction or by other influences. *Buffer solutions* are used generally to stabilize pH values. Buffer solutions contain similar amounts of both partners of a conjugate acid-base pair, which must be of low or medium strength. As an example, we consider a solution containing equal amounts of acidic acid CH_3COOH and its conjugate base sodium acetate. Utilizing the general Eq. (2.22), we get Eq. (2.25):

$$K_A(\text{acidic acid}) = \frac{c(CH_3COO^-) \cdot c(H_3O^+)}{c(CH_3COOH)} = 10^{-4.7} \, .$$

(2.25)

The pH of this mixture nearly exclusively depends on the concentration ratio of the acid and its conjugate base, assuming that acid as well as base concentrations are higher than hydrogen ion or hydroxyl ion concentrations. Thus, the pH can be set arbitrarily to a predetermined value. This function of buffer solutions can be seen clearly when considering the logarithmic form of Eqs. (2.22) and (2.25), the so-called *Henderson–Hasselbalch Eq.* (2.26):

$$pH = pK_A + \log \frac{c(B)}{c(A)} \, .$$

(2.26)

For equal concentrations of acetic acid and acetate, we get $pH = pK_A = 4.7$.

'Buffering' means to make a solution resistant to changes in hydrogen ion concentration caused by the addition of foreign substances. The limit for buffer action is given by the *buffer value (buffer capacity)* β. The latter is calculated by $\beta = d[B]/dpH$, where $d[B]$ is the increment (in moles) of a strong base required to produce a certain pH change of the buffer solution. The maximum value for a given total concentration of buffer agents is achieved with equal amounts of acid and base of the conjugate pair (e.g. for acetic acid/acetate the corresponding pH is 4.7). The buffer value depends on pH itself, as shown in Fig. 2.23 for the given example. If the total concentration of buffer agents (acetic acid plus acetate) is increased, β also increases.

For a long time, chemical indicator substances have been used to get information about acid-base systems. *Acid-base indicators* are dyes which change their colour with changing pH. This indicator function can be found if the colour of an acid differs from the colour of its conjugate base. A well-known example is methyl orange, where the acid is red and the corresponding base is yellow. A continuous colour change from red to yellow is visible when going from strong acidic towards strong basic pH. This behaviour can be used to

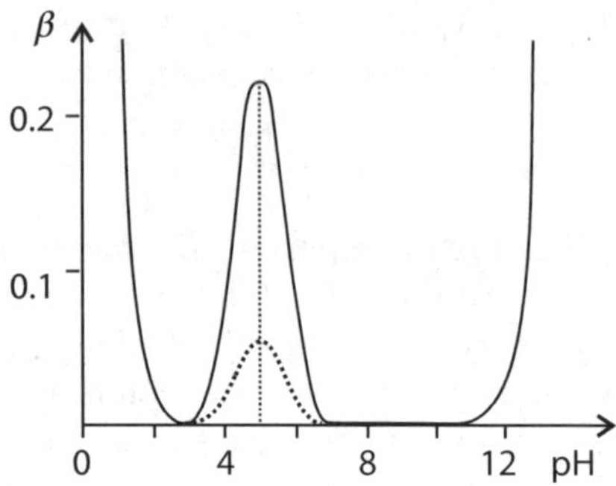

Figure 2.23. Buffer value β of acetate buffers depending on pH and total concentration of buffering agents. *Solid curve*: Total concentration 0.4 M; *dashed line*: $c_{\text{tot}} = 0.1$ M

develop a photometric pH measurement. In the world of chemical sensors, indicators are used e.g. in *pH-optodes*, where a thin layer of an acid-base indicator is immobilized at the front end of an optical fibre or some similar light conductor. The colour change in contact with a solution can be measured by means of a photometer.

Precipitation Processes and Complex Formation

For chemical sensors, the equilibrium between a sparingly soluble substance and its corresponding saturated solution is important. The corresponding equilibrium is the *solubility equilibrium*. If the solubility of the dissolved solid is very low, such equilibria are utilized to remove constituents by precipitation. As an example, consider the following reaction:

$$\text{Ag}^+ + \text{Cl}^- \rightleftarrows \text{AgCl}^- \downarrow \ .$$

In the above reaction, a solution containing silver as well as chloride ions is in equilibrium with the sparingly soluble solid silver chloride. In the corresponding form of the law of mass action, the concentration of the solid AgCl is a constant with the value 1, if given in terms of *molar fraction*. Consequently, we get the following Eq. (2.27):

$$K_{\text{sp}} = c(\text{Ag}^+) \cdot c(\text{Cl}^-) \ . \tag{2.27}$$

The *equilibrium constant K_{sp}* is often called the *solubility product*. One may derive from Eq. (2.27) the process by which a constituent can be removed from a solution as completely as possible. If, for example, silver ions must be removed, we should add an excess of chloride ions. The higher this excess, the lower the remaining silver ion concentration, due to the fact that the system tries to keep the product of the concentrations constant.

Precipitation processes are important if samples must be prepared for analytical measurements. In chemical sensors, precipitation processes are used to deposit (i.e. to *immobilize*) agents at solid surfaces.

Another equilibrium of interest is the formation of complexes. In this process, complex compounds are formed by the combination of *ligands* with a *central ion*. Characteristic for complexes is the *coordinate bond*. It forms when an electron donor donates an *electron pair* to an electron acceptor. Complex formation equilibria can be depicted by the following general scheme, where M stands for a central ion and L for a ligand:

$$M + L \rightleftharpoons ML$$
$$ML + L \rightleftharpoons ML_2$$
$$ML_2 + L \rightleftharpoons ML_3$$

etc., until

$$ML_{n-1} + L \rightleftharpoons ML_n \; .$$

The complex equilibrium constants express the stability of complexes vs. chemical attack. They are called *stability constants* (K_{stab}). As demonstrated in Table 2.4, such constants can be expressed in the form of *stepwise formation constants* or *overall stability constants*. Each type can be converted mathematically to the other one and vice versa.

A large variety of ligands are denoted here simply by the symbol L. Whether a ligand bonds more or less specifically to a metal ion depends mainly on its *hard* or *soft* character. These descriptive terms describe the behaviour of small, highly charged (*hard*), or big, easily deformable (*weak*) ions. Hard central ions preferably combine with hard ligands, and soft central ions with soft ligands. There are, however, other aspects which contribute to make a bond between a central ion and ligand more or less specific. Ligands can be synthetized 'taylor-made' in some cases to make an optimum fit for a special central ion.

Some ligands can be immobilized at particular surfaces and can act as 'traps' for specific analytes. Ligands can have molecular cavities which fit a specific ligand. In this way it is possible to 'disguise' cations so much that their normal

Table 2.4. Complex formation equilibria and stability constants

Reaction	Individual stability constant	Reaction	Overall stability constant
$M + L \rightleftharpoons ML$	$K_1 = \dfrac{c(ML)}{c(M) \cdot c(L)}$	$M + L \rightleftharpoons ML$	$\beta_1 = K_1 = \dfrac{c(ML)}{c(M) \cdot c(L)}$
$ML + L \rightleftharpoons ML_2$ etc. till	$K_2 = \dfrac{c(ML_2)}{c(ML) \cdot c(L)}$	$M + 2L \rightleftharpoons ML_2$ etc. till	$\beta_2 = \dfrac{c(ML_2)}{c(M) \cdot c^2(L)}$
$ML_{n-1} + L \rightleftharpoons ML_n$	$K_n = \dfrac{c(ML_n)}{c(ML_{n-1}) \cdot c(L)}$	$M + nL \rightleftharpoons ML_n$	$\beta_n = \dfrac{c(ML_n)}{c(M) \cdot c^n(L)}$

properties are no longer visible and the ions can no longer be recognized. An example of this function is the ligand valinomycin (Fig. 2.24), a natural product. In the molecular cavity of valinomycin, a potassium ion is trapped when a complex forms.

Normally, the ion K^+ is surrounded by a shell of water molecules that act as ligands. Like all cations, K^+ prefers an aqueous environment. A completely different behaviour is displayed if the potassium-valinomycin complex has formed. This complex appears to be a rather large organic molecule which easily dissolves in hydrophobic solvents, but only sparingly in water. In this way, potassium ions are 'lured' into non-aqueous solvents. In living organisms, this is a way to transfer potassium through natural membranes. In chemical sensors, the highly selective interaction with K^+ is very important for detection and quantitative determination of potassium.

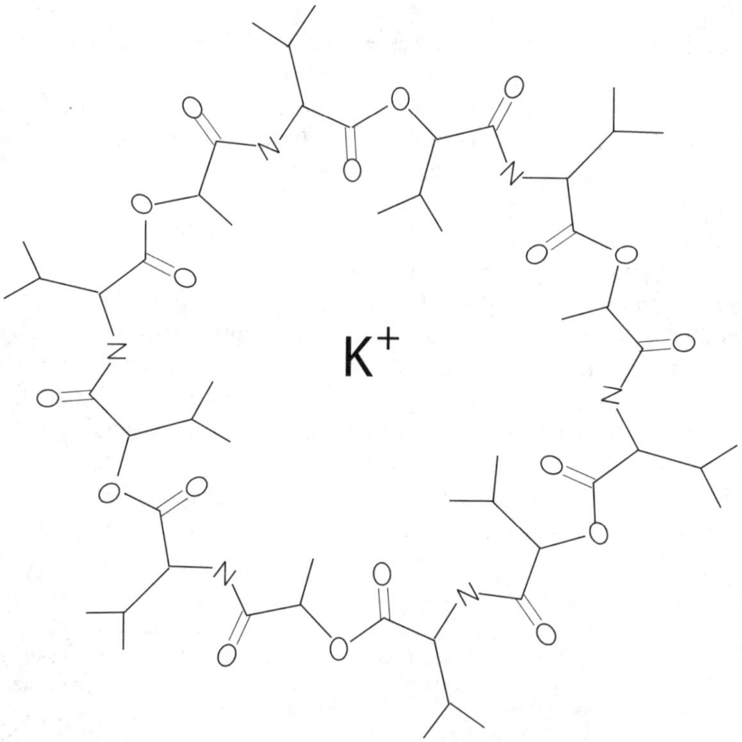

Figure 2.24. Valinomycin

2.2.5
Redox Equilibria

Redox equilibria, in some respects, make up the interface between fields of chemical and electrical phenomena. In such equilibria, electrons act as chemical reaction partners. Free electrons do not exist normally in solutions, but their charge can move in the form of ions. Conditions in redox processes are similar to those in acid-base reactions, where a proton acceptor (a base) must always be present to make possible an acid-base reaction where another particle (an acid) releases a proton. The same condition must be met in a redox reaction where one partner (the *oxidizing agent*) must be able to accept an electron when the other partner (the *reducing agent*) donates this electron. However, in contrast to protons, electrons can move freely in the 'electron gas' of metallic conductors. This results in a unique feature of redox reactions. Partners of such reactions are now able to cross a phase boundary and to pass over to a solid phase. This means a transfer of charges across an interface which is connected with voltage formation at the interface. Such phenomena are not found with other chemical equilibria. This is the reason for the historical development of theoretical concepts that seem quite different from the concepts of all other chemical equilibria.

Redox equilibria can be *heterogeneous* (with different phases participating as sketched above) or *homogeneous*. For the latter, the similarity to acid-base equilibria is obvious. An *oxidizing agent* is defined as a substance which is able to accept electrons, whereas a *reducing agent* is an electron donor. A redox reaction involves the transfer of electrons from an oxidizing to a reducing agent. Such reactions can be written in a schematic equation:

$$Red_1 + Ox_2 \rightleftarrows Ox_1 + Red_2$$

as an example

$$Fe^{2+} + Ce(IV) \rightleftarrows Fe^{3+} + Ce^{3+}$$

Again, we have conjugate pairs, this time *conjugate redox couples*, e.g. the pair Fe^{2+}/Fe^{3+}.

To classify redox couples with respect to their strength, a trick is used that is similar to that used with acid-base equilibria. Consider what would happen if the redox couple studied was brought to react with a 'standard partner'. Long ago, the couple hydrogen gas/hydrogen ion (H_2/H_3O^+) was chosen to act as the standard partner. To solve the problem of comparing the strength of redox couples, we simply initiate the reaction and, after equilibrium has been established, determine the actual equilibrium constant (as with acid-base equilibria). This time, however, the classification is done in a somewhat different way.

Indeed, the strength of a redox couple is measured by its reaction with H_2/H_3O^+, but the quantity for comparison is not the equilibrium constant. Instead, an *electric potential* is used. In fact, it would be better to speak of a *voltage* rather than a *potential*, but there are historic reasons to retain the term potential. In contrast to the common procedure with other equilibria, the potential values discussed here can be measured directly in many cases. This is a good reason to prefer an electric quantity. In the following considerations, a more detailed motivation will be given.

Optional redox couples in reaction with the standard couple results in equations like the following:

$$2Ce(IV) + H_2 + 2H_2O \rightleftharpoons 2Ce^{3+} + 2H_3O^+ \,,$$

$$Zn^{2+} + H_2 + 2H_2O \rightleftharpoons Zn + 2H_3O^+ \,,$$

$$I_2 + H_2 + 2H_2O \rightleftharpoons 2I^- + 2H_3O^+ \,.$$

For each of the reaction schemes given above, a definite value for its driving force $\Delta_R G$ can be determined, as e.g.

$$\Delta_R G = \Delta_R G^\theta + RT \ln \frac{c^2(Ce^{3+}) \cdot c^2(H_3O^+)}{c^2(Ce^{IV}) \cdot p(H_2)} \,. \tag{2.28}$$

For our 'standard redox couple', we must set standard concentration values. Units for dissolved substances are the usual ones, for gases like H_2, the partial pressure $p(H_2)$ is used. Standard concentration values are made dimensionless and set to 1, following some general conventions. In this way, Eq. (2.28) is reduced to

$$\Delta_R G = \Delta_R G^\theta + RT \ln \frac{c^2(Ce^{3+})}{c^2(Ce^{IV})} \,. \tag{2.29}$$

In Eq. (2.29) it seems that the driving force depends only on the special redox couple just considered. But nevertheless, the partner, i.e. the couple H_2/H_3O^+, did not disappear! Its effect is just standardized to a constant value, equal for every redox couple to be classified.

In the course of a redox reaction, a certain *amount of electric charge* ('*quantity of electricity*') is transferred from one partner to the other. For standardized conditions, this amount of charge can be calculated if we know the number of moles of electrons z which are transferred in the reaction. This number multiplied by the *molar charge* (or Faraday number) $F = 96\,500\,A \cdot s \cdot mol^{-1}$ yields the desired amount of charge q:

$$q = z \cdot F \,. \tag{2.30}$$

The driving force of the reaction $\Delta_R G$ was calculated in Eq. (2.30). This quantity divided by q would then mean something like 'driving force of the electrons transferred by the considered redox couple'. This quantity has the unit volt (V), i.e. it is a *voltage* or a *potential difference*. It is denoted traditionally by the symbol E (derived from the historic term *electromotive force*). By convention, E has the opposite sign of $\Delta_R G$. For our example discussed above, the number z has the value 2. We can set

$$-\frac{\Delta_R G}{2 \cdot F} = -\frac{\Delta_R G^\theta}{2 \cdot F} + \frac{RT}{F} \ln \frac{c(Ce^{IV})}{c(Ce^{3+})} \,. \tag{2.31}$$

Table 2.5. Electrochemical series

	Redox couple	Standard potential E^θ/V
1	$Ce(IV) + e^- \rightleftharpoons Ce^{3+}$	1.713
2	$PbO_2 + SO_4^{2-} + 4H^+ + 2e^- \rightleftharpoons PbSO_4 + 2H_2O$	1.685
3	$MnO_4^- + 8H^+ + 5e^- \rightleftharpoons Mn^{2+} + 4H_2O$	1.51
4	$Cl_2 + 2e^- \rightleftharpoons 2Cl^-$	1.36
5	$MnO_2 + 4H^+ + 2e^- \rightleftharpoons Mn^{2+} + 2H_2O$	1.23
6	$Fe^{3+} + e^- \rightleftharpoons Fe^{2+}$	0,7704
7	$CrO_4^{2-} + 4H_2O + 3e^- \rightleftharpoons Cr^{3+} + 8OH^-$	0.72
8	$O_2 + 2H^+ + 2e^- \rightleftharpoons H_2O_2$	0,682
9	$H_3AsO_4 + 2H^+ + 2e^- \rightleftharpoons H_3AsO_3 + H_2O$	0.58
10	$I_3^- + 2e^- \rightleftharpoons 3I^-$	0.536
11	$[Fe(CN)_6]^{3-} + e^- \; [Fe(CN)_6]^{4-}$	0.36
12	$Cu^{2+} + 2e^- \rightleftharpoons Cu$	0.346
13	$Cu^{2+} + e^- \rightleftharpoons Cu^+$	0.170
14	$2H^+ + 2e^- \rightleftharpoons H_2$	0.0000
15	$Zn^{2+} + 2e^- \rightleftharpoons Zn$	−0.7628

Finally we get

$$E_{Ce^{IV}/Ce^{3+}} = E^\theta_{Ce^{IV}/Ce^{3+}} + \frac{RT}{F} \ln \frac{c(Ce^{IV})}{c(Ce^{3+})} \; . \tag{2.32}$$

If we do a last step and postulate that the considered couple should also given standard conditions, then the potential E changes to the *standard redox potential E^θ*. This potential is useful for classifying redox couples with respect to their oxidizing or reducing capability. Now we can arrange redox couples in the order of their individual standard potentials. The resulting tables are called *electrochemical series* or *electromotive series*.

Table 2.5 lists homogeneous as well as heterogeneous redox couples. In heterogeneous couples, in the course of a redox reaction, electrons cross a phase boundary. As a result, electric quantities like potential or current are directly related to chemical quantities. They can be measured easily by *electrochemical methods*, since *electrochemistry* deals with charge transfer processes. Consequently, the group of *electrochemical sensors* is one of the most important groups of chemical sensors.

2.2.6
Electrochemistry

Electrodes in Equilibrium

Electrodes and Electrochemical Cells. Some redox couples, like Zn/Zn^{2+} (no. 15 in Table 2.5) a priori consist of two different phases, since the partners of the couple exist in different aggregate states. Both partners in contact with each other form a special kind of an *electrode.*

Electrodes are defined to be systems of at least one electron conducting phase in contact with at least one electrolyte (a phase with ionic conductance).

The quantities *redox potential* and *standard redox potential* have been introduced so far in an abstract fashion. Now, we can combine electrodes to establish an *electrochemical cell,* and this can be used to measure directly the potentials mentioned. For every redox couple, we must find a way to represent this couple by an electrode. The Zn/Zn^{2+} couple already is a special sort of electrode, namely a *metal/metal ion electrode,* also called an *electrode of the first kind.* Other types are *gas electrodes,* which can be designed with couples like nos. 4, 8 and 14 in Table 2.5 by combining the redox couples with a body made of an inert metal. A similar combination is useful for couples where both partners are dissolved species. The resulting electrode type is the *redox electrode.* A symbolic notation for electrodes is given for the examples listed in Table 2.6. A phase boundary is symbolized by a vertical dash or just by a slash.

A highly important gas electrode is the *hydrogen electrode.* The latter represents the redox couple H_2/H^+. Combining this electrode (internally designed to fulfil the requirements for standard conditions) with any other being studied, we can measure a voltage between both electrodes. This voltage is not only related but is in fact *identical* to the standard redox potential, which may then alternatively be called a *standard electrode potential.* Standard conditions mean that all *concentrations* (more exact *activities*) have a value of 1, and partial gas pressures are below atmospheric pressure (101.3 kPa). Combining electrodes to give an electrochemical cell means making a connection between the electrolyte phases either via a *diaphragm* permeable for ions or something similar. The arrangement of such cells can be denoted by cell symbols like the following example: $Zn/ZnSO_4(aq)//HCl(aq), H_2(g, p = 101.3\,kPa)/Pt$.

A diaphragm or similar liquid–liquid junction is symbolized by a double slash.

The electrode potentials discussed above can be measured directly. They depend on concentrations. Consequently, we can determine concentration values by potential measurements. Obviously, this type of situation presents a good opportunity to apply a chemical sensor. The basis of such measurements is Eq. (2.32). If the relationship is expressed in a more general way, the equation

Table 2.6. Examples of electrode sorts

Metall–metal ion electrode	Ag/Ag^+
Redox electrode	$Pt/Fe^{3+}, Fe^{2+}$
Gas electrodes	$Pt/Cl_2, HCl\,(aq, 0.1\,mol \cdot l^{-1})$
	$Pt/H_2, H_3O^+\,(aq, 1\,mol \cdot l^{-1})$

assumes the form given in Eq. (2.33):

$$E = E^\theta + \frac{RT}{nF} \ln \prod a_i^{\nu_i} . \tag{2.33}$$

This is a fundamental relationship, widely known as the *Nernst equation*.

In the concentration-dependent term of the Nernst equation, *activities a_i* are written. This fact may be useful for determining activity coefficients. More important, however, at this point is the fact that we can determine analytical concentrations if we set the activity coefficient equal to one, or at least give it a constant value by addition of a background electrolyte. It can be quite easy to make such measurements. A simple electrode can be set up simply by dipping a metallic body into an aqueous solution. For a silver wire, e.g., the Nernst equation gets the following form (2.34):

$$E = E^\theta + \frac{RT}{F} \ln a(\text{Ag}^+) . \tag{2.34}$$

Clearly, a simple silver wire can be used as a chemical sensor for silver ions.

Potentiometry. Determination of concentration by means of potential measurements at electrodes is an analytical method called *potentiometry*. To apply this method successfully, some preconditions must be met. One of them is that the potential (i.e. voltage) measurement should be done currentless, i.e. no electric current should flow through the electrode when measuring. If this condition is not fulfilled, we could not guarantee that the electrode is in *equilibrium with the surrounding solution*. Alternatively, we can illustrate the necessity of currentless measurement if we imagine the consequences of a current flow through the electrode. If this occurs, then we conduct an electrolysis, and this could generate, or consume, just the ion which we intend to determine.

All the standard potentials in lists like that in Table 2.5 refer to the standard hydrogen electrode. Working with this electrode is a technical problem. The electrode contains an activated platinum body, an additional acid solution with an activity exactly equal to 1, and finally a stock of hydrogen with a partial pressure equal to atmospheric pressure. The latter condition is fulfilled usually by means of an open gas cylinder. The streaming gas automatically assumes atmospheric pressure as soon as it has left the pressure cylinder. Altogether, the construction of such gas electrodes is not very user-friendly. For experimental work, we should look for a reference electrode that can be handled easier. It is not necessary to look for an electrode with the same potential as the standard hydrogen electrode, but it would be sufficient to find a design which delivers a stable potential which is referred exactly to the potential of the hydrogen electrode, since this is defined as the general standard. Very useful reference electrodes are *electrodes of the second kind*. A good example is the so-called *saturated silver chloride electrode*, made by combining a silver wire with a saturated potassium chloride solution which contains a certain amount of solid silver chloride. This combination provides a stable, precise potential difference of +0.198 volt vs. a standard hydrogen electrode. This electrode is written symbolically as $Ag/AgCl_{(s)}$, $KCl_{(sat)}$. Saturated silver chloride electrodes are commercially available. An example is presented in Fig. 2.25.

Potentiometry is one of the most important measurement techniques for chemical sensors. A closer look reveals that potential differences can appear at each phase boundary inside the system. Each potential difference (a so-called *Galvani potential difference*) is actually a type of voltage (sometimes named a *Galvani voltage*) and contributes to the overall electrode potential. The electrode potential could be interpreted as the sum of all voltages at phase boundaries. A single Galvani voltage cannot be measured, since we always just measure the voltage between two electrode terminals. What we can do is keep constant all the voltage contributions except the one which gives us the desired information about chemical composition. The location where the interesting voltage forms is not necessarily the interface between the metal and the electrolyte solution. It might also be that a concentration-dependent potential difference appears, e.g. at the interface of two electrolyte phases. The reason for such electric effects can be an *extraction equilibrium*, a *solubility equilibrium*, or, of course, an *electrochemical equilibrium*. Regardless of the special chemical origin of the concentration, in every case where ions are included, the quantitative concentration dependence is given by the Nernst equation.

Most important for the construction of useful electrodes for potentiometry is the design of the sensor interface with the sample solution. This interface should be designed such that the maximum possible degree of selectivity with the interesting constituent is achieved. As a result of such efforts, the field of *ion-selective electrodes* (ISEs) has been established. IESs are not a priori chemical sensors. Normally, they do not fulfil the condition of being small and cheap. Nevertheless, ISEs are a very important preliminary stage on the way to highly useful chemical sensors.

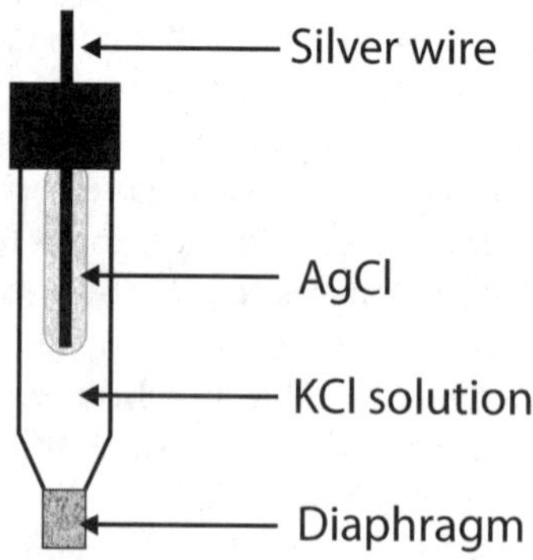

Figure 2.25. Silver/silver chloride reference electrode

Electrolytic Processes

The equilibrium at the electrode surface is the basis of potentiometry, but it is not identical to an equilibrium of an electrochemical cell. Such a cell, also called a *galvanic cell*, is capable of converting chemical energy into electric energy. Consequently, galvanic cells are transducers based on the *energy conversion principle*. A typical galvanic cell is the Daniell cell, this is the combination of a copper electrode with a zinc electrode. This galvanic cell, symbolically written as $Zn/Zn^{2+}//Cu^{2+}/Cu$, was used formerly as a source of electric energy for small appliances. If we connect both electrode terminals via an electric resistor, a current will flow, caused by the redox process taking place in the cell: $Zn + Cu^{2+} \rightarrow Zn^{2+} + Cu$.

At the zinc electrode (the anode), oxidation takes place, i.e. zinc metal is oxidized to give a zinc ion. At the second electrode (the cathode), all the copper ions are equally reduced to give metallic copper. Electrons released by zinc oxidation flow into solution where they discharge copper cations. Prior to the onset of the reaction, we can measure a voltage that is a measure of the driving force $\Delta_R G$ available during this initial state. The equilibrium state of the cell is achieved when this driving force has been consumed completely, i.e. when $\Delta_R G = 0$. To reach this state, we must allow current flow and then wait until the current approaches zero. This means that a discharged battery is in equilibrium state.

The electrolytic current flowing between the terminals of an electrochemical cell can be considered an expression for the *reaction rate*. We should expect that the magnitude of this current depends on the concentration of reacting substances. Hence it should be possible to design chemical sensors on the basis of a measurement of the electrolytic current. Under certain conditions, the transducer principle can change from *energy conversion* to *current limiting*.

Processes at Electrodes. Different processes take part in current flow through an electrochemical cell. Among them are the following which are essential:

- Transport of reactants towards the electrode by diffusion or ion migration in an electric field
- Charge transfer through the electrode-solution interface
- Transport of reaction products away from the electrode

No other processes are discussed here, although they are important also. Among them are *adsorption* of reactants or products at the electrode surface, and *nucleation*, i.e. the formation of crystal nuclei, where a new phase is generated by deposition of a solid or by gas evolution. However, only the three specially emphasized processes are really essential.

Each one of the processes mentioned above can be the critical one which controls the reaction rate and, consequently, the current amplitude. If charge transfer is the slowest process, then it is the controlling process, and the reaction is said to be *kinetically controlled*. An alternative term for such processes is *irreversible*. Although the expression '*an irreversible electrochemical reaction*' is not an exact denomination, it is used frequently. If such kinetic hindrances play a role, then we cannot find a simple relationship between the current and the concentration of the reacting species. This is one of the reasons why analysts always try to make kinetics as fast as possible, mainly by application of catalytic layers at the electrode surfaces, or e.g. by homogeneous catalysis when enzymes are used.

If the kinetics of a reaction is fast (a *reversible* electrode reaction), then transport processes are the slowest ones in the series of consecutive partial processes. To get a clear relationship between current and concentration, it is useful to organize the cell such that only one of the transport processes is in function, namely *diffusion*. We must suppress the *migration*. This is achieved by addition of a large amount of an inert *supporting electrolyte* which enhances the overall conductance to such an extent that no significant electric field strength can arise in the bulk of the solution.

With suppressed migration, the only way to transport ions to and away from the electrode is *diffusion*. For a *reversible electrode reaction*, the overall reaction rate is then said to be *diffusion controlled*. The laws of diffusion are valid for ions as well as for neutral particles. Particles diffuse in the direction where a lack of substance exists, i.e. towards a negative concentration gradient. As soon as a substance is consumed at an electrode surface (e.g. by electrochemical reduction of ions), particles start to move in order to compensate the deficiency.

There exist two laws of diffusion. The most important one for actual consideration is Fick's First Law of Diffusion, given in Eq. (2.35):

$$\frac{\mathrm{d}n}{\mathrm{d}t} \cdot \frac{1}{A} = -D\frac{\mathrm{d}c}{\mathrm{d}x} \, . \tag{2.35}$$

The law expresses that the flow of particles ($\mathrm{d}n/\mathrm{d}t$, measured in moles per second) through an area A is proportional to the concentration gradient $\mathrm{d}c/\mathrm{d}x$ (where x is the position, or length, in metres). The *diffusion coefficient D* acts as the factor of proportionality.

A simple relationship exists between the flow of particles $\mathrm{d}n/\mathrm{d}t$ and the electrolytic current I. This relationship is a consequence of Faraday's law [Eq. (2.36)]. This law defines the amount of charge q corresponding to the amount of substance n which has been converted in the course of an electrolytical process (with F the Faraday constant and z the number of electrons transferred for one molar reaction):

$$q = zF \cdot n \, . \tag{2.36}$$

Electrolytic current I and flow of particles $\mathrm{d}n/\mathrm{d}t$ are aligned together via the following equation (since $I = \mathrm{d}q/\mathrm{d}t$):

$$I = z \cdot F \cdot \frac{\mathrm{d}n}{\mathrm{d}t} \, . \tag{2.37}$$

A *concentration gradient* at the electrode is the origin as well as the precondition for a permanent electrolytic current, but also it is, eventually, the reason for current limitation. This fact can be utilized to design chemical sensors following the principle of *current limiting transducers*.

Current-Potential Curves at Macroelectrodes. Assuming that none of the chemical reactions taking part in the overall electrolytic process is inhibited, the reaction is then *diffusion controlled*. The shape of the curves $I = f(E)$ is determined by the laws of diffusion. It should be possible to extract from the curves analytical information about the electrolysed sample.

To record curves $I = f(E)$, also called *voltammograms*, preferably the potential is varied arbitrarily either step by step or continuously, and the actual current value is measured as the dependent variable. The opposite procedure is possible also but less common. The shape of the curves depends on the speed of potential variation and on whether the solution is stirred or quiescent. Two basic shapes are found (Fig. 2.26). The right curve in Fig. 2.26 is *sigmoidal* in shape. Such curves appear if a continuous convection is stimulated, either by stirring the solution or by movement of the electrode vs. solution. An important group are the *hydrodynamic electrodes*. They have in common a laminar, steady flow of solution at the electrode surface. This flow is generated by mechanical devices.

Figure 2.27 presents some examples for classic (i.e. macroscopic) electrodes which are meaningful for voltammetry. Among them is the famous *dropping mercury electrode* (DME), which was extremely successful for the development of analytical chemistry during the second half of the last century. It contributed to the establishment of a field of analytical chemistry which was new at the time – *electroanalytical chemistry*. The relative movement of solution vs. the electrode leaves unaffected only a thin layer adhering to the electrode surface, the *hydrodynamic layer*. Inside this layer, an additional, much thinner layer is located which is completely quiescent. This layer, called the *Nernst diffusion layer*, can be crossed by ions or molecules only by diffusion.

A closer look at the diffusion processes during voltammetric measurements can give an explanation of how the different shapes of the curves emerge. Let us consider as an example the reduction of copper ions at the copper electrode

Figure 2.26. Current-potential relationship for electrodes in quiescent solution (*left*) and for electrodes with convection or for microelectrodes (*right*)

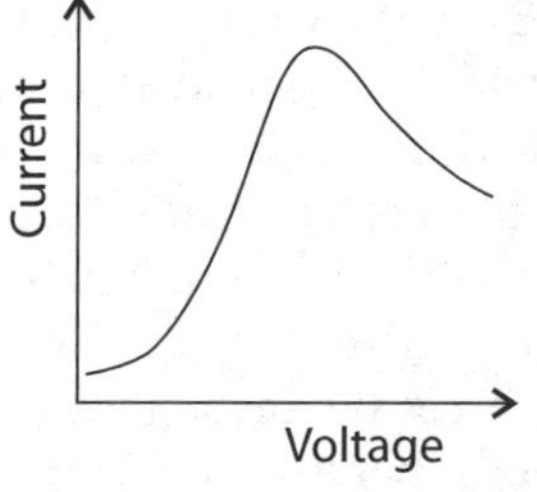

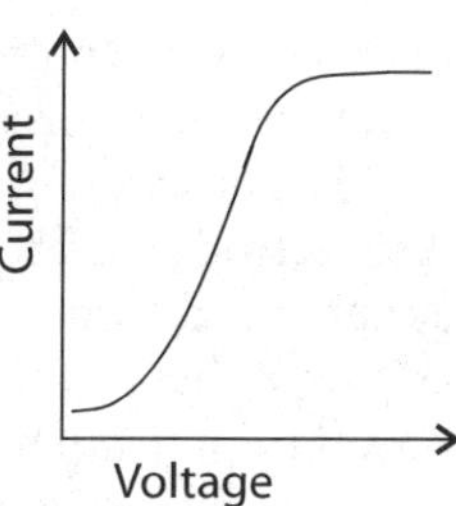

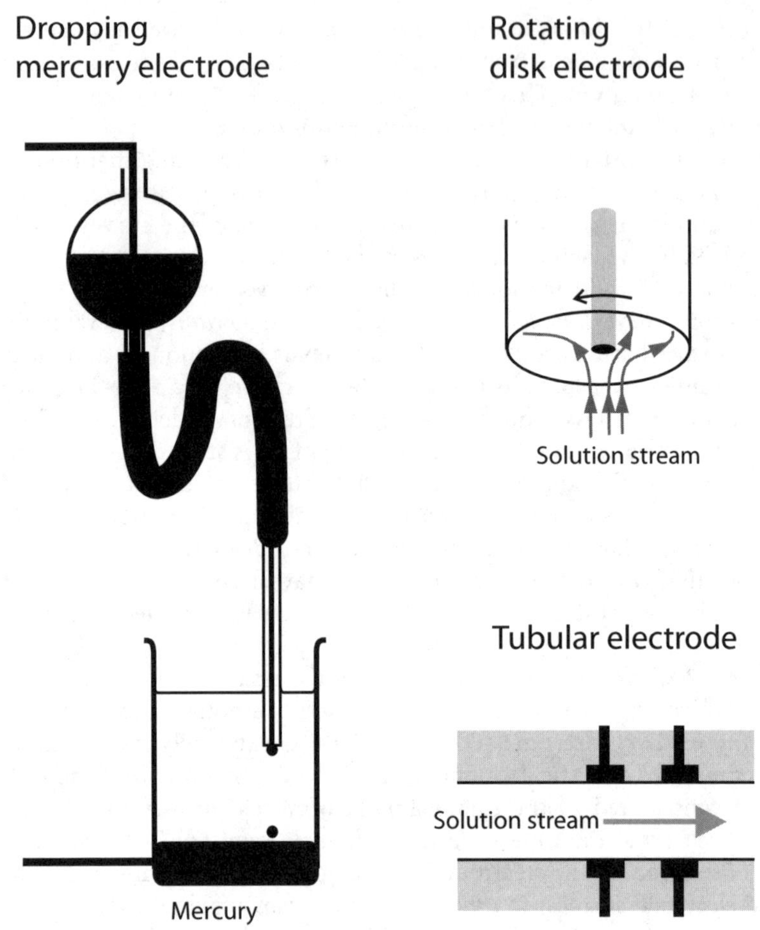

Figure 2.27. Classical hydrodynamic electrodes

in continuously stirred solution. In the concentration-dependent term of the
Nernst equation, only the copper(II) concentration must be written:

$$E = E^{\theta} + \frac{RT}{2F} \ln c(Cu^{2+}) \,. \tag{2.38}$$

In this case, the potential E is not measured, but it is an independent variable;
the electrode is forced to adopt it from an external source. The system reacts
to this distortion of equilibrium by generating the 'proper value' of Cu^{2+} at
the electrode surface. This local concentration, called now $c(Cu^{2+})_{surf}$, is made
either by reduction of copper ions or by oxidation of the electrode body to such
an extent that just the condition given by the Nernst equation is fulfilled. This
way, the applied potential forces the instantaneous value of a local concentration. The copper ions, either generated or disappearing, cause a concentration

gradient which gives rise to the diffusive transport according to Eq. (2.35), and finally to the actual value of electrolysis current. The resulting depletion layer near the electrode is identical to the aforementioned Nernst diffusion layer. A permanent, smooth convection establishes the constant thickness of this layer. Outside the hydrodynamic layer, the concentration is kept uniform as a result of stirring. Concentration changes generated at the electrode surface are homogenized in this region. If the applied potential is changed towards more negative values, then the point $c(Cu^{2+})_{surf}$ will shift downward in our example, towards lower values of the local concentration. The conditions are demonstrated in Fig. 2.28.

Since the diffusion layer is of constant thickness, the concentration gradient becomes steeper. Consequently, the electrolysis current increases with potential shifting in the negative direction. The maximum current is reached when the surface concentration approaches zero. This maximum current is called the *diffusion-limited current* I_D. With applied potential beyond this point, c_{surf} cannot become less than zero, the gradient does not increase, and consequently I_D remains constant and independent of further potential change. These interrelations are typical for current-potential curves of the sigmoidal type (Fig. 2.26).

Taking into account the conditions discussed above, a general equation for the diffusion-controlled reversible voltammogram can be derived [Eq. (2.39)]:

$$E = E^{\theta} + \frac{RT}{zF} \ln \frac{I_D - I}{I} .$$

$$(2.39)$$

A most useful property of the diffusion-limited current ID is its strict proportionality to the solution concentration of electrochemically active components. In the limiting-current region, i.e. for potential values negative enough to ensure that every ion arriving at the electrode is reduced, the concentration of copper(II) ions is zero. This means that the concentration gradient (and consequently the limiting current) is controlled exclusively by the homogeneous concentration in the bulk solution $c(Cu^{2+})$. Combining Eqs. (2.35) and (2.37), the relationship given in Eq. (2.40) results, where we have set the quotient of the

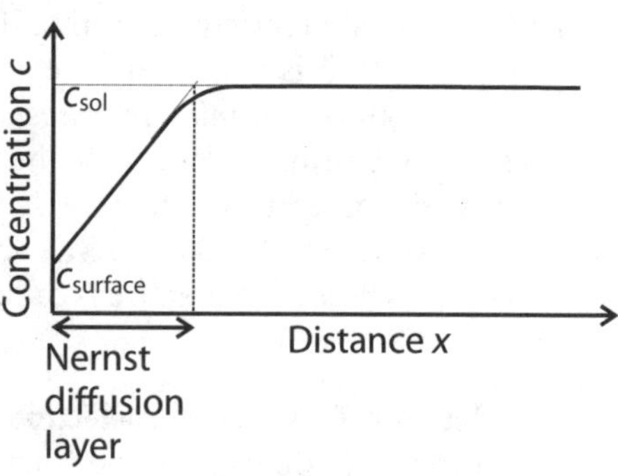

Figure 2.28. Concentration gradients at an electrode with convection

differences $\Delta c/\Delta x$ instead of the concentration gradient dc/dx. Furthermore, Δc is replaced by $c - c_{surf}$. For limiting-current conditions, we can set $\Delta c = c$. The quantity Δx is equal to the thickness of the stationary diffusion layer δ.

The concentration proportionality of the limiting current is not restricted to the example given. It is valid universally. All together, a voltammetric electrode used under current-limiting conditions really represents a *current-limiting transducer*:

$$I_D = -\frac{z \cdot F \cdot D \cdot A}{\delta} c \, . \tag{2.40}$$

We can also write

$$I_D = k \cdot c \, . \tag{2.41}$$

The shape of a current-potential curve also depends on the speed at which the potential range is swept through, i.e. the applied *scan rate*. The slower the potential sweep, the broader the extension of the diffusion layer into solution. Peak-shaped curves (as in Fig. 2.26, left) appear if the diffusion layer continues to grow during potential variation. Only if δ stays constant from start to end of the potential variation will we get a sigmoidal curve shape. Peak-shaped curves also contain analytical information; however, this cannot be extracted easily. On the other hand, we can derive highly informative diagnostic tools by means of peak-shaped curves.

Stationary diffusion can also be achieved by interposing a diffusion barrier, e.g. a semipermeable membrane (Fig. 2.29). An instructive example is the Clark sensor, a chemical sensor for determining dissolved oxygen (see also Sect. 7.2.2). In this device, a gas permeable membrane is located between the cathode and the sample solution.

Although not very well known, *electrically heated electrodes* are a very good source of sigmoidal voltammograms. The *thermal convection* at such electrodes indicates a highly efficient stirring effect. However, practical use of such systems was achieved only with the development of an arrangement for successful suppression of mutual interference of heating and measuring circuits (Gründler and Kirbs 1999). Surface temperature of heated microelectrodes can be controlled precisely, assuming that it is somewhat below the solution's boiling point. In this way, a thin heated layer close to the electrode surface acts as some kind of a *microthermostat*. Another variant of this technology utilizes short heating pulses with the result that convection plays no role, and the local temperature can increase far higher than the boiling point of the solution, currently up to 250 °C. As a solvent, in this case the metastable, superheated water is used. The method is called *temperature pulse voltammetry*, TPV (Gründler et al. 1996).

Current-Potential Curves at Microelectrodes . Microelectrodes have dimensions in the range of micrometres. The diffusion layer which forms at such electrodes is not of a planar but a bent or spherical shape. Thus the diffusional transport

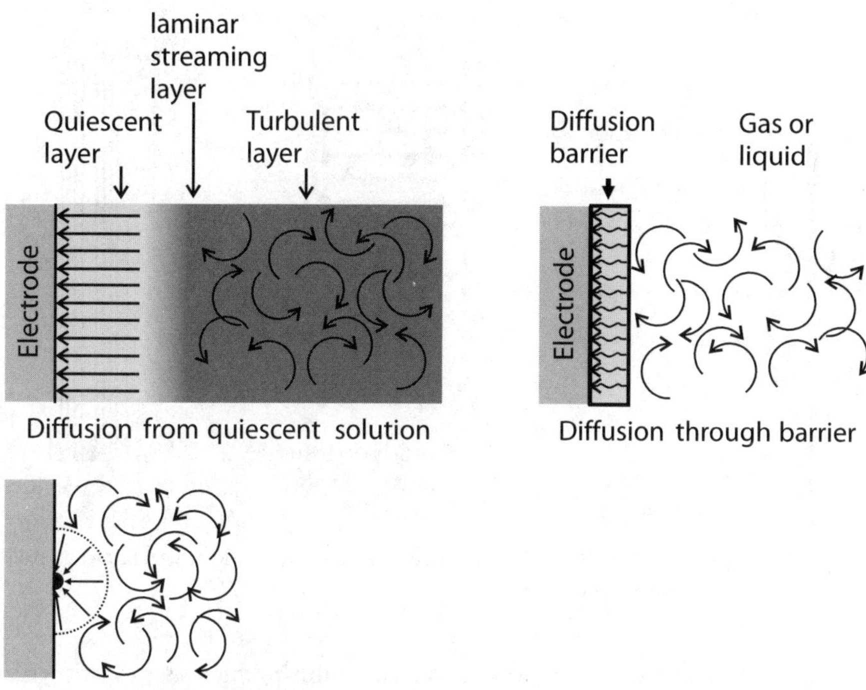

Figure 2.29. Stationary diffusion with planar and spherical diffusion layers. *Left*: diffusion layer of constant thickness in stirred solution; *centre*: diffusion barrier of constant thickness; *right*: semispherical diffusion region (*halo*) of constant thickness at a microelectrode

from the edges of the active area is more meaningful than with normal electrodes, where the active area is much larger than the thickness of the diffusion layer. As a result diffusion is more efficient with microelectrodes (Fig. 2.29). This is shown clearly if we compare the area of a given macroelectrode with a collection of microelectrodes where an identical surface area is fragmented into many tiny islands. The diffusion transport reaches a steady state after a certain time depending on the dimensions of the electrode. Overall, with microelectrodes much higher current densities ($j = I/A$) can be achieved in comparison to macroscopic electrodes.

Microelectrodes can be designed in different shapes (Fig. 2.30). Their response time vs. potential or concentration changes is much shorter than that of classical electrodes. Microelectrodes are not sensitive to convection from external sources, and they need a lower supporting electrolyte content; sometimes they even work without any background electrolyte. All these positive properties make them ideal chemical sensors. The problem is that they exhibit high current densities, however low total currents. Special low-amplitude current amplifiers are necessary to measure current values between femtoam-

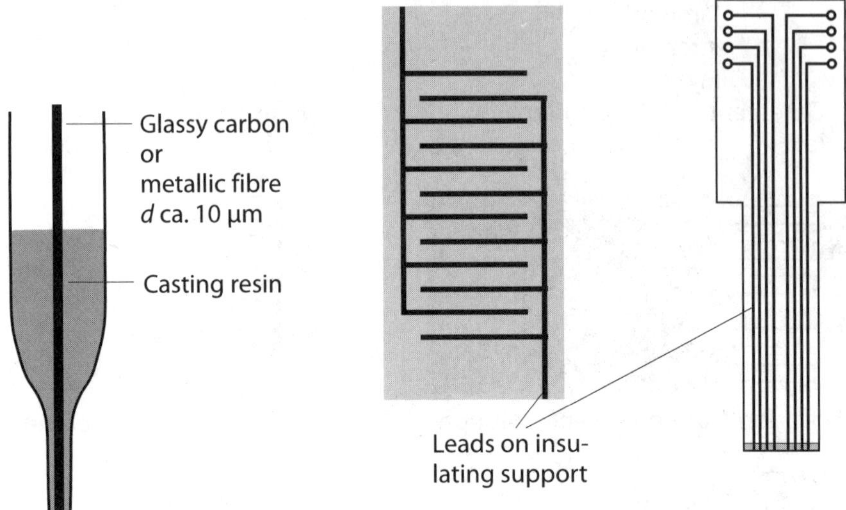

Figure 2.30. Design of microelectrodes. *Left*: needle with a microdisk; *centre*: interdigitated structure; *right*: stiletto shaped array

peres and nanoamperes. One way to overcome this problem is to work with microelectrode arrays, where many single-electrode areas are gathered to give a larger total electrolysis current. Microelectrode behaviour of such an array is achieved only if there is sufficient space between the single electrodes. This means e.g. that between electrode spots $1\,\mu m$ in diameter there must be a distance of $50\,\mu m$ at least.

Voltammetry. The collective term 'voltammetry' encompasses all methods based on the evaluation of current-potential curves. To record such curves, the electrode must be 'polarized', i.e. arbitrarily adjustable potential values must be imposed. This is done by means of an electronic circuit called the *potentiostat* (see also Sect. 2.4.1) and using three electrodes in an electrochemical cell (Fig. 2.31). The potentiostat is provided at its input with a linearly increasing reference voltage, i.e. with *voltage ramp*. The electrolytic current flowing through one of the electrodes, i.e. the *working electrode*, is recorded as a function of the applied potential. Microelectrodes can work without this electronic circuitry. Instead, the necessary polarization voltage is simply inserted between *working* and *reference electrodes*.

Voltammetry with quiescent macroelectrodes results in peak-shaped curves as sketched in Fig. 2.26. If analytical information is desired, then quiescent electrodes are useful only in the form of microelectrodes. For diagnostic purposes, however, macroelectrodes in quiescent solution are very useful. The corresponding method, the *cyclic voltammetry* (CV), allows one to obtain information about the studied electrochemical system very fast. The measure-

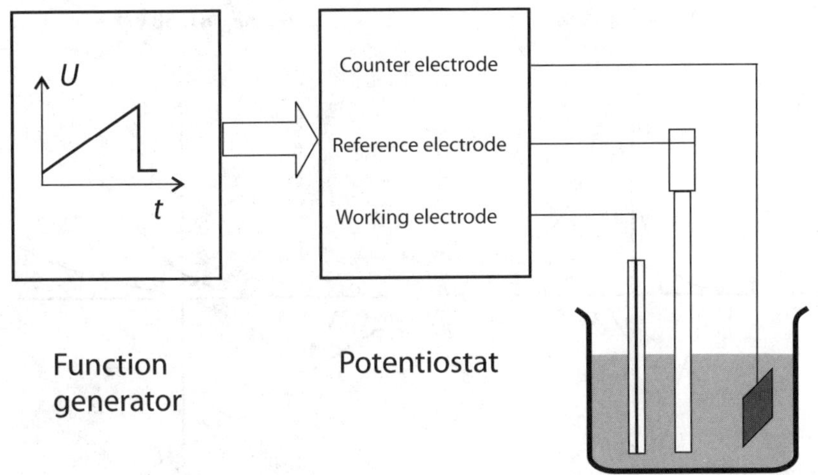

Figure 2.31. Voltammetric set-up. The potentiostat is an electronic controlling circuit imposing a predetermined voltage between terminals of working and reference electrodes. The function generator provides the reference voltage as a function of time

ment set-up is identical to that given in Fig. 2.31. However, a reference voltage signal with a triangle-shaped time function rather than a ramp-shaped one is imposed. The electrode response is recorded as a function of potential E, not of time t. The resulting cyclic voltammograms commonly have a shape like that given in Fig. 2.32. Very often two opposite peaks appear. One of these peaks (for the 'forward' scan) is known as the reaction of the substance initially present in solution, the second peak as the product of this reaction when the voltage scan is reversed. Hence, by cyclic voltammetry substances can be studied which have been generated by the method itself.

Cyclic voltammograms are rich in important and concise information (Fig. 2.33). The peak potential difference ΔE_p can be used to answer the

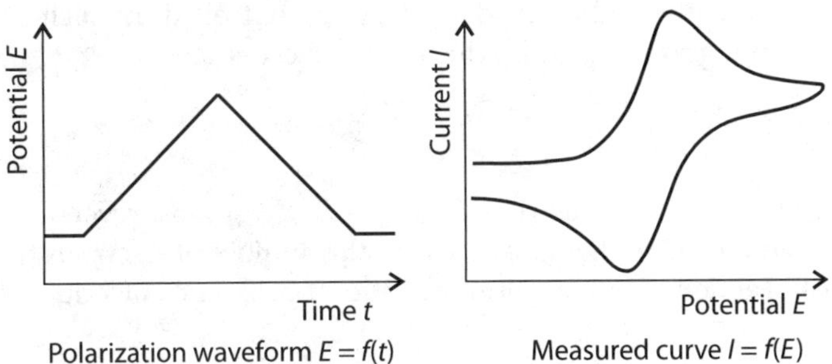

Figure 2.32. Shape of the reference potential for cyclic voltammetry (*left*) and electrode response (*right*) as $I = f(E)$

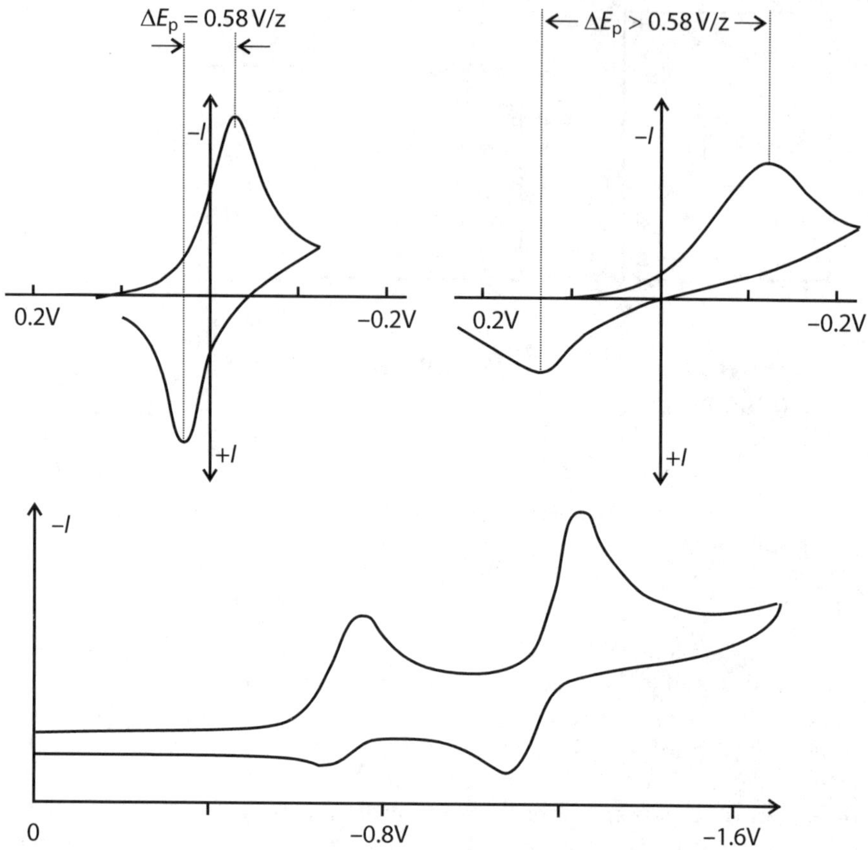

Figure 2.33. Information extractable from cyclic voltammograms. *Top*: ΔE_{peak} as a criterion for reversibility; *bottom*: multiple peaks as an evidence for generation of more than one product

question whether the electrochemical reaction of the substance studied is *reversible*, i.e. which kinetic hindrances of the charge transfer exist. An *ideally reversible reaction* would be totally diffusion controlled. In such a case, the cyclic voltammogram must fulfil the following criteria:

- $\Delta E_{peak} = E_{p\,anod} - E_{p\,cathod} = 57.0\,\text{mV}/z$
- $I_{p\,anod}/I_{p\,cathod} = 1$

The difference of peak potentials $E_{p\,anod} - E_{p\,cathod}$ for a reversible reaction should amount to 57 millivolts divided by the number of electrons transferred in a molar reaction. Furthermore, the ratio of peak current values should be ca. 1.

A cyclic voltammogram also provides information about the reaction mechanism. If multiple peaks appear, then electrons are transferred in consecutive steps.

If analytical information is desired, i.e. if the composition of a solution must be studied, the sigmoidal curves are useful since they can be evaluated easily (Fig. 2.34). The *diffusion-limited current* I_D is proportional to the analyte concentration over a wide concentration range. This high degree of linearity is characteristic for *amperometric sensors* which are based on voltammetry. Further analytical information is given by the *half wave potential* $E_{1/2}$. This special potential value is independent of concentration. It belongs to the current value $I = I_D/2$ and is characteristic for the *kind of substance studied*. This can be understood if $I_D/2$ is introduced in Eq. (2.39). As a result, we get $E_{1/2} = E^\theta$. In this way, the standard electrode potential of many substances can be determined (of course, since we usually do not use a standard hydrogen electrode, the experimental results must be corrected). Since E^θ is characteristic of an individual chemical substance, we can consider the half wave potential as a tool to identify substances. Of course, this can be only a rough estimation. In contrast, I_D yields highly precise values and is an important tool of quantitative analysis.

It is not necessary in every case to record a complete voltammogram. In many cases it is sufficient to choose a potential in the diffusion-limited potential range and to measure the corresponding value of I_D. The measured value will then follow all the concentration changes. Such a procedure is called *amperometry*. Amperometric sensors are preferred for indicating the equivalence point of titration and also for detecting concentration changes in flowing streams.

Voltammetric techniques for classic (macroscopic) electrode shapes have been improved strongly by the introduction of *pulse methods*. In these techniques, not a simple voltage ramp is imposed on the electrode. Instead, pulse sequences are superimposed on the exciting ramp signal. The response of the electrode vs. such pulses is evaluated separately. In this way, signals can be extracted with a highly improved signal-to-noise ratio. Most important are *differential pulse voltammetry* (DPV) and *square-wave voltammetry* (SWV). In chemical sensors, such up-to-date technologies are hardly used.

Stripping voltammetric methods are two-step procedures. In a first step, electrolysis is performed with the intention of accumulating the material of interest at the electrode surface. In this step, the electrode is rotated or the solution is stirred to allow a good transport of the analyte towards the electrode. With

Figure 2.34. Useful pieces of analytical information in sigmoidal voltammograms. $E_{1/2}$ half-wave potential, I_D diffusion-limited current

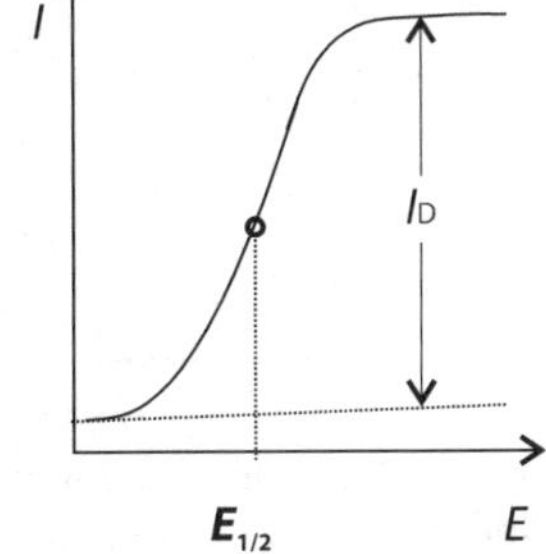

microelectrodes, the spherical diffusion zone gives rise to an improved transport without further precautions. As the deposition potential, generally a value is chosen in the diffusion-limited region. In a second step, the accumulated material is 'stripped', i.e. it is re-oxidized (if deposited cathodically in the first step) or brought to reaction otherwise. During this step, an analytical signal is achieved. The resulting signal appears to be 'amplified' since the local concentration of the accumulated analyte is higher than its homogeneous solution concentration. In Fig. 2.35, the temporal programme of such a determination is given.

The accumulation of a substance at an electrode surface also may be achieved by precipitating a sparingly soluble compound. An example is the determination of lead traces by anodic depositing lead dioxide at a platinum electrode: $Pb^{2+} + 6H_2O \rightarrow PbO_2 + 4H_3O^+ + 2e^-$. Traces can be accumulated also by *adsorption*. This is done preferably by adding an excess of an adsorbable ligand, which forms strong complexes with the ion to be determined. The potential dependence of adsorption equilibria [Eq. (2.45)] is used to find optimum conditions for adsorption. In stirred solution, the optimum deposition is imposed, so that first a layer of adsorbed ligands will form. The latter binds ions from solution by the formation of chemical bonds. In the next step, the adsorbed layer of cations is made the object of an electrochemical reaction. This electrochemical reaction is the source of an analytical

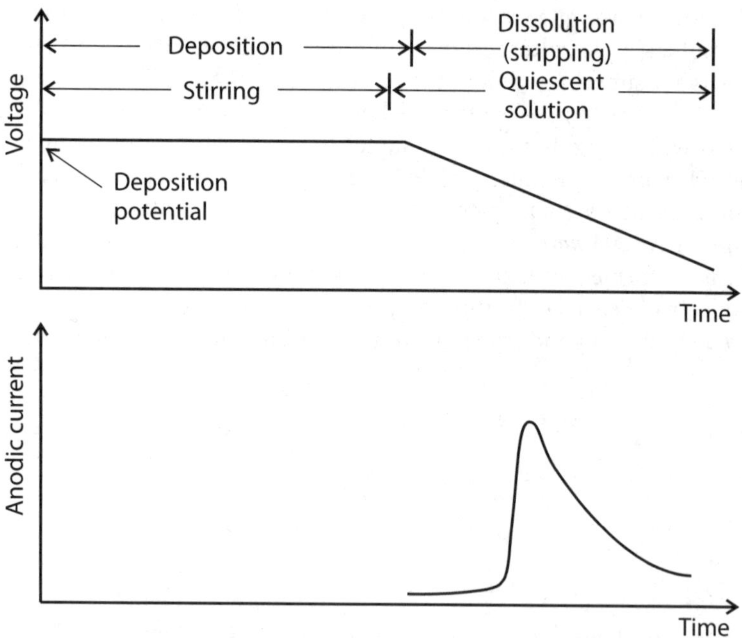

Figure 2.35. Controlling waveform of a voltammetric stripping analysis. *Top*: polarization voltage reference as function of time. *Bottom*: recorded signal

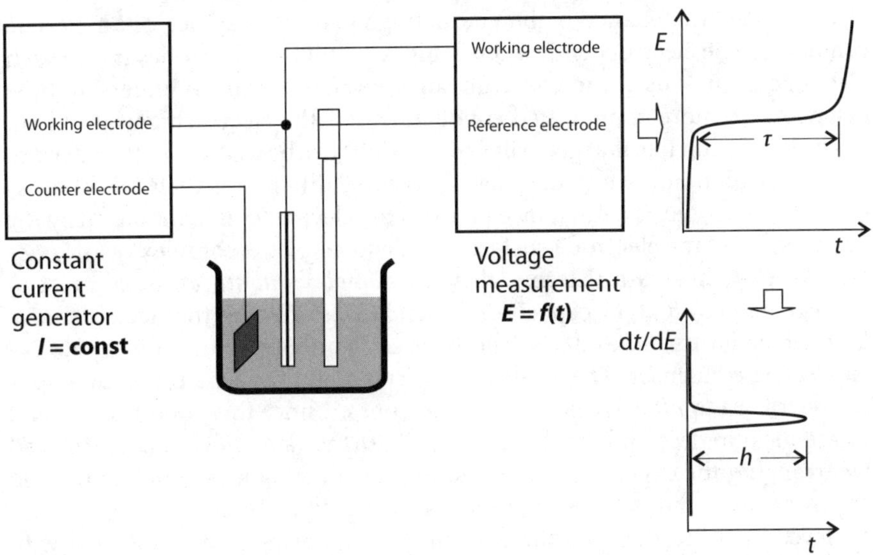

Figure 2.36. Experimental set-up for *constant-current stripping analysis* (also *chronopotentiometric stripping* or *potentiometric stripping analysis* (PSA))

signal. For all kinds of stripping analysis, in the second step the accumulated analyte layer is removed (i.e. 'stripped'). The accumulation step must be done under precisely defined conditions and with an accurate deposition time.

There are different ways to yield a stripping signal. Most common is *voltammetric stripping* (e.g. Fig. 2.35), where a voltage ramp is imposed on the electrode. A typical procedure is to start with the accumulation potential and to vary the voltage until anodic oxidation takes place. An anodic current peak appears whose height depends approximately linearly on analyte concentration. Alternatively, a constant current can be imposed to perform the stripping process. Such procedures are called *chronopotentiometric stripping* (also 'potentiometric stripping', PSA). This procedure results in staircase-shaped potential-time curves, where the time interval τ (Fig. 2.36) is a concentration proportional analytical signal. Instead of τ, the derivative dt/dE can also be evaluated. Peak height linearly depends on concentration.

Stripping analysis belongs to the most efficient trace determination methods in analytical chemistry. A detection limit as low as 10^{-10} M has been achieved in many cases.

Measuring Conduction and Impedance

Conductance measurements provide information about ionic concentration (Chap. 2, Sect. 2.3). Further information can be extracted from the *impedance* of electrodes. Impedance is the alternating current resistance of an electric conductor. The electrochemical cell equipped with two electrodes is also such a conductor. It is useful to construct an *equivalent circuit* to understand the behaviour of more complicated conductors. For the electrochemical cell, an equivalent circuit like that given in Fig. 2.37 (left) can be written. At the interface electrode/solution, we find the galvani potential differences (galvani voltages) g_1 and g_2. The solution resistance between the electrodes is symbolized by R_L. Impedances of the electrode surfaces are denoted by the complex resistors Z_1 and Z_2. In mathematical terms, they are *complex quantities*, in contrast to the *real quantity* R_L. The capacities C_{D1} and C_{D2} symbolize the electric *double layer* which forms at the interface between electrode and solution. They act like an electric condenser. The equivalent circuit simplifies when measurements are restricted to *alternating current* (AC) ones. Under this condition, g_1 and g_2 are meaningless and can be omitted. A further simplification is achieved by arranging the experimental design in such a way that only one of the two electrodes responds to the imposed electric excitation signals. This can be achieved by utilization of a three-electrode arrangement in connection with a potentiostat. The resulting simplified equivalent circuit is shown in Fig. 2.37 (right).

The *electrochemical double layer* mentioned above exists at every interface between an electronic conductor and an electrolyte. It is a special case of the phenomena appearing when phases of different conduction types come into contact. As discussed in the preceding chapter, a parallel arrangement of charge carriers with opposite signs is formed along the interface. A special feature of the electrochemical double layer appears if the electrolyte is an ion-containing solution. In this case, the strict parallel order of the double layer is permanently distorted by the heat-induced movement of ions on the solution side. In addition to the *static double layer* (also called the *Helmholtz layer*), which is also a diffuse part, the *diffuse double layer* forms on the solution side. This layer extends somewhat into homogeneous solution. Ions in solution are surrounded by a layer of solvent molecules, i.e. they are *solvated* (or *hydrated* in water). This effect plays an important role primarily in aqueous solution. Following the generally accepted model, we find inside the *Helmholtz layer* mentioned an inner part, the *inner Helmholtz plane* (IHP). The latter mainly contains adsorbed water molecules and anions, as indicated in Fig. 2.38.

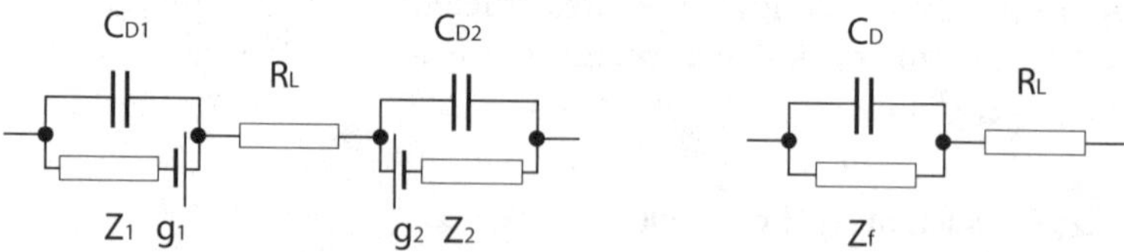

Figure 2.37. Equivalent circuits of an electrolytic cell. *Left*: complete circuit; *right*: simplified for AC measurements, restricted to response of working electrode alone

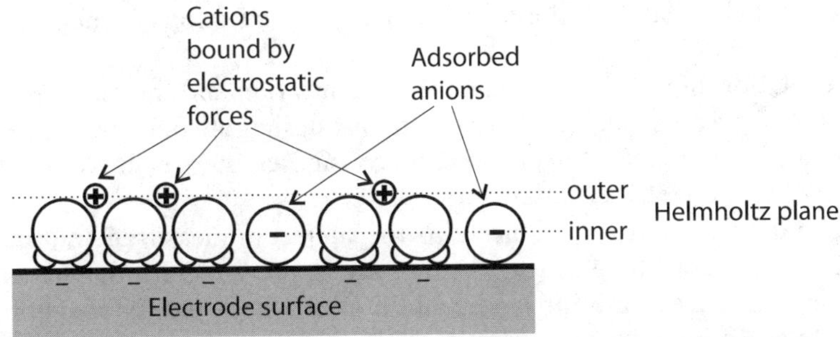

Figure 2.38. Double-layer structure at an electrode surface in aqueous solution

The charge centre of water dipole molecules, or of adsorbed ions, respectively, constitute the plane of the inner Helmholtz layer. The *outer Helmholtz plane* (OHP) is limited by the radii of adsorbed anions or molecules.

The capacity of the double-layer condenser can be determined by AC measurements. Adsorption of foreign particles from solution can change strongly the actual capacity. This can become a source of analytical information. The diffuse part of the double layer tends to disturb electrochemical measurements. By adding a large excess of supporting electrolyte, this distortion can be suppressed.

In the simplified equivalent circuit given above, the symbol Z_f stands for a special complex resistance, the so-called *faradayic impedance*. The latter reflects the behaviour of the electrode surface when current is flowing. The resistance R_L is the homogeneous solution resistance between electrodes. It is reciprocal to the conductance and is of pure *ohmic* (*real*) character, i.e. its actual value does not change if we go from AC to DC measurement. The impedance Z_f, by contrast, depends on frequency as well as on time.

Analytical information can be extracted from R_L as well as from Z_f. The experimental conditions must be designed to ensure that only one of both quantities is measured.

If an *electrolytic conductance* solution is to be measured, the influence of phenomena at the interface electrode/solution must be minimized. Either the electrode surface must be prepared in a proper way or the conductance measurement must be performed in a region far from the electrode surfaces. For preparing the electrode surface, large metallic electrodes with a high degree of roughness are used. The double-layer capacity C_D of such electrodes is very high. Since the capacitive part of AC resistance decreases with capacity and frequency, already by measuring in the kilohertz region, the capacitive resistance approaches zero, so that the double layer acts like a short circuit, and consequently R_L is active alone. Thus the solution resistance can be measured by means of the well-known Wheatstone bridge as with any other ohmic resistance. In the *four-point technique*, which is better suited for small sensors, the conductance is measured between two points located along a distance between the external electrodes. A closer look at technical questions of measurement is given in Chap. 5.

The relationship between total ionic concentration and the measured conductance is given by Eq. (2.19). The technical design of conductance sensors will follow in Chap. 5. Not in all the cases considered there is the conductivity measured an *electrolytic conductivity*.

When the faradayic impedance Z_f is the subject of measurement, the processes occurring at the electrode surface are in the foreground, whereas the solution resistance is minimized by addition of a large excess of supporting electrolyte. An impedance measurement is performed by means of a potentiostat which has been modified by superimposing the reference voltage with a low-amplitude AC voltage. The potentiostat forces the working electrode (WORK in Fig. 2.39) to attain a definite potential difference in relation to the reference electrode (REF). This potential is disturbed periodically by the AC signal. The response of the system is analysed in such a way that the alternating current part is separated by an electronic filter. This AC response can be plotted as a function of exciting frequency. The device described, the *frequency analyser* (Fig. 2.39), also ensures that the signals studied are related to processes at the working electrode alone. The amplitude of the exciting AC voltage is low (in the range of a few millivolts), so that the equilibrium state of the working electrode is not disturbed markedly. In a *frequency spectrum analysis*, the superimposed AC is varied in a broad range, commonly between 10^6 and 10^{-2} Hz, starting with the higher value, ca. 50 to 100 steps.

Extraction of information from faradayic impedance Z_f follows a scheme quite different from that in voltammetry. Z_f is separated into two partitions, namely a *real part* and an *imaginary part*: $Z_f = Z' - jZ''$. In this equation, j is the imaginary unit $j = \sqrt{-1}$. Plotting the imaginary part of impedance, Z'', as a function of the real part Z', the so-called *Nyquist diagram* results (Fig. 2.40).

In the Nyquist diagram, a pure ohmic resistance would appear as an isolated point at the Z' axis of the diagram. A pure capacity C would result in a vertical line at the Z'' axis when the frequency is varied. An RC circuit (a resistor and a capacitor, either in parallel or in series) would give a semicircle with

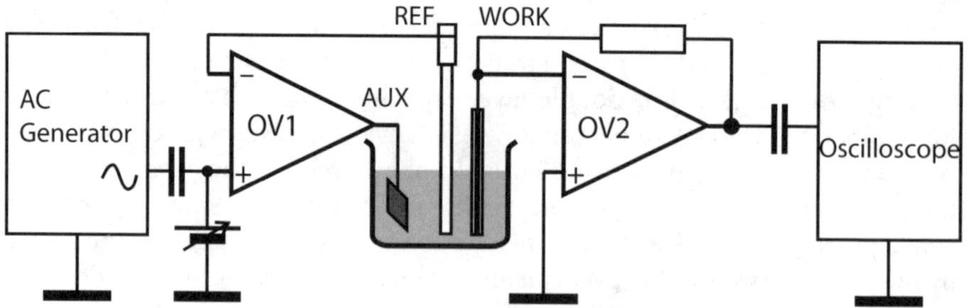

Figure 2.39. Simplified scheme of a frequency analyser

frequency variation. The behaviour of an electrode indeed can be symbolized by an equivalent circuit consisting of an RC circuit with a resistor in series. However, this equivalent circuit would not be sufficient to reflect all properties of an electrode. Hence the equivalent circuit must be extended by subdividing the faradayic impedance into two partitions which are symbolized by two complex resistances Z_T and Z_W. Z_T expresses the behaviour of the partial process *charge transfer*, and Z_W (the *Warburg impedance*) is an expression of the contribution of *transport processes* (mainly *diffusion*). In the majority of cases, Z_T is independent of frequency, hence it can be symbolized by R_T. The resulting equivalence circuit is given in Fig. 2.41.

In the high-frequency range, the effect of kinetic hindrances (symbolized by R_T) dominates. The lower the frequency, the stronger will be the effect of diffusion, denoted by the Warburg impedance Z_W. The behaviour of the latter brings about a straight line with a 45° slope in the Nyquist diagram. Thus the Nyquist diagram of many electrodes appears like that in Fig. 2.42. Such diagrams present a good deal of information; however, interpreting the results is a problem since they are not presented in a descriptive manner.

Modern frequency analysers automatically impose the correct distortion signal $U^\sim$ and record the corresponding response $I^\sim$. For each value, the phase shift θ is determined. The faradayic impedance Z_f is calculated by interpretation of these pieces of information. The analyser automatically plots

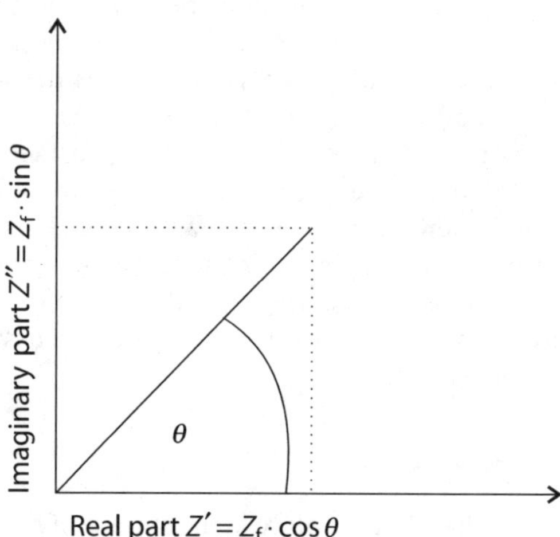

Figure 2.40. Nyquist diagram

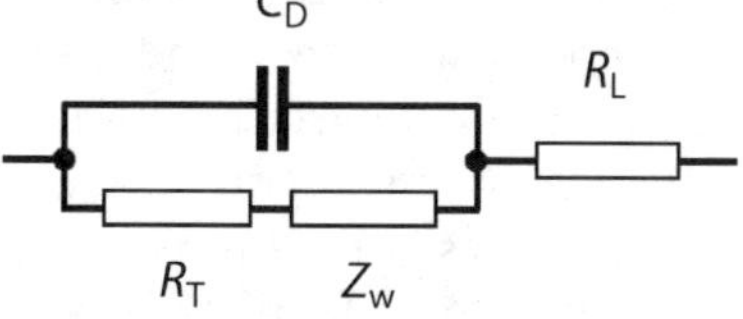

Figure 2.41. Equivalence circuit of an electrode. R_T charge transfer resistance (real), Z_W Warburg impedance (complex), R_L solution resistance (real)

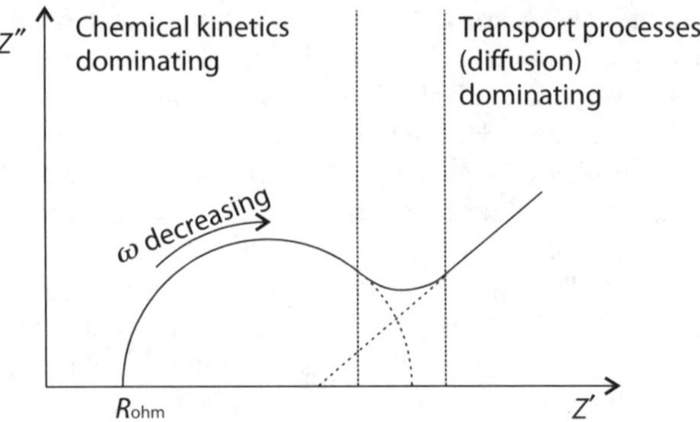

Figure 2.42. Nyquist diagram for typical electrode

the interrelationship of Z' and Z'' in the form of a Nyquist diagram. Modern devices for frequency analysis have reached such a high degree of automation that they are attracting increasing interest for sensors. On the other hand, so-called *impedimetric sensors* are often nothing more than simple *conductometric* devices.

2.2.7
Ion Exchange, Solvent Extraction and Adsorption Equilibria

The partition of molecules between two phases can be based on different sorts of equilibrium. Meaningful are equilibria concerning the processes of ion exchange, partition of substances between immiscible solvents (solvent extraction), and accumulation of substances at solid surfaces (adsorption). In some cases, real chemical bonds are formed, but sometimes only weak forces control the process. These equilibria generally are reversible, and they are mobile, i.e. they tend to react fast to concentration changes. This is a valuable property for sensor applications. Furthermore, they contribute to the accumulation of traces at surfaces, and they are important in manufacturing ordered structures at surfaces. The following discussions are dedicated to equilibria of particular interest.

Ion Exchange

Ion exchange takes place at the surface of *polyelectrolytes*. Some natural minerals are polyelectrolytes, among them the *zeolites*. More common are synthetic organic resins, which contain, at their surface, sites able to trap or release ions. Other sorts of phase boundary with ion exchange are e.g. an oxidized graphite surface or the interface between water and an organic solvent with dissolved *amphiphilic* (*amphipathic*) substances that have accumulated at the interface. In the process called ion exchange, one ion type is released, and another is trapped by loose binding forces.

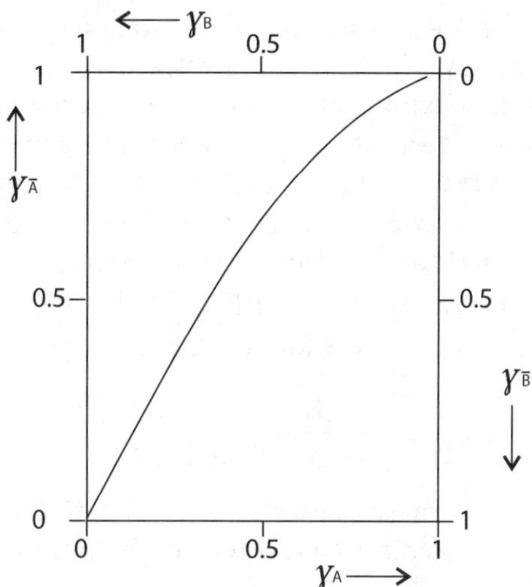

Figure 2.43. Exchange isotherm for ions A and B

Figure 2.44 gives a schematic view of the basic structure of two types of ion exchangers. Both types, the cation exchanger and the anion exchanger, belong to the group of *strong electrolytes*. Ions to be exchanged are bound by pure electrostatic forces. We can consider the anion exchanger as *strongly basic* and the cationic exchanger as a *strongly acidic* resin. The exchange process with such resins is *non-specific*, i.e. the extent of exchange does not depend on the chemical properties of the electrolyte but only on their concentration. If an excess of sodium chloride in solution is brought into contact with an anion exchanger covered with OH ions, then chloride is bound and OH^- is released, so that a solution of NaCl is converted into sodium hydroxide solution. Similar processes will proceed with a cation exchanger when sodium ions are bound and H_3O^+ is released instead. Ion exchange resins are subject to swelling and possess a very large *internal surface* as a result of their very high water content. Their capacity to bind ions is extremely high. This is important for their application in water purification to produce soft water.

Weakly acidic resins commonly contain COOH groups instead of SO_3H groups. In weakly basic resins are found groups like $-NHR-NH_3OH$. In such polyelectrolytes, the interaction with ions is no longer purely electrostatic but partially covalent. Thus ion exchange is more *selective*.

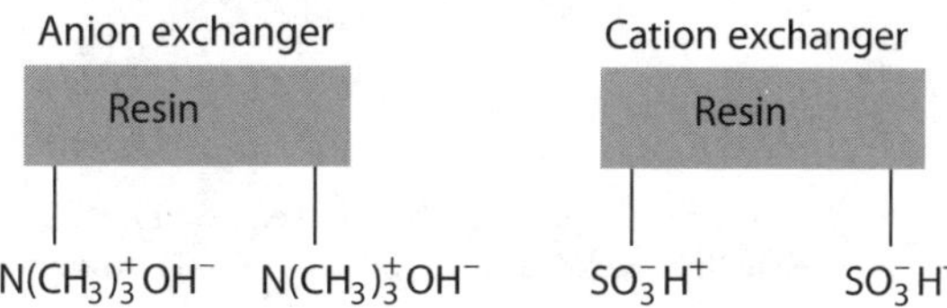

Figure 2.44. Ion-exchange resins with typical surface sites for non-specific ion exchange

Table 2.7 lists examples of ion exchangers, among them also exchange membranes, which are characterized by their permeability for only one type of ion, either cations or anions. The material of such exchanger membranes can be used as an active layer for electrodes in electrochemical sensors. Ion exchange layers are a good basis for further chemical modification.

The exchange equilibrium can be written as follows: $\overline{H^+} + Na^+ \rightleftharpoons H^+ + \overline{Na^+}$ with $\overline{H^+}$ and $\overline{Na^+}$ ions that are fixed at the resin. To characterize the equilibrium, an *exchange constant* $K_{Na^+,H^+}^{\dagger}$ is defined, which is a thermodynamic constant according to the following expression:

$$K_{Na^+,H^+}^{\dagger} = \frac{a(\overline{Na^+}) \cdot a(H^+)}{a(Na^+) \cdot a(\overline{H^+})}. \tag{2.42}$$

An alternative description is given by the *selectivity coefficient* K_{Na^+,H^+}, defined by concentrations rather than activities:

$$K_{Na^+,H^+} = \frac{\gamma(\overline{Na^+}) \cdot \gamma(H^+)}{\gamma(Na^+) \cdot \gamma(\overline{H^+})}. \tag{2.43}$$

In the above equation, a special type of concentration quantity with the symbol γ is applied, since the usual concentration unit mol/L is not useful for a hydrated resin. This concentration quantity is defined by $\gamma_{Na^+} = \dfrac{c(Na^+)}{c(Na^+) + c(H^+)}$ and $\gamma_{\overline{Na^+}} = \dfrac{c(\overline{Na^+})}{c(\overline{Na^+}) + c(\overline{H^+})}$.

A common graphical representation of the equilibrium is the exchange isotherm. An example of the equilibrium $A + \overline{B} \rightleftharpoons \overline{A} + B$ is given in Fig. 2.43.

Table 2.7. Examples of ion exchangers

Inorganic ion exchanger: zeolites (alumosilicates) and apatites		
Organic cation exchanger resins		
Basic organic polymer	Functional groups	
Phenol formaldehyde resin	$-OH$	weakly acidic
	$-COOH$	weakly acidic
	$-SO_3H$	strongly acidic
	$-PO(OH)_2$	weakly acidic
Sulfonated cellulose	$-SO_3H$	strongly acidic
Organic anion exchanger resins		
Amino resin	$-NH_2$	weakly basic
	$-NHR$	weakly basic
	$-NR_2$	strongly basic
Ion exchanger membranes (e.g. NAFION®)		
Selective permeability for ions of only one type		

Solvent Extraction

Solvent extraction equilibrium or *partition equilibria* arise when a substance is partitioning between two immiscible solvents in contact with each other. The corresponding law is the *Nernst partition law*. According to this law of nature, the ratio of concentration (better activity) values in both phases is a constant:

$$K_{\text{Extr}}^{\dagger} = \frac{a^{\text{II}}}{a^{\text{I}}} \, . \tag{2.44}$$

The associated constant is the *partition coefficient*. Numbers I and II in the equation are phase numbers denoting liquid phases in contact. Equation (2.44) is valid approximately also when written with concentration rather than activity values. The *distribution ratio* $D = \dfrac{c_{\text{tot}}^{\text{II}}}{c_{\text{tot}}^{\text{I}}}$ gives an alternative description of the equilibrium using total concentration c_{tot}, where all the existing species are summarized, including products of association and dissociation.

As in other equilibria discussed above, the mutual interdependencies of partition equilibria can be visualized in the form of an *isotherm*. An example is given in Fig. 2.45. The plot is a straight line only if no association or dissociation is connected with a phase transfer of the dissolved species.

Different kinds of partition equilibria are found in chemical sensors. In sensors on the basis of ion-selective electrodes with liquid membranes, specific ligands are dissolved which form complexes with sample ions. In this way, at the sensor surface a concentration-dependent galvani potential difference is formed which can be measured potentiometrically. The principle of *extraction photometry* is applied, in modified manner, in *optical sensors*. Extraction photometry is a classical method of trace analysis, where the sample solution is extracted using a water-immiscible solution of a ligand which is able to

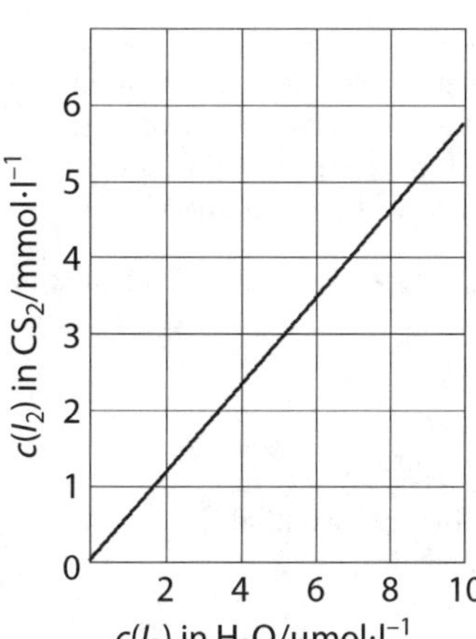

Figure 2.45. Partition isotherm of iodine between water and carbon disulphide

form coloured complexes with the analyte. The organic phase containing the coloured complex is studied photometrically. This measurement scheme can be adapted for optical sensors based on optical fibres.

Adsorption

Generally, the term *adsorption* is used to describe accumulation, i.e. a *concentration rise close to a surface*, or to an interface between neighbouring phases. In principle, the opposite of accumulation, i.e. the *negative adsorption* (*depletion*) of substances at a surface, also exists. However, this case is considered infrequently.

The excess concentration Δc (Fig. 2.46) divided by the surface area A yields a 'two-dimensional concentration' Γ (in mol/cm^2), also called the *surface concentration* or *adsorbed amount* $\Gamma = \frac{\Delta c}{A} = f(T, c)$.

Adsorption equilibria can be classified into two subgroups (Table 2.8). Both subgroups are meaningful for chemical sensors.

A general mathematical description of *physisorption equilibria* is given by *Gibbs' law*, which relates the adsorbed amount Γ to the decrease in surface tension σ if the concentration c of an adsorbable substance in solution increases. In its complete form Eq. (2.45), Gibbs' law displays σ as a function of the *chemical*

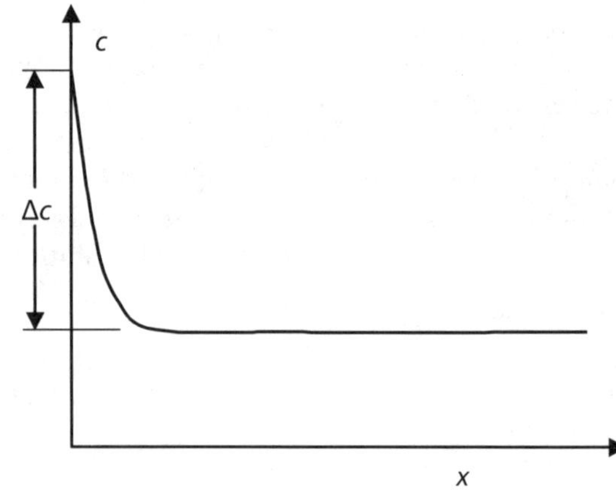

Figure 2.46. Adsorption at an interface. x distance from the interface, Δc concentration rise caused by adsorption

Table 2.8. Two kinds of adsorption equilibria

	Physisorption (capillary condensation; van-der-Waals adsorption)	Chemisorption (spezific adsorption)
Adsorption enthalpy	8–25 kJ/mol	over 40 kJ/mol
Chemical bond	weak, non selective	strong, selective
Equilibration speed	fast	slow

potential μ (this is the free Gibbs energy ΔG of one type of particle) and of the electrode potential given by the potential difference between metal phase ϕ_M and solution phase ϕ_L. The quantity q in Eq. (2.45) denotes the surface charge on the solution side:

$$\mathrm{d}\sigma = -\sum \Gamma \mathrm{d}\mu - q \cdot d(\phi_M - \phi_L) \,. \tag{2.45}$$

The consequences of Gibbs' law are well known to everybody. When surface active agents are dissolved in water, they tend to accumulate at the water/air interface, and this accumulated layer drastically decreases the surface tension. That is, accumulation of foreign substances normally means decreasing surface tension. The second term in Eq. (2.45) means that the extent of adsorption is potential dependent. This is a very important feature for analytical chemistry, since certain substances can be accumulated from solution by application of a defined potential at an electrode surface. The potential dependence of Γ is the basis for an analytical determination procedure for surface active agents. Their accumulation also changes the double-layer capacity, so that the amount of substance in the adsorbed layer can be determined by capacity measurements. Such measurements belong to the tasks of electrochemical impedance spectroscopy (EIS), which is being used more and more with chemical sensors and sensor arrays.

Adsorption isotherms are very important tools for experimental applications. They characterize the dependence $\Gamma = f(c)$ for definite conditions. There are different types of such isotherms. Most important is the *Langmuir isotherm* given by Eq. (2.46). An impression of the graphical representation is given in Fig. 2.47. A linear rise of Γ for low concentration values is followed by consecutive flattening, and finally by approaching a saturation value Γ_∞ in the range of high concentration. This curve size is typical for *monomolecular adsorption layers* (*monolayers*), which are the prevalent type. There exist numerous modifications of the Langmuir isotherm that consider further effects like mutual interaction of adsorbed molecules.

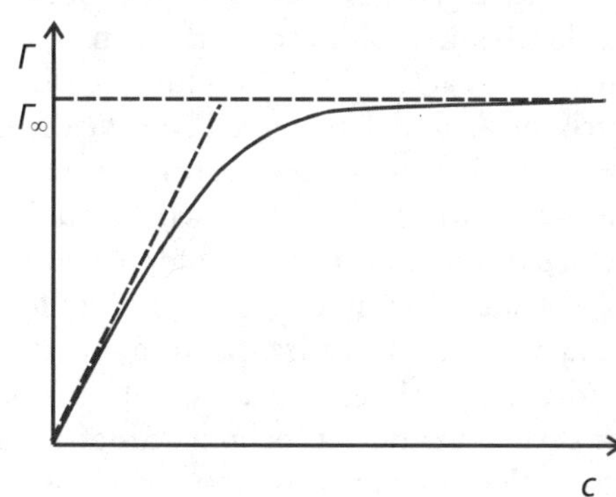

Figure 2.47. Langmuir's adsorption isotherm

Further types of adsorption isotherms are better suited for multiple-layer adsorption:

$$\Gamma = \Gamma_\infty \frac{c}{k + c} \; . \tag{2.46}$$

Adsorption processes play an important role in chemical sensors. In many cases, formation of an adsorbed monolayer is the first step in functionalizing the sensor surface. This is meaningful for electrochemical as well as for optical sensors.

2.2.8
Special Features of Biochemical Reactions

Although biochemical reactions are only a small part of the countless variants of chemical reactions, their special features are unique enough to discuss at least some outstanding aspects. Otherwise, it would be difficult to understand the functionality of modern biosensors.

A characteristic feature of a biochemical reaction is the participation of large organic molecules, polyelectrolytes in most cases. Preferably such molecules belong to the group of proteins, but other molecules also play an important role, e.g. the nucleic acids. These natural molecules are not randomly formed heaps of molecular building blocks but well-structured, highly complex molecules which perform many complicated operations very precisely. Some of these functions are of particular interest for biosensors. They will be discussed briefly in the following sections.

Enzymatic Reactions

Enzymes are biocatalysts with an extremely high selectivity. Their molecules are protein molecules with a molecular mass between 10^4 to 10^5 Da. Enzymes work under mild conditions, i.e. at room temperature or slightly above and at near-neutral pH. Biosensors with enzymes generally contain a layer of enzyme molecules immobilized at the sensor surface. This layer is able to catalyse just one reaction with a definite biologically active substance. The latter is recognized and determined specifically in this way.

The most important feature of enzyme molecules is their *specific three-dimensional configuration* with a molecular cavity including an *active site* which is suitable for a special sort of substrate molecule. At this site, the reaction of the substrate molecule during the formation of a *product* takes place. The enzyme molecule recognizes the substrate sterically, i.e. by following the lock-and-key principle. The active site makes up only a small part of the overall molecular volume. Its primary function is to stabilize the *activated complex*, i.e. the molecular transition state which is formed between enzyme and substrate

(ES) in the course of a reaction. In this way, the activation energy of the process is decreased, as with every other catalyst.

Enzyme-catalysed reactions follow the general scheme

$$E + S \underset{\overleftarrow{k_1}}{\overset{\overrightarrow{k_1}}{\rightleftharpoons}} ES \overset{k_2}{\longrightarrow} E + P ,$$

where E denotes the enzyme, S the substrate, P the products and ES the enzyme-substrate complex. $\overrightarrow{k_1}$, $\overleftarrow{k_1}$ and k_2 denote the reaction rate constants of the participating reactions. As soon as all the active sites of the enzyme molecules present are occupied, *saturation* is established. The reaction rate assumes a constant value if the substrate is present in large excess.

If the backward reaction of E and P is negligible, the rate of ES formation can be written as follows:

$$v = \frac{d(ES)}{dt} = \overrightarrow{k_1} \cdot c(E) \cdot c(S) - \overleftarrow{k_1} \cdot c(ES) - k_2 \cdot c(ES) = 0 . \tag{2.47}$$

An important quantity of enzymatic reactions is the *Michaelis–Menten constant* K_M, which is defined by Eq. (2.48):

$$K_M = \frac{\overleftarrow{k_1} + k_2}{\overrightarrow{k_1}} . \tag{2.48}$$

Since the total concentration of enzyme $c_{tot}(E)$ is equal to the sum of the free enzyme concentration $c(E)$ plus that of the enzyme in the complex $c(ES)$, the following equation holds true:

$$c(ES) = \frac{c_{tot}(E) \cdot c(S)}{K_M + c(S)} . \tag{2.49}$$

The rate of an enzyme-catalysed reaction then follows an important relationship, the *Michaelis–Menten equation*:

$$v = -\frac{dc(S)}{dt} = k_2 \cdot c(ES) = \frac{k_2 \cdot c_{tot}(E) \cdot c(S)}{K_M + c(S)} . \tag{2.50}$$

For the limiting condition, when all active sites are occupied [i.e. $c(S) \gg K_M$], and for high substrate concentration, the maximum reaction rate is given by $v_{max} = k_2 \cdot c_{tot}(E)$. For very low substrate concentration [$c(S) \ll K_M$], we get however

$$v = \alpha \cdot c(S) , \tag{2.51}$$

with the constant $\alpha = \frac{v_{max}}{K_M}$. Equation (2.51) is the basis of the majority of enzymatic biosensors. The reaction rate can be measured e.g. in terms of electrolysis current. It is proportional to the substrate concentration and, hence, to the analyte.

Immunochemical Reactions

Antibodies are produced by living organisms in order to attack disturbing substances (*antigens*), preferably foreign proteins. Antibodies also belong to the group of proteins. They are serum proteins. In the course of antigen-antibody reaction, both antagonists form a stable complex, which prevents the foreign substance from doing damage. The bonding in this complex is reversible to some extent, i.e. the complex can be decomposed by exposure to certain substances. This does not mean, however, that the complex formation is a mobile equilibrium.

Simple antibody molecules are Y-shaped (Fig. 2.48). Mainly they consist of four polypeptide chains, which are held together by covalent bonding and intermolecular forces.

The majority of reactions relevant for biosensors are based on antibodies of the *immunoglobuline G* class. They are large molecules with a molar mass of ca. 15000 Da. Small molecules (the so-called *hapten molecules*) may act as antigens, in particular if they are adsorbed at protein molecules. Such *hapten-protein conjugates* commonly are utilized in *immunosensors* in order to recognize low-molecular compounds.

The reaction of antibody (Ab) and antigen (Ag), generating the complex Ab–Ag, can be expressed by the following rate laws:

$$v_{\mathrm{Assoz}} = k_{\mathrm{Assoz}} \cdot c(\mathrm{Ab}) \cdot c(\mathrm{Ag}) \,, \tag{2.52}$$

$$v_{\mathrm{Dissoz}} = k_{\mathrm{Dissoz}} \cdot c(\mathrm{Ab\text{–}Ag}) \,. \tag{2.53}$$

Combining these equations results in the equilibrium constant

$$K_{\mathrm{Assoz}} = \frac{k_{\mathrm{Assoz}}}{k_{\mathrm{Dissoz}}} = \frac{c(\mathrm{Ab\text{–}Ag})}{c(\mathrm{Ab}) \cdot c(\mathrm{Ag})} \,. \tag{2.54}$$

The association reaction, i.e. the formation of an antibody-antigen complex, is a second-order reaction. The opposite reaction, i.e. dissociation of the complex, is a first-order reaction.

Immunosensors are biosensors that utilize the antibody-antigen reaction. They are commonly manufactured by immobilizing a layer of immunoglobuline G at the sensor surface. The complex is formed by interaction with the antigen in aqueous solution. The concentration of the latter at the surface depends

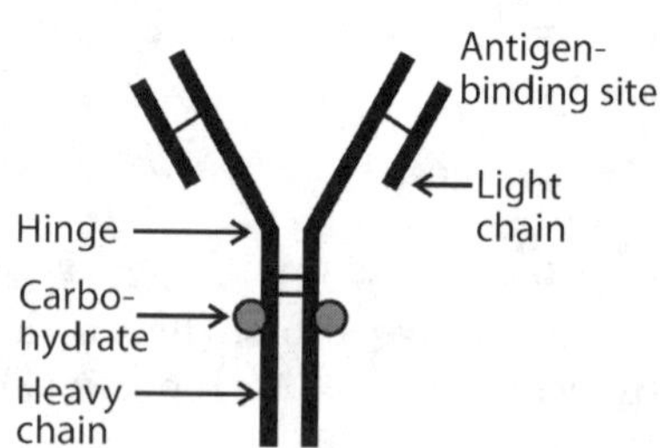

Figure 2.48. Constitution of antibody molecules (simplified)

on concentration of the free antigen, according to the equilibrium constant. The complex formation changes those properties of the surface which can be measured, e.g. the double-layer capacity. The latter can be measured electrochemically. Also, mass change and variation of the sensor's optical properties of the depend on free antigen concentration and can be measured. With optical immunosensors, commonly it is necessary to mark one of the reactants, i.e. to attach an electrochemically or optically active functional group. This could be, e.g. a fluorescing group.

Which measuring procedure with immunosensors is useful depends on the transducer type. Common groups of measuring techniques are the *direct indication* (either Ag or Ab is electrochemically active or luminescing, respectively) and the *competitive indication*. The latter is based on a competition between the sample (not active in indication) and an active (marked) derivate. If for example a sensor surface is prepared with an antigen-antibody layer containing fluorescing antigen molecules, then the sample (an antigen without fluorescence) would displace a certain part of the fluorescence-marked molecules in the layer. The decrease in fluorescence would provide information about sample content.

Living organisms can synthesize a precisely fitting antibody for nearly every existing substance. The antibody can be isolated, purified and immobilized at a sensor surface. In principle, immunosensors are extremely selective. However, in practice distortions may occur as a result of non-selective adsorption.

Reactions with Nucleic Acids

Nucleic acids are biopolymers of prime importance for life. Deoxyribonucleic acid (DNA) is the natural tool for storing and transmitting genetic information. The structure of a DNA molecule is sketched in Fig. 2.49. DNA is usually *double-stranded*. It consists of two *hydrogen-bonded single strands*. A molecule of a single strand contains a 'backbone' built from pentose sugar and phosphate groups. Heterocyclic base groups are attached to this backbone. Four bases exist, namely adenine, guanine, cytosine and thymine. Two single strands combine to give a double strand, the famous *double helix*. The strands are held together by hydrogen bonds between pairs of bases. The pairing is not arbitrary, but only two pairs exist: adenine-thymine and guanine-cytosine. An important consequence of this pairing scheme is that a single-stranded DNA (ss-DNA) molecule can combine only with its precisely fitting complement. This fact can be utilized to recognize individual DNA molecules and, hence, individual organisms. A characteristic part of a single strand is immobilized at a surface and acts as a *probe* searching for its complementary strand. In the course of *hybridization*, the complementary units form the double strand.

DNA sensors utilize specific properties of the molecule. Electrochemical reactions are possible since the guanine base is electrochemically active. It can be oxidized at different electrodes. The determination process commonly starts

Figure 2.49. Simplified representation of DNA structure. A 'backbone' of phosphate and sugar groups carries the *bases* A, T, and G. By combination of complementary bases via *hydrogen bonds*, the *double helix* is formed

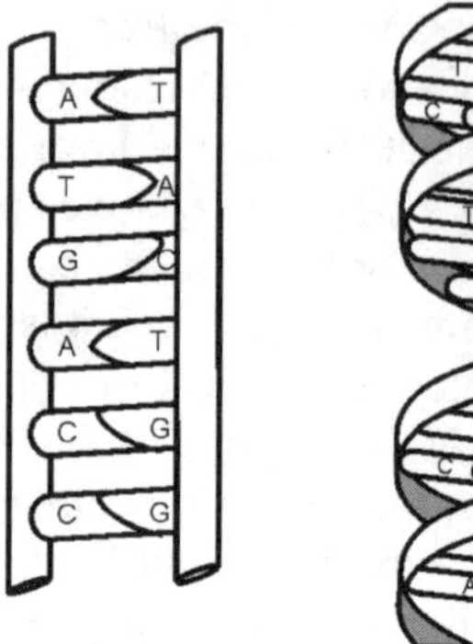

with adsorptive accumulation of DNA followed by anodic oxidation, which yields an analytic signal. Hybridization probes for detection of individual strands need additional substances to indicate completed hybridization.

2.3
Sensor Technology

Chemical sensors are products of quite different fields of science and technology. Consequently, there exist quite different manufacturing techniqes. On the other hand, the field of sensors has formed in the course of the rapid development of microtechnologies, and these technologies have strongly influenced the design of chemical sensors. In this way, construction details have emerged with sensors which cannot be found in any other field of analytical chemistry. It is useful to have a closer look at design details characteristic of chemical sensors.

The common design of a chemical sensor includes an interface in direct contact with the sample, the *receptor*. Very often the receptor represents a thin layer located at the surface of an inert carrier. The next element cannot be described by a simple rule. There are electric contacts, devices for signal processing and many other types of units, depending on the kind of sensor. The transducer, i.e. the most important element for sample recognition, can be manufactured by different means. In *thick-film technology*, different layers are screen-printed on the surface of a carrier. Thin-film technology, the predominant technique in microelectronics, also plays an important role for sensors. Thin layers are generated by vapour deposition, sputtering or chemical vapour deposition (CVD). A combination of thick-film and thin-film technologies can also be found in sensors. Examples of multilayer sensors are polycrystalline semiconductor gas sensors, where the sensitive tin dioxide or titanium dioxide layer is spread by sputtering on the surface of a conductive layer designed to be a heater, consisting either of platinum or of ruthenium dioxide. The latter can also be produced by sputtering. Another example is a gas sensor type where a paste of fine oxide particles is deposited on a ceramic substrate. The paste is fired and sintered in the next manufacturing step. Layers of equal material and in equal order can be fabricated by thin-film as well as by thick-film techniques.

Techniques for generating structures have reached a high degree of perfection and are very important for microelectronics. The efficiency of such techniques can be illustrated by considering modern integrated electronic circuits where millions of transistors are located on a single silicon chip. Microelectronic structures can be two or three dimensional. They strongly vary with respect to resolution. With thick films, only rough structures can be made. With thin films on silicon wafers, extreme resolution is attainable and complex three-dimensional structures can be realized. The latter are fundamental for new technological fields like micromechanics and microfluidics. Chemical sensors can be integrated into such devices. As a result, complete miniature laboratories (*lab-on-the chip*) can be assembled. Highly complex instruments like liquid chromatographs also are becoming available in miniaturized form.

2.3.1
Thick-Film Technology

Thick-film technology is fully developed so that in recent years no considerable new contributions have been made. Thick films are made almost entirely by screen printing. In this technology, a fine sieve is pressed tightly onto the substrate, which must be covered by a print. The sieve is prepared prior to its application so that some part of its area is covered by impermeable regions forming a structure. Commonly, stainless steel sieves with a mesh size of 50 to 200 μm are used. The material to be spread onto the substrate normally is a fine powder which has been worked up with binders, solvents etc. to form a paste (the so-called *ink*). The ink is applied on the upper side of the sieve in the form of a string. In the next step, the ink is spread over the sieve by means of a rubber squeegee. When the sieve is lifted, a structured layer showing the contours of the sieve pattern remains on the substrate. This layer is finished either by drying or by an additional burning process after drying. In the burning process, binders are ashed and mineral particles are sintered. The processes are illustrated in Fig. 2.50.

Ceramic materials are the preferred basis for thick films produced with higher temepratures. Among them are substances like aluminium oxide, glass and quartz. For lower-temperature manufacturing, the assortment of substrate materials is more versatile. Plastic foils as well as specially prepared paper or cardboard are widespread materials.

The pattern on the screen is made mainly by photolithography. The screen is covered by a layer of light-sensitive paint illuminated through a negative (or positive, depending on the actual procedure) which contains the pattern to be generated. After removing the paint at the places which should be left open (the *developing process*), the sieve is ready for use.

Screen-printing inks are commercially available in large variety. Inks for generating metallic leads contain noble metal powders, mainly of gold, platinum and palladium. Carbon inks contain glassy carbon particles, often spherical in form and of a defined particle size range. Ruthenium dioxide is a common material for inks which are used to print electric resistors. Inks designed to be sintered on a ceramic substrate commonly contain glass powder as a binder additive.

The resolution of thick-film structures is restricted. Three-dimensianal arrangements have been achieved only in a later stage. They are produced by preparing thin flexible foils made of a plastic membrane with embedded particles of starting material for ceramics. Holes are punched through the membranes with high precision in such a way that after stacking foils to form a package, the holes (so-called *vias*) are arranged on top of each other. The resulting continuous hole is filled with a paste containing noble metal particles. In this way, a 'vertical' electric connection between different 'levels' can be achieved. The technology has been developed to construct complex electronic circuitry with printed components like resistors, condensors, coils etc. In this way sensors can be equipped with an electronic adaptation circuit, e.g. a preamplifier. A highly sophisticated variant of this stacking technology is the *low temperature cofired ceramics* (LTCC) technique. In LTCC, after stacking, the foils are fired and converted into thin ceramic plates.

For thick-film technology, a broad range of materials is available. Costs for materials and production are comparatively low. Small series can be produced with reasonable effort on the laboratory scale. On the other hand, production of a large number of pieces also is not a problem since automatic screen-printing machines are available. Disadvantages are the low resolution and the high surface roughness of the screen printed layers, in particular after thermal

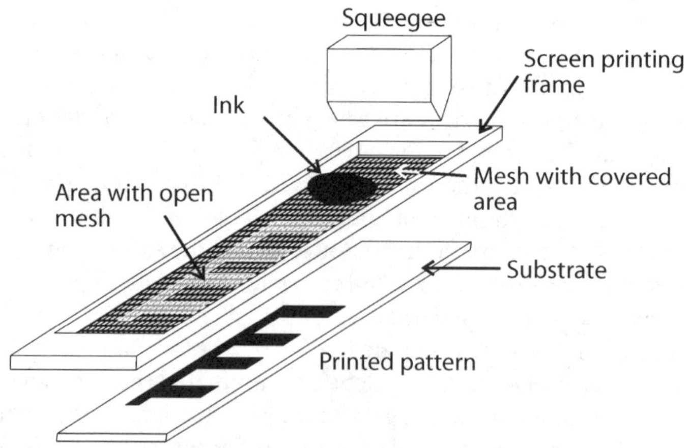

Figure 2.50. Manufacturing of thick-film structures by screen printing

treatment. This is a problem mainly for electrochemical sensors when the active electrode area has been made by screen printing.

2.3.2
Thin-Film Technology and Patterning Procedures

Thin films are of interest here only in connection with the techniques that evolved from microelectronics. The preferred material of such techniques is highly purified crystalline silicon. This material is not only a universally applicable semiconductor but is also characterized by very useful mechanical properties. This is the reason why silicon is preferred also in micromechanics and microfluidics. Even porous layers can be manufactured.

Semiconductor components commonly are fabricated in the course of a precisely organized sequence of a few technical unit operations. Major steps are covering by metallic layers, making etching masks by photolithography, etching processes, and attaching of insulating layers. Insulating layers can be produced analogously to metallic layers, or they are formed by oxidation in the oxygen-containing gas phase. Overall, the unit operations mentioned ensure a high resolution.

Thin films of inorganic materials on a substrate are deposited by *high vacuum deposition*, by *sputter deposition* or by chemical procedures like CVD or MOD.

Vapour deposition and sputter deposition are similar procedures. They utilize different methods for generating the metallic particles in the gas phase. Vapour-deposition procedures are thermal processes where the material is heated electrically or by electron bombardment to raise the vapour pressure to an appropriate value. Such procedures work under high vacuum. Sputter deposition (Fig. 2.51) operates by means of plasma, mainly generated in an argon atmosphere. Between a heated cathode and an anode, a high voltage of 2 to 5kV is applied. The resulting glow discharge generates argon ions which are accelerated by the electric field and shot against the *sputtering target* on the cathode where they knock out a few atoms at a time. Atomic clusters generated inthis way tend to deposit on all surfaces in the chamber. The geometric arrangement ensures that the substrate is deposited sufficiently.

Film deposition can be enhanced by addition of small amounts of reactive gases like oxygen, nitrogen or hydrogen (reactive sputtering). Nitride layers, e.g. tantalum nitride, are produced by sputtering a metallic target in a nitrogen-argon atmosphere.

Sputtering is able to produce layers of elements or alloys which are not volatile. The deposited films have the same composition as the target material. If films of insulating materials must be deposited, DC sputtering will not work. It can be performed, however, by RF sputtering, where a high-frequency field is applied between the poles. The equipment shown schematically in Fig. 2.51

Figure 2.51. Equipment for sputter deposition

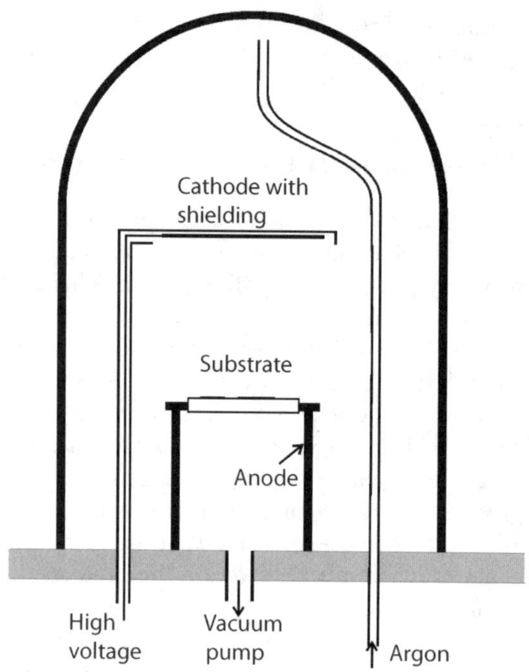

is completed then with coils generating a concentric magnetic field. A further improvement is magnetron sputtering. A magnetic field in parallel to the target surface causes a strong intensification of the gas discharge. In this way, even films of silicon dioxide can be generated.

By *chemical vapour deposition* (CVD), volatile compounds of the material to be deposited are evaporated. The vapour is decomposed thermally, sometimes supported by gas plasma. At the substrate surface, the products react with components of the gas phase, generating a non-volatile film. The process runs at a relatively low temperature. The stoichiometric composition of the resulting amorphous films can be controlled widely.

The *metal-organic deposition* (MOD) *process* needs neither vacuum nor sputtering equipment. It is based on easily decomposed metal-organic compounds. Such compounds consist of a central atom with ligands that are bound by *coordinative* bonds (ligands possess a free electron pair which coordinates with a corresponding gap of the central atom). The compounds are slightly soluble in certain non-aqueous solvents. The resulting solutions are called *inks*, as with screen-printing procedures. The ink is spin-coated onto the substrate in a centrifuge or similar device. In the next step, the film is decomposed thermally. A large variety of films can be produced in this way, e.g. zirconium dioxide ZrO_2 for electrochemical oxygen sensors (*lambda probe*), barium titanate $BaTiO_3$ for pyroelectric gas sensors and tin dioxide SnO_2 for so-called *Taguchi sensors*.

Generally, films must be patterned to get useful devices. The simplest case is conducting structures which are nested like two combs. Such *interdigitated* structures are useful to measure conductance or to perform electrochemical reactions. Microelectronics makes use of much more complex structures, which meanwhile are utilized also in highly efficient modern electrochemical sensors. A highly complex example is discussed later in connection with *ion-selective field effect transistors* (ISFET) in Chap. 7.

Thin-film patterns can be generated in the course of film deposition, when, e.g. the sputtered film is deposited through a mask lying on the substrate. More common is structuring by removing parts of the even film. With masks, only rough structures are available. The common procedure consists in removing parts of a photosensitive polymer layer by illumination and development, and etching the underlying film. Alternatively, the *lift-off technique* has become popular in recent years. With this technique, the polymer pattern is generated prior to film deposition. Next, the substrate plus polymer pattern is covered by the thin film. Finally, the polymer layer is removed, thereby also removing the thin film at the appropriate places.

2.3.3
Surface Modification and Ordered Monolayers

Surface Modification

According to the definition of chemical sensors (Chap. 1), they have an interface in direct contact with the sample. This interface is the preferred place where the critical interaction with the sample takes place. This interaction is connected with the functions of *recognition (receptor function)* and *signal transduction.*

Films on a substrate, either thick or thin films, cannot act generally as receptor or transducer immediately. An exception to this rule exists with layers of semiconducting metal oxide films like SnO_2, which can interact directly with reducing gases. It is much more common to *functionalize* sensor surfaces prior to use and to use metallic or ceramic layers as a support only for the sensitive film. Functionalizing is the role of *surface modification.* The latter has become a broad special field, important in particular for electrochemical sensors.

Surface Cleaning. The first step in surface modification is generally cleaning. Prior to cleaning, every surface is dirty, i.e. at the very least it is covered by adsorptive layers. But sometimes native adsorption layers are useful or can fulfil a receptor function, as with the oxide film on a platinum surface. In the majority of cases, adsorbed layers must be removed to establish a definite initial state for further modification.

Sensor surfaces can consist of most materials. Metals, oxide ceramics and carbon in its different modifications including diamond, silicates and organic polymers are the most important groups. Consequently, the cleaning operations are also extremely multifaceted. In what follows, a small selection for cleaning solid surfaces is presented.

- Chemical treatment:
  - Treatment with oxidizing solutions for metallic surfaces (preferably gold). Example: *'piranha solution'*, a mixture of sulfuric acid with hydrogen peroxide (35%) in the ratio 3:1.
  - Strongly alkaline and surface active solution for non-metallic inorganic surfaces (silicates, boron-doped diamond). Example: 40% ethanolic potassium hydroxide solution.
- Electrochemical treatment of metallic surfaces. Example: Oxidation of carbon surfaces by imposition of a highly positive potential; *electrochemical cycling* of gold surfaces (the potential is switched periodically between the values for oxygen and for hydrogen evolution in acid solution).
- Thermal treatment under vacuum or in inert gas atmosphere.
- Application of gas plasma, e.g. oxidizing oxygen plasma or argon plasma to enhance desorption of loosely bound substances, sometimes ablation of a surface layer of the base material.

Some of the substances listed above are highly aggressive cleaning solutions. The piranha solution mentioned will destroy the exposed metal if it acts upon the surface for too long a time. Organic parts, insulating materials and materials used to encapsulate details are subject to attack of this solution. Ethanolic potassium hydroxide solution acts similarly. It mainly converts organic substances into surface active compounds, but to some extent even glass is attacked and partially dissolved. Nevertheless, this aggressive solution has been applied successfully to clean boron-doped diamond (BOD).

Immobilizing Functional Groups. Sensors should interact *selectively* with groups of substances, or even *specifically* with a single molecule type. Selective interaction is achieved generally by the appropriate groups which must be *immobilized* at a surface. Different immobilization procedures are in use. Some of them are not generally considered to be surface-modification procedures, since they have been developed in another context. Such a procedure is *adsorptive stripping*, a common method in the field of electrochemical trace analysis. An excess of ligands is added to a solution containing metal traces to be determined. The ligands form an adsorptive monolayer at the electrode if the proper potential is applied. For a certain time we have a modified surface. This surface can bind selectively the interesting metal cations which are accumulated in this way, supported by stirring. When sufficient metal ions have been collected, they are removed (*stripped*) by electrochemical reaction (oxidation or reduction). In the course of stripping, a measurable signal is achieved. In this procedure, the adsorbed monolayer is designed primarily to catch metallic traces for consecutive analytical determination.

A common objective is to stabilize functional surfaces. The ideal would be to have a surface which can be used repeatedly for an indefinite length of time without regeneration or new preparation. Immobilized molecules should fulfil their function without damage and should be accessible for the sample. On the other hand, they should not become subject to leakage or bleeding. Current immobilization methods are preferentially dedicated to large organic molecules. This is a consequence of the strong interest in biosensors.

The most popular immobilization methods are

- Adsorption (mainly physisorption)
- Covalent attachment at a surface
- Inclusion in polymer layers, gels and conductive pastes.

A very elegant method of attaching functional groups is linkage to *ordered layers*. This is a special case of covalent attachment mentioned in the above list.

Large protein molecules, among them many enzymes, frequently can be attached to surfaces simply by non-selective adsorption. This works best with carbon surfaces that had been oxidized slightly. *Specific adsorption* (*chemisorption*) in monolayers normally is based on a covalent chemical bond.

Fixation of molecules via covalent chemical bonding generally yields robust functional surfaces. One way to direct bonding molecules is the interaction of their π-electron system with solid surfaces. This interaction occurs especially with carbon surfaces since the graphite lattice also contains aromatic π-systems. Dyes which act as *mediators* in oxidation processes of important biologically active substances can be attached to carbon surfaces safely in this way. The dye molecules are attached at one end with an aromatic substituent that is able to interact with the aromatic system of the graphite layer structure. Examples of such mediator dyes are given in Fig. 2.52 (Persson and Gorton 1990).

Dedicated synthetic attachment of substituents yields bifunctional molecules placed at one end of the linking group; at the other end is the group responsible for sensing. The art of synthesizing such molecules consists in establishing a strong bond as well as in arranging the molecule in the steric optimum, so that after immobilization the functional group can be active. Examples of bifunctional molecules are presented in Fig. 2.53. In these examples, molecules have the function of enhancing the electrochemical reaction of the important biomolecule cytochrome C (Allen et al. 1984).

If large organic molecules (proteins etc.) must be immobilized, the primary problem is to synthesize the molecular *linker* responsible for ensuring a stable bond with the sensor surface. Common linkers are e.g. groups such as the following:

$-NH-(CH_2)_n-NH-$

$-O-CO-(CH_2)_n-CH=N-$

$-O-(CH_2)_n-NH-$

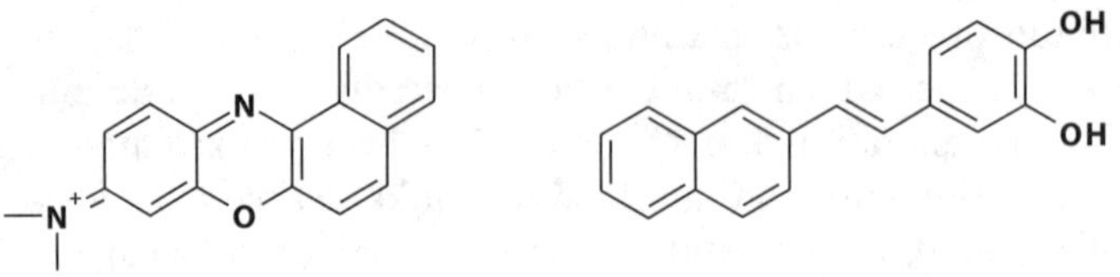

Figure 2.52. Dyes which can be bound to carbon surfaces by chemisorption via π–π interaction. *Left*: Meldola's blue, *right*: 4-[2-(2-naphthoyl)-vinyl]catechol

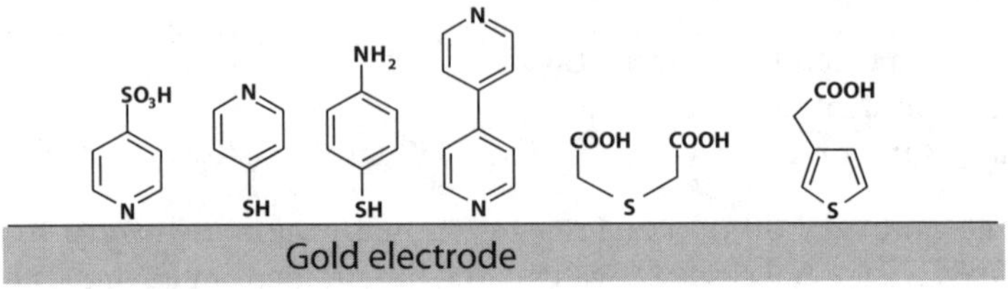

Figure 2.53. Bifunctional molecules bound to a gold surface

Obviously, an alkyl chain $(CH_2)_n$ is always located between the end groups. The reason is that a certain distance between immobilized molecule and solid surface must be provided. This is of particular interest for bulky enzyme molecules which should not lose their specific biocatalytic activity by immobilization.

The groups providing the bond with the solid surface vary depending on the kind of surface. For a gold surface, thiol groups are characteristic. These groups are highly efficient and easy to handle. The successful combination of gold with thiol linker groups resulted in the establishment of a special class of modified surfaces which contain a highly ordered monolayer. Such surfaces, the so-called *self-assembled monolayers* (SAMs) are discussed in the next chapter.

Not only metals are subject to surface modification. Silicate surfaces (e.g. front ends of optical fibres) can also be functionalized in a similar manner. The glassy surface is first silanized, i.e. the OH groups on the surface are allowed to react with dimethyldichlorosilane, so that $-Si(CH_3)_2Cl$ groups form on the surface. The latter can be linked with the functional group in the next step, e.g. by reaction with alcohols. After this treatment, the surface is covered by alkyl groups linked to silicon via an oxygen atom. Alternatively, for modification of glassy surfaces, RF plasma polymerization has been proposed. An RF discharge in a vapour atmosphere generates a reactive plasma which modifies the surface by a chemical reaction. In this way, perfluorinated polymers have been attached to optical fibres. Optical sensors based on such fibres can absorb volatile organic substances from the gas phase. This results in a measurable change in optical properties.

Polymers are giant molecules forming solid amorphous layers of disordered networks. In such layers, certain substances are soluble as in liquids.

Alternatively, polymers like PVC can be mixed with highly viscous liquids. The latter also can act as solvents for certain molecules that can act as receptors. The so-called *softeners* in classical polymers frequently have useful solvent properties. For preparing a selective sensor membrane, the polymer containing a softener is dissolved in a volatile solvent like dioxane, the active sensor substance is added, and then the solution is poured onto a flat surface. After evaporation of the dioxane, a thin membrane remains. In ion-selective electrodes, such membranes are used. They contain certain ligands (e.g. *crown ethers* or other *neutral carriers*) which are able to react selectively with cations to form complex compounds.

Polycarbonate, polyurethane and PVC are preferred for fabrication of thin polymer layers.

Most successful receptor-molecule carriers are conducting polymers. Among them are substances with metallic conductivity. These are e.g. polypyrrole, polyaniline and polythiophenes. Their molecules contain multiple conjugate double bonds which are the reason for electron mobility in the molecule. A different type of conductance exists in *redox polymers*, where redox centres are inserted into the polymer. Charge carriers can be exchanged between such centres.

A very elegant way to synthesize conducting polymers is electropolymerization, although this method is restricted to electrochemical sensors. The electrode is positioned in a monomer solution. The monomer is reduced or oxidized. Reactive radicals are formed which cause coupling reactions, creating finally a dense, adhesive film at the electrode surface. The molecular groups owning the receptor function can be attached to the monomer prior to polymerization. Polymerization then generates highly stable functional layers. A typical example is formation of polypyrrole layers. The simplified reaction scheme is given in Fig. 2.54.

Layers of conducting polymers promote electron transfer. This is useful if the critical receptor process is a redox reaction. Such layers are utilized preferably in electrochemical biosensors with enzymes which catalyse biochemical redox reactions. Conducting polymers possess the advantages of classical polymers (solvent function and compatibility with organic substances) as well as of semiconductors or metals (conductance). Examples are discussed in the chapters dealing with biosensors.

Gels are *colloidal solutions* with the appearance of a solid or semisolid aggregation state. Among them, *hydrophilic gels* (*hydrogels*) are of particular importance. They preferentially consist of giant molecules, mostly proteins, covered by an aqueous layer. They form a three-dimensional disordered molecular network which contains a considerable amount of solvent. The molecules

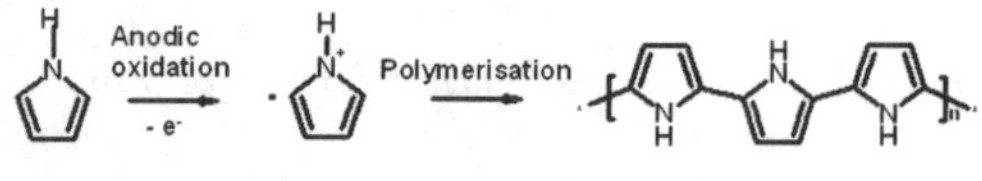

Figure 2.54. Mechanism of electropolymerization when polypyrrole is formed

to be immobilized are solved mainly in the solvent included in the gel matrix. Synthetic gels like polyacrylamide and inorganic hydrogels (silicagel) have been proposed. Gel layers are prepared easily. Normally, dipping the sensor body into heated (liquefied) colloidal solution is sufficient. Supporting gauze or other material is used frequently. Synthetic gels are made by mixing the initial substances (e.g. the monomer) with the functional substance plus auxiliary components. The mixture is painted on the sensor body and allowed to solidify by polymerization.

In addition to organic polymers, inorganic polymeric lattices are also used for coating sensor surfaces. An example is given by ferri/ferrocyanides and their analogues where iron is substituted by other elements like nickel. The complex formed must contain two ions of the same element but with two different oxidation numbers, e.g. $Fe^{3+}[Fe^{2+}(CN)_6]$, the so-called *insoluble Prussian blue*. Such compounds can be prepared in the form of thin layers on a surface. The layers are constituted by a cubic base unit (Fig. 2.55). Counterions securing charge neutrality are positioned in the centre of the cubic elements. Their mobility is restricted since they cannot leave their cages for steric reasons. Contact with a solution containing the counterions selectively brings about a potential difference at the interface.

Layers of inorganic redox couples act similarly to redox polymers; however, they have a higher thermal stability. They are useful only for electrochemical sensors.

Functional molecules can also be immobilized by preparing a paste containing these molecules together with binders and conductive solid particles. The paste is used to coat a solid surface. Carbon pastes prepared in this way are frequently applied in electrochemical sensors. A special variant is the use of a conductive organic salt substituting carbon particles as well as the binder. Sensors with pastes are well suited for laboratory use, in particular if tests must be conducted. When used in commercial applications, commonly the paste is substituted by a screen-printed mixture of similar composition. Screen-printing inks generally also contain binders, conductive particles and dissolved functional substances.

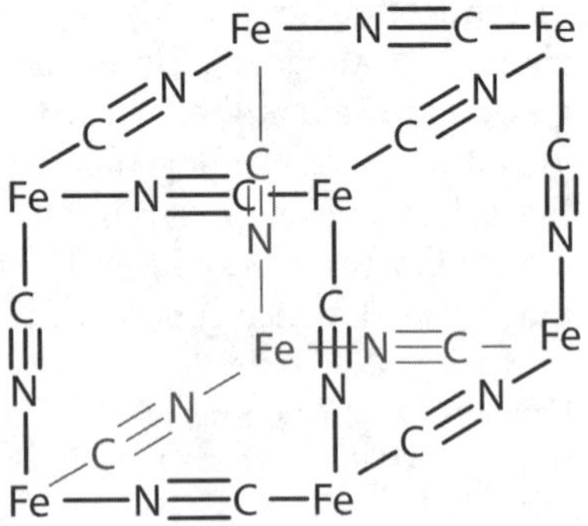

Figure 2.55. Structure of layers formed of *insoluble prussian blue*

Self-Assembled Monolayers (SAMs)

As mentioned above, with bifunctional molecules an efficient surface modification can be achieved. A further development of this idea resulted in *self-assembled monolayers*, abbreviated SAMs (Ulman 1996). They are characterized by a high degree of order. Between their terminating groups, alkyl chains of a certain length must be located. Between neighbouring adsorbed molecules, weak van der Waals forces are active. The result is a densely packed layer of molecules standing steeply at the surface. The molecules form some kind of a brush with only a little space between its 'bristles'. Even for small molecules it is impossible to permeate such a monomolecular film. SAMs are a highly efficient tool to design surfaces in a versatile way. An impression of a SAM is given in Fig. 2.56. The molecules in this example are alkylthiols attached to a gold surface. This combination is the most common one amongst SAMs. Further examples (Fig. 2.57) are long alkyl chains with alkyltrichlorsilane groups at the linking end, which are suited for binding at glass or quartz surfaces, and chains with carboxyle groups for combination with metal oxides or sulfides.

SAMs should be ordered perfectly and 'dense'. These requirements cannot be fulfilled always completely. Frequently, pinholes with molecular dimensions exist. These open a path for small molecules which might interact non-selectively with the solid surface.

Interfaces functionalized by SAMs can be useful for electron transfer in the direct oxidation of organic substances, and they may also act as ion-

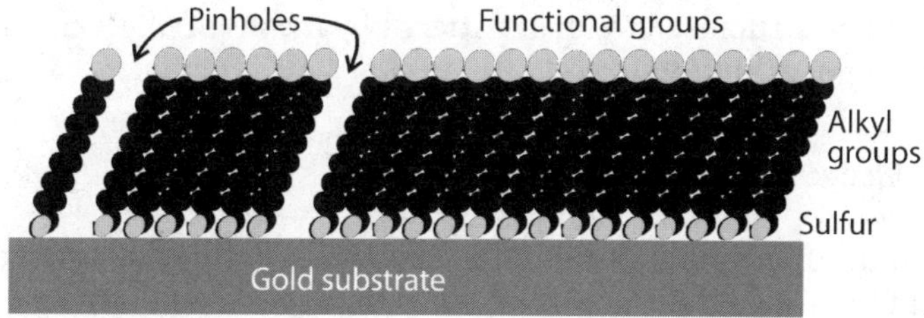

Figure 2.56. Structure of a SAM of alkylthiole molecules at a gold surface

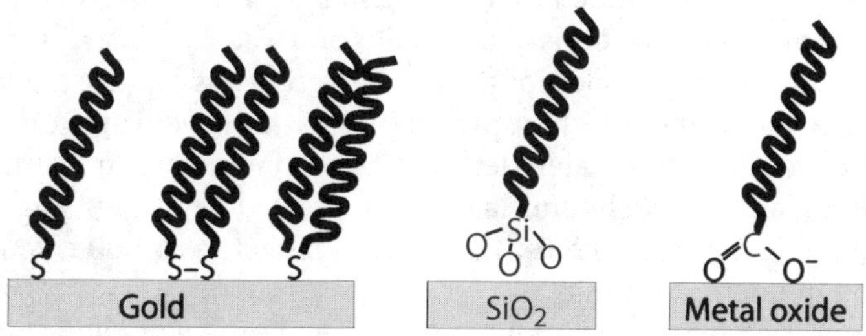

Figure 2.57. SAMs at gold, glassy and metal oxide surfaces, respectively

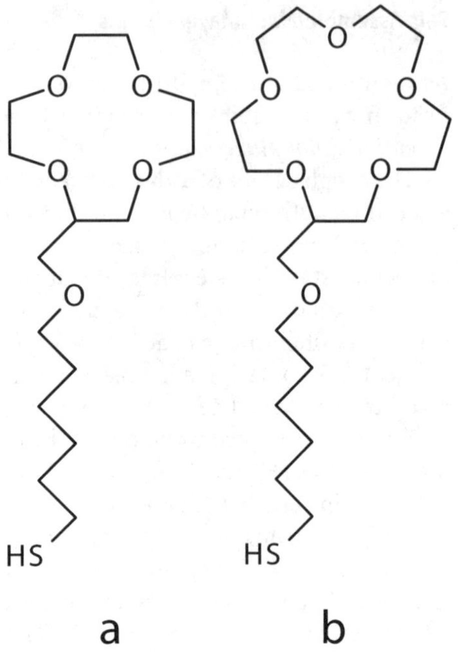

Figure 2.58. Molecules as functional groups for ion-selective SAMs bound via SH groups to a gold surface. **a** 12-crown-4 (selective for Na^+). **b** 15-crown-5 (selective for K^+)

selective surfaces. Such surfaces are formed e.g. when crown ether molecules are bound to a gold surface via chain molecules with an SH group acting as a tether (Fig. 2.58). The crown ethers coordinate selectively with certain metallic cations (Flink et al. 1998).

SAMs can be structured to form patterns by means of photochemical processes (Wollman et al. 1993).

Langmuir–Blodgett Layers (LBLs)

The Langmuir–Blodgett technique for generation of highly ordered monomolecular films on solid surfaces is based on the experience that thin molecular layers on a water surface can form spontaneously and that such layers possibly can be transferred to solid surfaces by dipping objects into the corresponding water volume. The technique has a long history. Agnes Pockels at the end of 19th century observed that films of fatty acids on a water surface in a tray can be compressed by shifting a slide along the surface. She was capable of measuring mechanically the changes in film properties that occur during compression. Lord Rayleigh and others later detected the monomolecular nature of the films. Manipulation of such films later was perfected by Irving Langmuir and Catherine Blodgett. In the 1970s, Langmuir–Blodgett techniques found their application in microelectronics.

A precondition for the technique is the existence of *amphiphilic* molecules, consisting commonly of a hydrophilic 'head' and a long, hydrophobic 'tail'. The classic example are *fatty acids* with the hydrophilic carboxyl group $-COOH$ as a head. Among them are e.g. *palmitic acid* $H_3C-(CH_2)_{14}-COOH$ and *stearic acid* $H_3C-(CH_2)_{16}-COOH$. Molecules of this kind spontaneously arrange on a water surface in such a way that the hydrophilic head dips into water and the hydrophobic tail is oriented towards the air, outside the water as far as possible. The degree of hydrophilicity should not be so strong that the compound is soluble. In this case, micelles could form where the hydrophobic ends agglomerate forming molecular clusters with hydrophilic ends outside. A choice of hydrophilic head groups is given in Table 2.9. The relative strength of the groups is indicated if they are bound C_{16} alkyl chain (Gaines 1966).

The films can behave on the liquid surface like two-dimensional analogues of the three states of aggregation *gaseous*, *liquid* or *solid*. The film is 'gaseous' if its molecules are far from each other and can move without mutual interaction. Analogous to the liquid state are *expanded films*. Their molecules can shift easily relative to each other and show certain viscosity. *Condensed films* are the counterpart to the solid state. In such films, the molecules are in contact with each other. Condensed film cannot be compressed further. By measuring the resistance of the surface against compression, the actual state of the film can be detected. Such measurements are done by means of the *Langmuir balance*. Plotting the measuring results vs. the molecular area yields a characteristic curve from which the desired information can be read. Further compression above the state of condensed film destroys the closed film. Molecules are then stacked in a three-dimensional manner.

Fatty acids and other constituents of SAMs normally are too soft and their melting point too low to allow practical use of their films. The films must be stabilized somehow. One way is to form SAMs of molecular groups which can be polymerized instead of linear alkyl chains. Once such a film has formed, it is cross-linked, yielding rigid, thermally stable monolayers.

Fabrication of Langmuir–Blodgett films follows the procedure outlined in Fig. 2.59. After adding the material to the solvent, the molecules are in two-dimensional '*gaseous*' or *expanded state* (Fig. 2.59, step 1). The film is pressed by moving the slide in the Langmuir trough until a pressure meter indicates that the state of a *condensed film* is attained (Fig. 2.59, step 2).

Table 2.9. Hydrophilic character of head groups at a C_{16} alkyl group

Very weak (no film formation)	Weak (unstable film)	Strong (stable film)	Very strong (molecule soluble)
$-CH_2I$	$-CH_2OCH_3$	$-CH_2OH$	$-SO_3H$
$-CH_2Br$	$-C_6H_4OCH_3$	$-COOH$	$-OSO_3H$
$-CH_2Cl$	$-COOCH_3$	$-CH_2COCH_3$	$-C_6H_4SO_4H$
$-NO_3$		$-NHCOCH_3$	$-NR_3^{\oplus}$

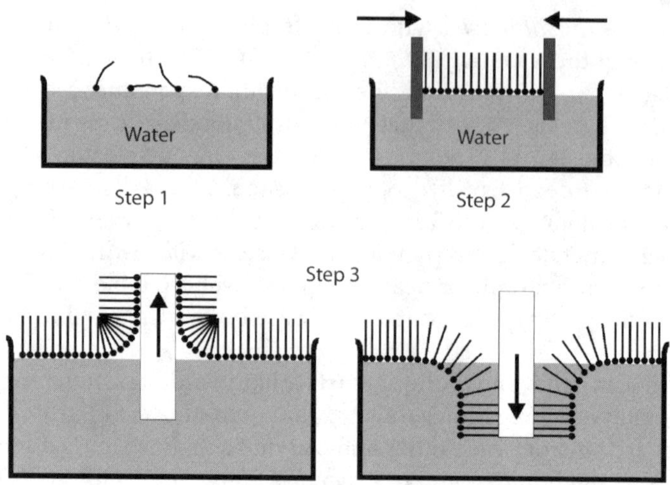

Figure 2.59. Procedures to create Langmuir–Blodgett films

Next, the body to be coated is dipped by a mechanical device slowly into the solution, from either the top or bottom, depending on which side of the molecules one would like to be facing the solid surface (Fig. 2.59, step 3). Of course, the character of the solid material must be considered, since it is important whether the surface is hydrophilic or hydrophobic. The solid surface must be cleaned carefully prior to film formation. The coating process is tedious; generally its speed ranges between a few millimetres to several centimetres per minute. Multiple layers are produced generally by repeated dipping. As a rule, the monolayers are then oriented either head-to-head or tail-to-tail.

Langmuir–Blodgett layers have a structure similar to that of natural cell membranes. They thus play an important role in biosensors. Enzymes and antibodies can be attached in this way to ISFET gates or to *surface acoustic wave (SAW) sensors.*

2.3.4
Microsystems Technology

Patterning techniques for semiconductors discussed in Sect. 2.3.2 have developed continuously with high speed since the invention of the transistor. An important milestone was the invention of *integrated circuits* (ICs). They are fabricated in bulk production style using large silicon wafers. Each technological step is executed simultaneously with thousands of single components, eventually resulting in highly complex electronic circuits of extreme efficiency. Mass production of integrated electronics requires high investment costs and can be done only by large companies. Reasonable prices are achieved since fully automated mass production is the standard in this industry. Nowadays, integrated circuits are widely used even in numerous everyday appliances.

The development of ICs went from single transistors over integrated analogue operational amplifiers up to digital circuits. An outstanding achievement is the microprocessor with connected electronic memory circuits. Engineers in this way became familiar with the properties of the preferred material, the highly purified single-crystal silicon. It became obvious that the established technological processes could also be used to solve problems in fields outside of electronics. In quick succession, new technological fields appeared, such as *micromechanics, microfluidics* and *micro-optics* (*integrated optics*). It appeared that highly purified silicon possesses very useful mechanical properties, allowing one to make pumps and other things of this material. On the other hand, the limits of ICs on a silicon chip showed up. The requirements of electronics frequently collided with the opposite requirements of optics and mechanics, so that integration on one and the same chip (*monolithic integration*) proved impossible. Problems of this kind must be solved by *hybrid technologies*, i.e. the device must be composed of complex subunits which have been manufactured by different technological processes.

As a generic term for all branches of integrated technology, the term *microsystems technology* is in use, however not in a uniform way. For chemical sensors, preferably two fields are important, namely microelectronics (integrated electronics) and micro-optics (integrated optics).

Integrated Electronics

Basically, there exist two families of integrated circuits, first the bipolar circuits (which is known from 'classical' transistors) and secondly the MOS circuits, for which examples will be presented later. Both technologies are based on the material silicon and both are the basis of mass production of highly complex circuits. The predominant technology up to now is photolithography working with photoresists, masks, oxide layers and etching processes. This established technology has reached a high degree of perfection. Today extreme package density is achieved with conductor dimensions in the nanometer range. Etching techniques with electron beams are emerging only slowly.

Integrated electronic circuits on one chip together with chemical sensing units are not common, since the requirements of both fields are more or less opposite. Some examples can be found with MOS sensors for gases. In the majority of cases, the electronic circuit is produced separately and completed by the sensor, so that a hybrid circuit results. On the other hand, it can be observed that highly complex electronic circuits are becoming increasingly meaningful for sensors. It proved very useful to place amplifying and conditioning circuits as close as possible to the signal source. The cheaper electronic circuits are becoming, the more calculator circuits are utilized to correct distortions mathematically. Increasingly, sensors and sensor arrays are combined with dedicated microcomputers. Such development lines brought about *intelligent sensor arrays* as well as *miniaturized total analytical systems* (μ-TAS). The results of this modern technology are discussed in Chap. 10.

Integrated Optics

The term integrated optics is not used uniformly. Mainly it means some kind of a loose analogue of integrated electronics. The idea is to combine optical components like light sources, light detectors, mirrors etc. on a common substrate to make miniature devices like, for example, spectrometers. In contrast to classical optical instruments, light beams in miniature devices do not cross the free space, but they are *conducted*. For that purpose, wave guides are used consisting of transparent materials with high refractive indices. In such waveguides, the light remains 'captured' by multiple total reflection.

The requirements of optics cannot be fulfilled by only one material. Silicon is a good carrier for optical components, and waveguides can be manufactured in the form of thin silicium dioxide layers. Additionally, glasses, polymers and special materials with high refractive indices are used. Currently, a complete monolithic integration of complex optical instruments is not possible, but in the future this could be managed by means of the lithium niobate ($LiNbO_3$). Monolithic integration would require many incompatible process steps. As an example, a typical instrument like a spectrometer should have a laser, a monochromator, and a light detector on only one chip. Thus, the state of the art today is to solve the problem by hybrid arrangements.

Integrated optics on a glass substrate has reached a relatively high degree of perfection. One requirement is to make very thin waveguides which are mandatory for *single-mode light conduction* (for details see Sect. 8.1). One way is to insert foreign materials with a high refractive index along a certain path of the glass surface. An interesting procedure is based on thermal ion exchange. As shown in Fig. 2.60, a silver ion containing melt is allowed to interact with parts of the glass surface through a metallic mask. At the contact sites, sodium ions of the glass matrix are substituted by silver ions. The resulting silver silicate has a much higher refractive index than the sodium glass. The waveguide produced

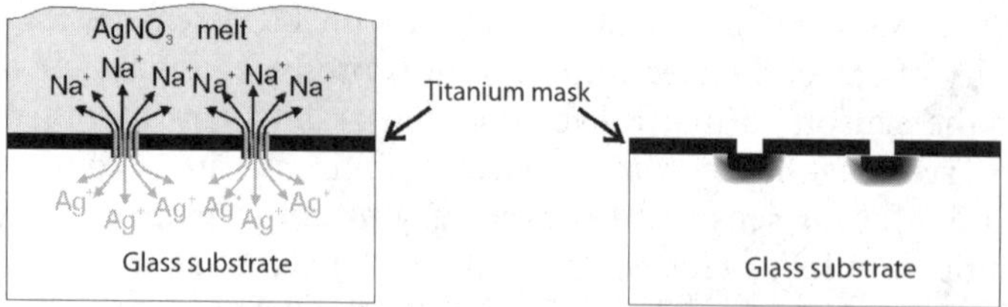

Figure 2.60. Generation of optical waveguides on a glass substrate by ion exchange

in this way may be 'buried' in the glass substrate by a further ion-exchange step. Finally, half cylinders are generated which can be used like classical optical fibres.

For chemical sensing, interferometers in integrated optics became particularly useful. Of these, the *integrated Mach-Zehnder interferometer* type predominates. This instrument is considered in Sect. 2.4.

2.4
Measurement with Sensors

Measurement with sensors and for sensors is not fundamentally different from techniques of classical instrumental chemical analysis. In contrast to classical laboratory instruments, however, sensors are designed to achieve maximum mobility and minimum dimensions. Hence, signal processing must consider these requirements as well. Furthermore, some primary electronic circuitry is mandatory. Working with chemical sensors requires some elementary knowledge of electronics. The field of chemical sensors cannot be considered a part of classical instrumental analysis alone.

2.4.1
Primary Electronics for Sensors

Sensing elements in nearly all cases require primary signal processing close to where the primary signal appears. Preferably, amplification of low-voltage or current signals is required. Also, time integration or formation of the time derivative may be necessary to obtain measurable signals. Of course, all these functions can be accomplished in a classical manner, e.g. using a measuring card in a PC. In this way is not the best one since the signal on its way to the instrument always catches distortion signals which are measured together with the quantity of interest. Thus, the signal-to-noise ratio is downgraded. Consequently, there are good reasons to locate the amplifier as close as possible to the signal source. Thus, classical analogue amplifiers have remained indispensable.

Amplification by Operational Amplifiers. Although operational amplifiers (OAs) contain numerous electronic components, they can be handled themselves like a single component. Working with integrated OAs is so easy that untrained persons can solve their problems in this way. Sometimes, some do-it-yourself electronics is a better solution than to look for specialized commercial instruments. It is sufficient to understand the graphic symbol of the OA and its inherent functionalities.

Experiments with OAs can start already with the basic circuit in Fig. 2.61, where the terminals $+U_B$ and $-U_B$ are connected with two 9-volt batteries to

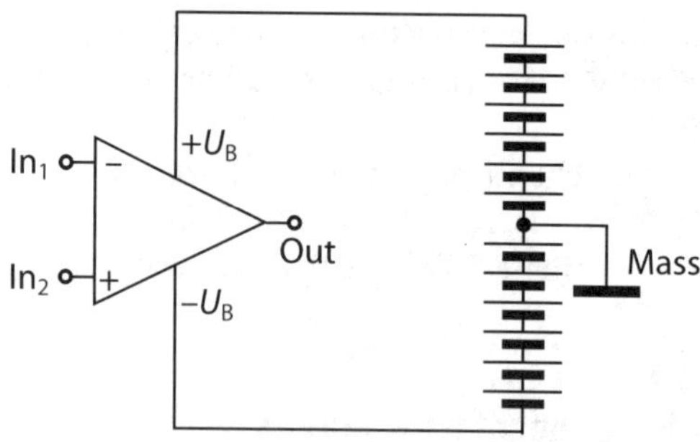

Figure 2.61. Graphical symbol of an operational amplifier and its basic arrangement

ensure a symmetric power supply. The outer connection between both batteries is the reference for all input and output voltages, defined to have the potential zero. This point can be connected to ground. The expression '1 volt at point x' means '1 volt imposed between the point x and the reference'. In all the circuits discussed later, the power supply is omitted. For further simplification, it is assumed that the OA has *ideal* properties listed as follows (this is a good approximation):

- No current can flow into inputs In1 and In2. They have *indefinite input resistance*.
- The output OUT can deliver all the current we need for the following units, i.e. its *output resistance is low* (near zero).
- An external voltage at non-*inverting input In1* (the one with a positive sign) generates at Out a highly amplified voltage of identical polarity. We should keep in mind that in practice the output voltage cannot be higher than the voltage of the power supply.
- An external voltage at the *inverting input In2* generates at Out a highly amplified voltage with the opposite sign.
- If there is a *negative feedback*, i.e. if the output signal is fed into the inverting input, then we can state that the OA always sets the potential difference between its inputs In1 and In2 to zero.

Indeed, commercially available OAs correspond to the criteria given above to such an extent that the circuit can be considered an ideal component. For most of the applications which deal with chemical sensors, the series 080 OA types are useful. They are produced by many different manufacturers. Since they are inexpensive, some loss by electric shock can be tolerated. Experiments can be done easily by means of commercially available OA sockets mounted on a piece of printed board material.

In common OA applications, negative feedback is necessary. In the *voltage follower* (Fig. 2.62, left), a connection is made between Out and In2. As a result of feedback, all points Out, In2 and In1 assume equal potential in relation to the reference. No voltage amplification can be stated; however, now we have a voltage source (of a magnitude equal to the input voltage studied) that can

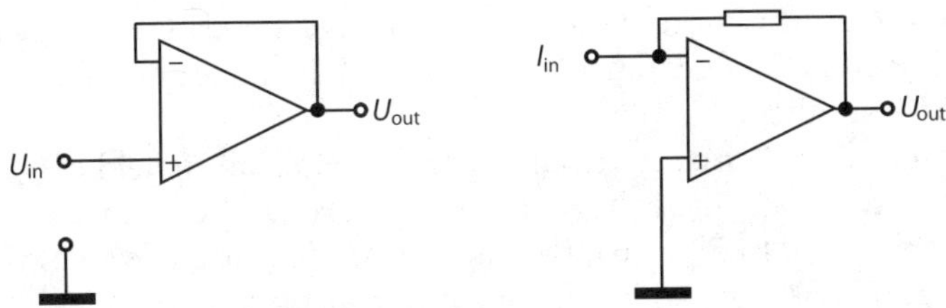

Figure 2.62. Circuits with operational amplifiers. *Left*: voltage follower, *right*: current follower

be loaded, i.e. we can connect it to some power consuming equipment like an old-fashioned voltmeter. To measure pH, we could connect a glass electrode with In2 and the associated reference electrode to mass (the reference). A direct connection between the electrodes and the voltmeter would not work since the line resistance of the electrodes would be much higher than the input resistance of an ordinary voltmeter. The latter would short close the galvanic cell consisting of glass and reference electrodes. Hence, the very simple circuit *voltage follower* is very useful for measuring voltages *without current flow*, as this is a requirement for *potentiometry*.

The *current follower* (Fig. 2.62 right) seems to solve an unsolvable problem. It can measure the current flowing in a short-closed circuit! The current to be measured would flow to mass (the reference) in the circuit given in Fig. 2.62. However, the current is redirected to flow over the resistor R towards output Out. Since In1 is directly connected to mass, In2 also has assumed mass potential (zero). For the current I the condition is as if it were connected to mass. Indeed it flows via R and generates there an IR drop of the magnitude $I \cdot R$. Consequently, the output voltage $U_{Out} = -I \cdot R$. This value can be measured by any voltage meter. Overall, one can measure a current in a virtually short-closed circuit.

By means of operational amplifiers, many different amplification circuits can be assembled. A further example is the *inverting amplifier* (Fig. 2.63 right). This voltage amplifier can be given an arbitrarily adjustable amplification factor given by R_2/R_1. The input of this amplifier does not work load free. If an indefinitely high-input resistance is necessary, a voltage follower can be used as preamplifier.

Operational amplifiers can fulfil more tasks than just voltage amplification. Sometimes important in analytical chemistry is *signal integration* (*summing*) over a certain time period. This can be done by an integrator (Fig. 2.63 right). Here also the current seems to flow to mass; indeed the charge carriers are redirected to charge the capacitor C, which consequently assumes the voltage $U_C = q/C$. Since q (the charge of the capacitor) is equal to the integral of I over time, the output voltage of the OA assumes the value $U_{Out} = -\frac{1}{C} \int_0^t I\, dt$.

An integrator can be used to summarize voltage values over a certain time (considering every changes during this time), but instable or oscillating voltage can also be smoothed.

Electronic control circuits are an important application field of operational amplifiers. Quantities like voltage, current or pH can be controlled and held at a constant value. In Fig. 2.64, the current flowing through a light-emitting diode is stabilized to keep the light emission constant.

The *potentiostat* is one of the most important circuits of electrochemical instrumentation. The development of electrochemistry is closely connected with this device. Several methods of electrochemistry became practicable only after the invention of the potentiostat. They became popular when potentiostats were available at reasonable prices. The simplest circuit of a potentiostat is shown in Fig. 2.64 (right). The problem to be solved is to keep constant the voltage between two points, the terminal *REF* (connected to a *reference electrode*) and the terminal *WORK* (connected to the *working electrode*, the measuring electrode). This voltage should stay constant, independent of the current magnitude flowing through the electrodes. To fulfil this task, a third electrode, the auxiliary electrode (*AUX*), is necessary. Between WORK and REF, a predetermined potential should be imposed, identical to the reference voltage U_{ref}, applied at the inverting input of the OA. The control circuit in a way 'reproduces' the reference voltage, since again its inputs must be at the same potential due to negative applied feedback. Consequently, at the output of the circuit, the current is always controlled to such a magnitude that the voltage between REF and

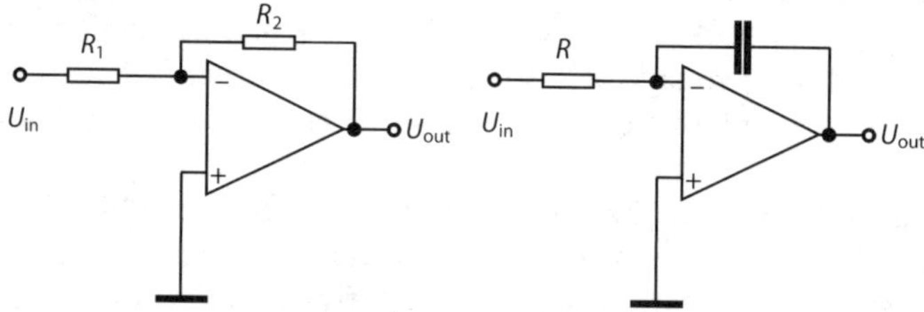

Figure 2.63. Inverting amplifier (*left*) and integrator (*right*)

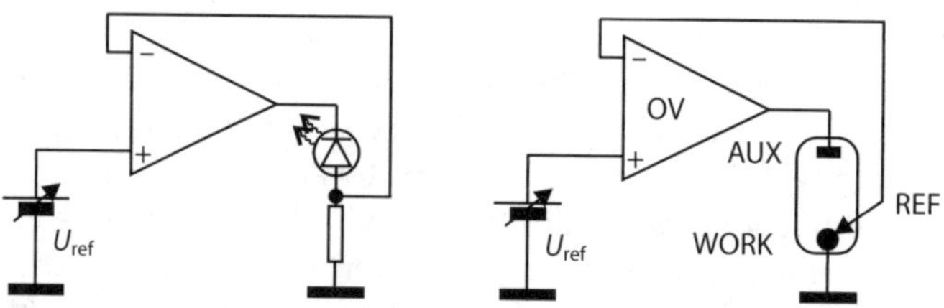

Figure 2.64. Electronic control circuits with operational amplifiers. *Left*: stabilization of a current flowing through a load (e.g. a LED). *Right*: potentiostat for control of electrolysis potential

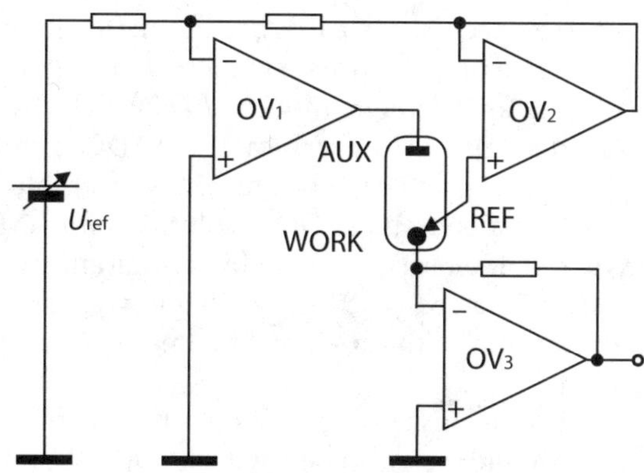

Figure 2.65. Potentiostat circuit of inverting type

WORK remains at the preset value. As a result, the current flowing through the electrolysis cell between AUX and WORK merely depends on chemical processes in the cell, in particular on the concentration of an electrochemically active species. This current is the quantity to be measured by the instrument.

An extended and improved potentiostat circuit is shown in Fig. 2.65. This circuit, the so-called inverting type, is widely used in practice as it is very reliable. One of the three OAs in Fig. 2.65, OA3, is connected to act as current follower. It keeps the working electrode input WORK at zero potential and outputs a voltage signal proportional to the electrolysis current. OA2 ensures that the reference electrode is kept free from current flow. OA1 is the controlling amplifier. In this special case, it makes the potential at WORK equal to the inverted reference voltage U_{ref}.

At present, it is possible to design control circuits like the potentiostat in integrated form on one chip together with the sensing element.

The counterpart of a potentiostat, the *galvanostat,* also exists. In comparison to the potentiostat, it is less meaningful in electrochemistry.

2.4.2
Instruments for Electric Measurements

Sensor signals can be measured by means of standard instruments, of course. Indeed, for testing it is useful to have an oscilloscope and a digital voltmeter. On the other hand, it is better to do sensor measurements by means of a somewhat modified computer. Signal processing facilities would be available, and graphical representation would be included.

In this section, an average PC is assumed to be the basis of instrumentation. The interface between preamplified sensor signal and computer is given by a *data acquisition card* which is a plug-in unit for a free slot of the PC bus. Such an interface card should contain at least one analogue input, one analogue output and some standardized digital input/output terminals. Data acquisition cards are commercially available in great variety. There are cards with a proprietary microcomputer and others which are equipped with just an *analogue-to-digital converter* (ADC) and a *digital-to-analogue converter* (DAC). In every case, an ADC is indispensable to convert sensor signals (mainly voltage or current values) into numbers which can be interpreted by a computer. The resolution of an ADC is the preferential performance characteristics. A 10-bit resolution means that the ADC can distinguish 1024 steps (2^{10}). An analogue voltage of 1 V at the ADC input could be measured with an error of somewhat less than ± 1 mV, according to ca. $\pm 0.1\%$.

A data acquisition card should be capable of providing analogue signals, i.e. it should be equipped with a DAC. Again, the resolution is an important characteristic. An ADC as well as a DAC should be sufficiently fast, i.e. the conversion of a signal into a digital number and vice versa should not need much more than several microseconds. Contemporary cards easily fulfil these requirements. The high resolution and precision of a modern DAC are good enough to provide reference voltage values by means of the DAC output. Such reference values are necessary e.g. for working with potentiostats.

Digital input/output terminals are the basis for cooperation of sensor and data acquisition cards. A typical process of cooperation is to start the measurement of an external instrument by a digital *trigger pulse* from the PC.

2.4.3
Optical Instruments

Spectrometers

In analytical instrumentation, optical signals are measured by *photometers* or *spectrometers*. Signals can be extracted evaluating either *absorption* or *emission* of the sample. To obtain selectivity for the analyte studied, measurement must be done within a defined wavelength range. Three fundamental configurations of spectrometers can be distinguished: (1) the sample emits light caused by thermal excitation, (2) the sample emits light by exposure to an exciting light beam and (3) the sample absorbs light. In Fig. 2.66, the basic configurations are depicted.

Spectrometers can be arranged either as single-beam or as double-beam instruments (Fig. 2.67). In both cases, the optical properties of the sample are compared to those of a *reference*. In double-beam spectrometers, the light beam is divided so that one partial beam interacts with the sample, the second one with the reference. The light detector accepts both beams in parallel, and their intensity is compared. In single-beam spectrometers, the comparison of sample and reference effects is done consecutively, i.e. two consecu-

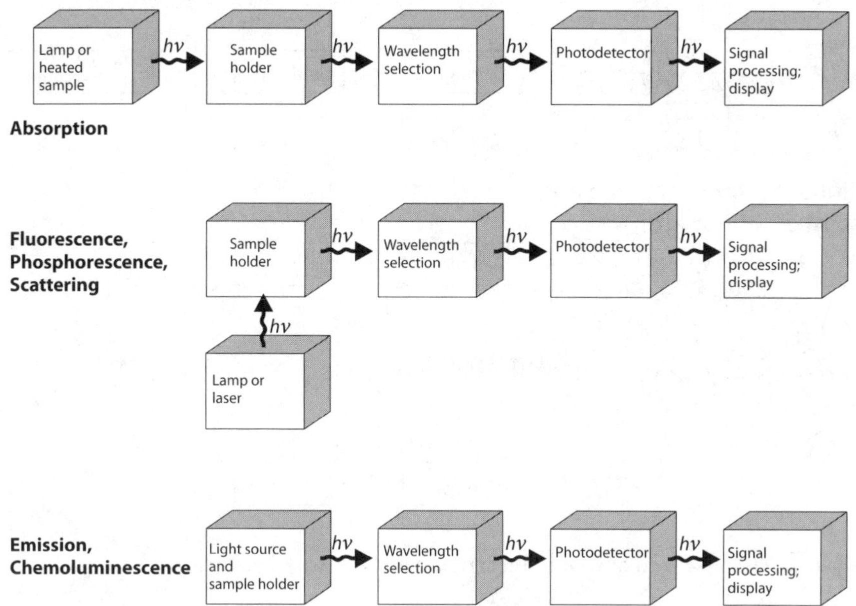

Figure 2.66. Configuration of different spectrometers

tive measurements are performed, one with the sample in the light path, the other with the reference. Both arrangements have advantages as well as disadvantages. Double-beam instruments are capable of compensating short-time fluctuations of the light source or the background. This is not possible with single-beam spectrometers; however, the latter commonly have the advantage of a larger light intensity. With single-beam spectrometers, as a rule, a lower detection limit can be achieved.

Light sources for spectrometers emit either continuous light or a line spectrum. Line spectra are useful if the emitted lines fit well to the analytical task. Wave selection in this case is simple, and requirements for *monochromator* resolution are not demanding. It is sufficient to select a certain wavelength range covering the chosen emission line. Mercury vapour lamps are typical line-emitting light sources (Fig. 2.68). Their emission spectrum can be modified by variation of lamp pressure and addition of other gas components. Such additives increase the number of available spectral lines.

Gas lasers became available to date thanks to reasonable prices. Their restrictions are the same as those of gas discharge lamps. Generally, with lasers the emission lines are more narrow-band. Tunable dye lasers would be the ideal light source. However, they are too sophisticated and too expensive. Furthermore, the dyes used are not sufficiently stable due to their light sensitivity.

Among discontinuously emitting sources, LEDs are particularly attractive for sensor applications. The reasons are their small dimensions, their low energy demand and their manufacturing technology, which is compatible with

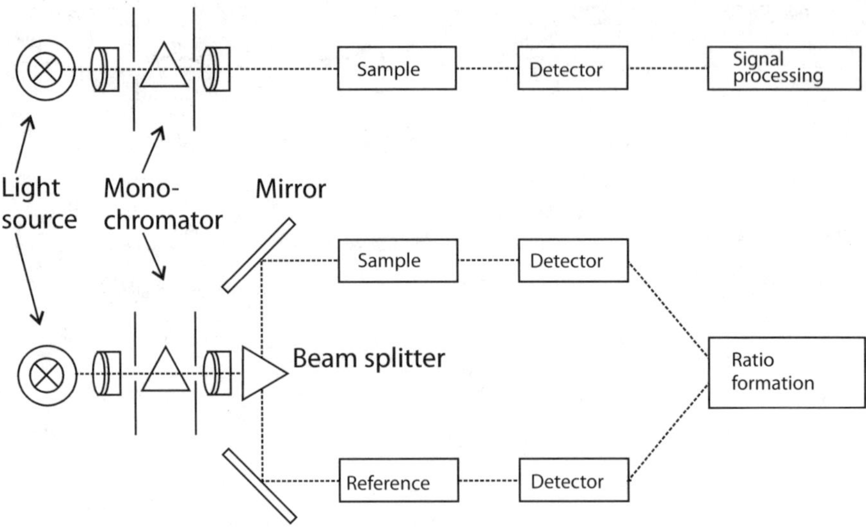

Figure 2.67. Single- and double-beam arrangements of spectrometers

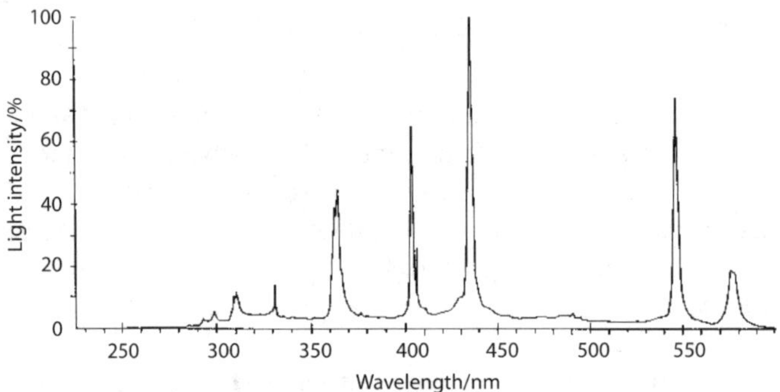

Figure 2.68. Emission spectrum of a high-pressure mercury vapour lamp

sensor production. Light emission of LEDs is based on a semiconductor effect, when electrons recombine at the pn-junction following their movement from conductance band to valence band. Excess energy is emitted in the form of light irradiation during recombination. Energy for this process is supplied continuously by the electric current crossing the pn-junction. LEDs for visible light are manufactured mainly from so-called $A^{III}B^{V}$-semiconductor materials. Such materials are e.g. gallium arsenide GaAs, gallium phosphide GaP, or a combination of both. LEDs have a spectral band width of ca. 40 nm. They are available for all colours of visible light.

Laser diodes (Fig. 2.69, bottom) are made from the same materials as LEDs, but their structure is somewhat more complex, and they are processed differently. Low current density at the pn-junction results in non-coherent emission,

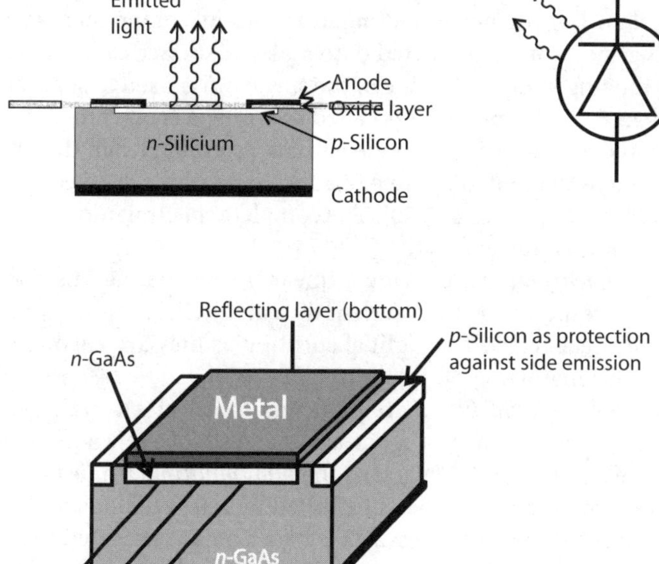

Figure 2.69. *Top*: LED, *top left*: semiconductor assembly, *top right*: graphical symbol. *Bottom*: assembly of laser diode

i.e. in ordinary LED function. From a certain threshold value, the emission becomes coherent and the diode acts as a laser source. Laser diodes normally operate in pulse mode.

Monochromatic light can be obtained alternatively by filtering the white light of continuous light sources. Monochromators for such purposes must fulfil stringent requirements. They must have a very high resolution, i.e. they must be capable of separating a narrowband section of the continuous spectrum. Many continuous-light sources are of the *black-body-emitter* type. Black bodies are solids which are excited thermally to emit broadband radiation. The well-known tungsten lamp is an example of this group of radiation sources. In the special form of halogen lamp, the emitted light range is shifted somewhat towards shorter wavelengths. Nevertheless, the emission of black bodies in the UV range is insufficient for spectroscopy. A more useful quasi-continuous light source including a large part of UV is the *deuterium lamp*. This lamp does not belong to the group of black-body emitters; however, its properties are similar to theirs.

Wavelength Selection and Light Detectors. There are two ways to separate a certain range of wavelengths, either by means of optical *filters* or by *monochromators*. Optical filters absorb parts of a wavelength mixture in such a way that only a certain part is transmitted. Monochromators, working either with a *prism* or an *optical grating*, disperse the radiation into its component wavelengths. The resulting spectrum is projected onto a planar surface called a *focal plane*. An exit slit moving along the image of the spectrum is used to separate the desired wavelength. This operation is called *scanning* a spectrum. Alternatively, a spectroscopic measurement is done without scanning when the light detector is not a single element, but a line of a thousand or more detectors forming a linear *photodiode array* (Fig. 2.70). The complete spectral information is thus obtained in one go and very fast.

Grating monochromators provide a linear dispersion, i.e. distances at the focal plane correspond to the wavelength scale. Reflection gratings are free from adverse effects caused by light absorption as they are encountered with prism monochromators.

In some cases, *optical filters* are sufficient to select a certain wavelength range. *Interference filters* have a spectral bandwidth of a few nanometres.

Among the different *light detectors*, the *photomultiplier tube* (PMT) is currently the best and could not be substituted by a less tedious and expensive device. The overview given in Fig. 2.71 is based on an estimated relative value for spectral sensitivity (Malmstadt et al. 1981). The photomultiplier is a result of further development of the vacuum phototube, where electrons are released by illumination with photons from a photocathode. These electrons are collected ('sucked off') by means of an electric field and can be measured as an electric current. In the case of the multiplier tube, electrons are guided to other areas which are able to emit secondary electrons, the so-called *dynodes*. In this way, the primary current is amplified in an avalanche manner. Among the advantages of the multiplier tube is not only its high sensitivity, but also its flexibility regarding adaptation to different wavelength ranges. Its fast response also is a very useful property. CdS photoresistors, for example, are comparable in sensitivity but too slow to be used in spectrometers.

A most desirable requirement of modern sensor technology is to find a real substitute for the PMT. Current semiconductor components do not fulfil all the demands, but there are strong efforts to minimize the disadvantages of semiconductor devices by mathematical correction procedures.

Classical light detectors surely can be used with optical sensors, but there arises a discrepancy with the sensor characteristics, namely they are small, mobile and cheap. Semiconductor photodiodes fulfil these requirements much better. Among them, the silicon photodiode is the most common, but its spectral sensitivity is not optimal, since it is extremely sensitive in the infrared region. Consequently it suffers strongly from temperature changes. Photodiodes based on GaAs, GaP and InP have better spectral properties. All the different photodiodes have in common the *internal photoelectric effect*. If a pn-junction is *reverse biased*, a depletion zone is formed which blocks current conduction from the n-type region to the p-type region. The depletion zone absorbs photons, resulting in the release of charge carriers. The electron-hole

Figure 2.70. *Bottom*: spectrometer with wavelength scanning; *top*: Photodiode array spectrometer

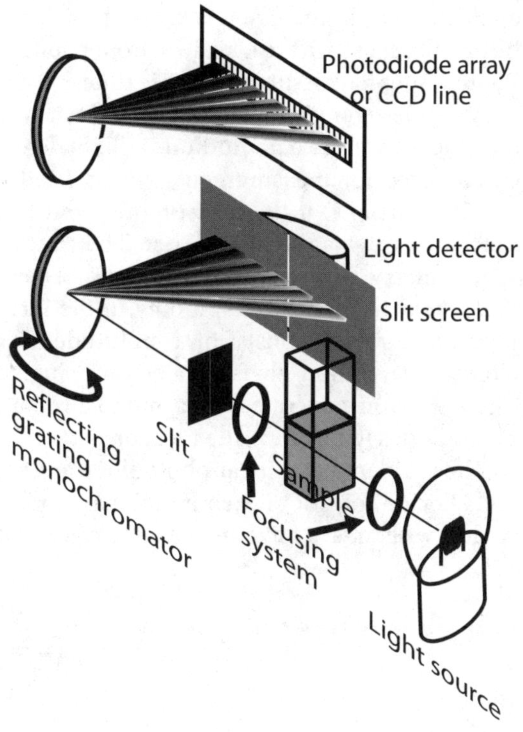

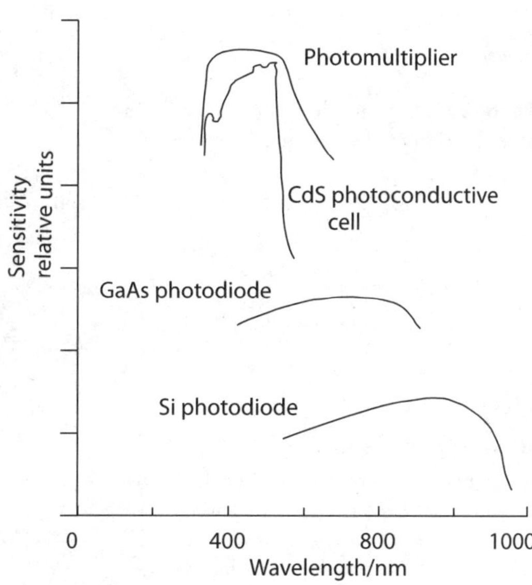

Figure 2.71. Relative spectral sensitivity of different photodetectors

pairs formed in this way may engender a photocurrent if an electric field is applied. This photocurrent is proportional to the number of photons per time hitting the sensitive area, i.e. it is proportional to the intensity of incident light. Photodiodes have a short response time, but their contribution to noise is high.

A simple arrangement for an optical sensor consists of a LED as light source, an optical fibre and a photodiode as light detector. This 'single-channel' set-up can be used for measurements with a fixed wavelength. If there are higher requirements, i.e. if the measurement must be extended to a broad spectral range, then two approaches exist. The classical one is wavelength dispersive spectrometry with a PMT at the end. The second one is using an *optical multichannel analyser*, commonly in the form of a *photodiode array*. *Linear photodiode arrays* consist of a multitude of identical photodiodes arranged along a line, commonly fabricated on a single semiconductor chip (Fig. 2.72). The spectrum generated by a monochromator is projected on this line of photodiodes lying along the focal plane.

The problem with linear photodiode arrays is to make electric connections and allocate control to each individual diode of the line. This problem is less critical with arrays forming a *charge coupled device* (CCD). Only a few single

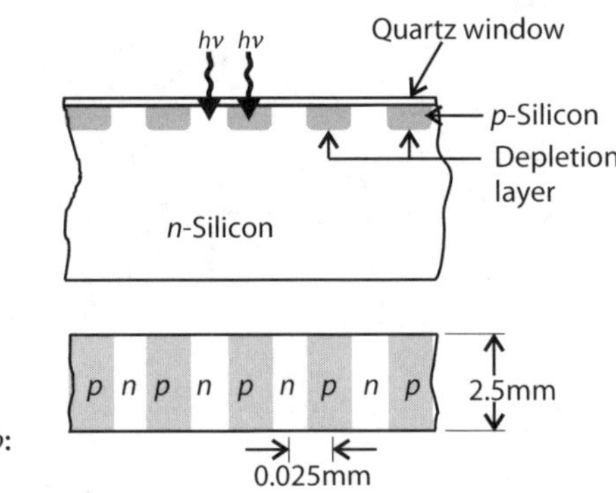

Figure 2.72. Linear photodiode array. *Top*: cross section of chip; *bottom*: *top view*

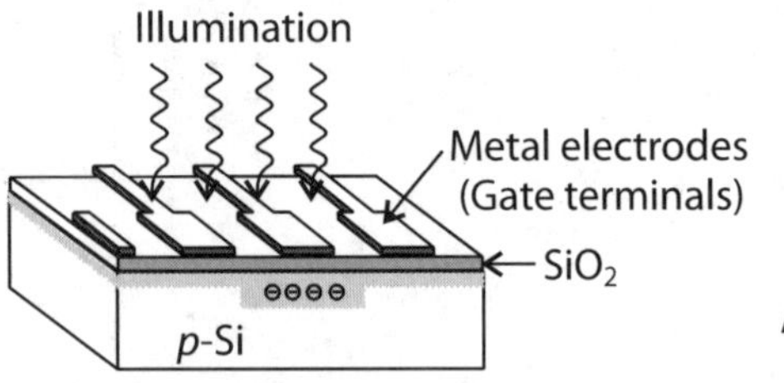

Figure 2.73. Function of CCD array. A Free charges are formed by illumination and fixed by an imposed electric field. B Synchronous charge transfer in shift register in order to read information

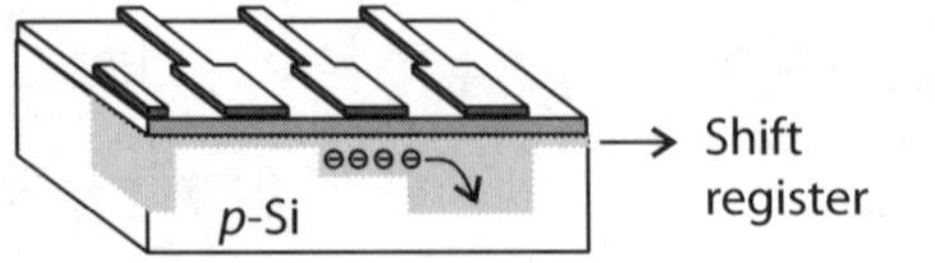

contacts are necessary with such structures (Fig. 2.73). The space charges formed by illumination in the semiconductor material are pushed by a pulsed electric field into a shift register where the information is read digitally. CCD arrays are very common nowadays since they find broad application in digital cameras. Thus they became available at reasonable prices.

Fibre optic elements play an important role in optical chemical sensors. They are discussed in detail later.

Interferometers and Fourier-Transform Spectrometers

Non-dispersive spectrometry is an alternative to the classic, dispersive technique where monochromators are used, as depicted schematically in Fig. 2.67. Non-dispersive techniques, used mainly in *infrared spectroscopy*, are based generally on *interferometry*. The sample is irradiated by *polychromatic radiation*. After transmitting the sample, the light forms an *interference pattern*, which contains all pieces of information that could be obtained alternatively by dispersive methods. Several types of interferometers can be built in *integrated optics* having miniature dimensions. They are useful for chemical sensors.

The function of an interferometer can be explained best if considering the Michelson-type interferometer shown schematically in Fig. 2.74 (left). Polychromatic light is separated into two beams by means of a semitransparent mirror. Both beams are unified at the detector after being reflected at the surface of a second mirror. If both beams are of equal length, then there will be no difference between the beams at the input and the output of the interferometer. However, one of the mirrors moves back and forth with an amplitude Δx. Consequently, in each period interference is encountered. If the actual shift

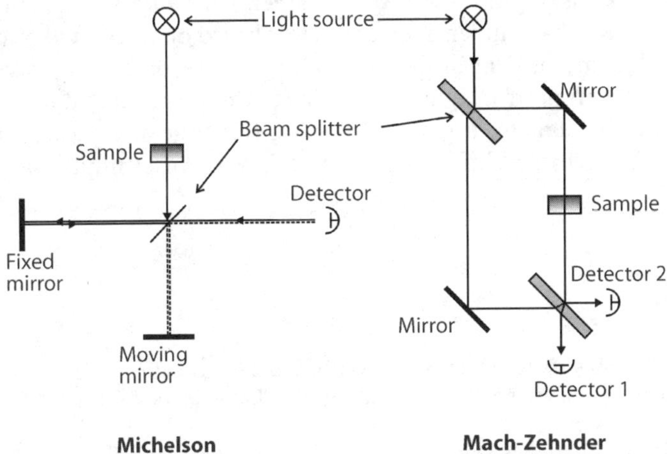

Figure 2.74. Functional principle of interferometers. *Left*: Michelson type, *right*: Mach-Zehnder type

of the mirror is equal to a multiple of a certain wavelength, i.e. if peak meets peak, then the light intensity is amplified. The opposite case occurs with an odd-numbered multiple of half the wavelength. This means that wave troughs meets wave troughs, and consequently light is extinguished. A characteristic pattern results when the polychromatic light intensity is plotted as a function of actual mirror position x. Utilizing the relationship between mirror position and wavelength, intensity can be plotted as a function of frequency or wave number. The result is a spectrum. An equal result can be obtained by *Fourier transformation* of a time-dependent intensity function which is a result of many physical measurements. The intention is to transform the function $I = f(t)$ into a function $I = f(\bar{v})$. The spectrum determined by *Fourier transform spectroscopy* contains all information obtainable by spectroscopy using monochromatic light of varied wavelength.

Devices manufactured by means of the technology of *integrated optics* are of particular importance for sensors. The preferred type of interferometer for this technology is the *Mach-Zehnder* type (Fig. 2.74, right). Again here is the light beam divided into two paths. If we assume first that both beams are of equal length, then all wavelengths at the detector are *in phase* equal to the state at the input. No interference is observed. Interference is encountered as soon as the path length of the light beam is changed in one of the paths, e.g. by insertion of a sample. The resulting interference pattern contains all optical information about the sample material. A special variant useful for sensor application is the Mach-Zehnder interferometer equipped with optical fibres as light paths. One of the branches is exposed to a sample solution with the effect that the angle of internal total reflection is changed. Hence the total light path length in one of the branches is modified. The resulting interference pattern can be evaluated to yield analytical information.

Sensors for gases and dissolved components have been constructed with miniaturized Mach-Zehnder interferometers based on integrated optics.

A complete instrument in this technology can be built on a space of only a few square centimetres. Light-conducting paths are generated commonly by allowing foreign ions to diffuse into a glass substrate along a line, resulting in a thin channel of a different refractive index. An example was given in Sect. 2.3.4.

2.5
References

Allen PM, Hill HAO, Walton NJ (1984) J Electroanal Chem 178:69

Flink S, Boukamp BA, van den Berg A, van Veggel FCJM, Reinhoudt DN (1998) J Am Chem Soc 120:4654–4657

Gaines Jr GL (1966) Insoluble monolayers at liquid-gas interfaces. Wiley, New York

Gründler P, Kirbs A (1999) The technology of hot-wire electrochemistry. Electroanalysis 11:223–228

Gründler P, Kirbs A, Zerihun T (1996) Hot-wire electrodes: voltammetry above the boiling point. Analyst 121:1805–1810
Malmstadt HV, Enke CG, Crouch SR (1981) Electronics and instrumentation for scientists. Benjamin/Cummings, Reading, MA
Persson B, Gorton L (1990) J Electroanal Chem 292:115
Ulman A (1996) Chem Rev 96:1533–1554
Wollman EW, Frisbie CD, Wrighton MS (1993) Langmuir 9:1517

3 Semiconductor Structures as Chemical Sensors

Semiconductor structures in diodes, transistors, gates etc. are the basis for the extraordinary success of microelectronics. Initially, such structures were designed mainly to amplify low signal amplitudes. Influences of the environment by temperature changes or light irradiation were first considered to be distortions. Obviously, it could be worthwhile to make the best of it and to utilize the strong sensitivity of semiconductor devices simply to measure 'parasitic' quantities. A priori semiconductor structures are 'sensors' for temperature and electromagnetic waves. Chemical sensors can make use of this sensitivity after converting analytical quantities into heat or light emission. Alternatively, amplifying semiconductor units are useful for chemical sensors since the amplifier can be brought very close to the receptor.

Metal-oxide-semiconductor (MOS) structures are of particular interest for chemical sensors. Such structures are widespread in millions of memory chips and CPUs in our computers. The *MOS field effect transistor* (MOSFET) was presented in Chap. 2. Commonly, this device is designed to amplify low voltages since such a voltage imposed between the gate (G) and source (S) terminals brings about a high field strength due to the very thin isolating SiO_2 layer covering the gate. The electric field of the gate voltage acts as controlling quantity for the current through the drain terminal.

The ability to amplify low-voltage signals is the basis for different applications of MOSFETs in the field of chemical sensors. This development started with a more peripheral property of the device. A gas-sensitive MOSFET has been designed more or less as a by-product of microelectronics (Lundström 1975). Hydrogen gas sensors on this basis played a certain role. The chemical sensitivity of MOSFETs generally implies some voltage signal at the gate input which must reflect concentration changes. In the case of the hydrogen sensor, this signal is based on a thin palladium layer on the gate. Hydrogen dissolves homogeneously in palladium. At a temperature of ca. 150 °C, hydrogen molecules start to dissociate into free atoms. The latter diffuse through the metallic layer and are adsorbed at the interface with silicon dioxide (see schematic representation in Fig. 3.1). Atoms are polarized at the interface forming a layer of dipoles. The electrical asymmetry of this layer corresponds to a partial charge separation, i.e. formation of a potential difference at the interface. This voltage gives rise to a current change in the output circuit (source-drain) of

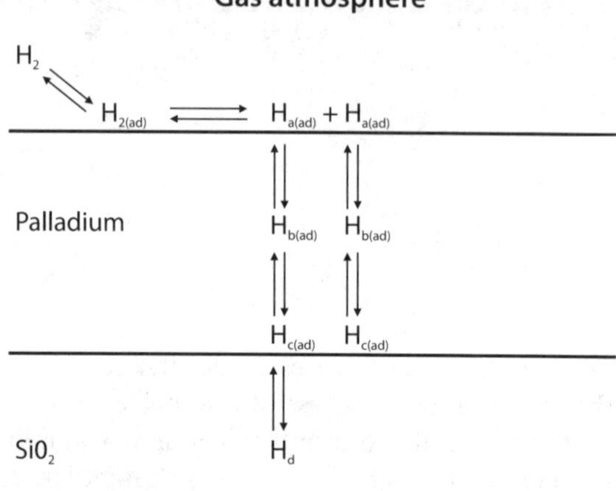

Figure 3.1. Formation of potential difference at interface between metallic gate layer and isolating SiO₂ layer. $H_{a(ad)}$ and $H_{b(ad)}$ are adsorbed hydrogen atoms, H_d hydrogen atoms mobile in SiO₂

the MOSFET, i.e. the drain current is modified. The relationship between the measurable current change and the concentration of hydrogen is logarithmic. Such a logarithmic relationship is characteristic of galvanic cells. Hence the interface between metal and isolator can be considered an electrode which follows the Nernst equation [Eq. (2.33)].

Gas-sensitive MOSFETs have been designed also for ammonia and other hydrogen-containing gases like arsine and hydrogen sulphide. Gas sensors for fluorine, hydrogen fluoride and oxygen have been mentioned in the literature. They contain a thin sputtered platinum layer covering a lanthanum fluoride film on the gate. The function of these sensors cannot be explained completely. It seems clear, however, that exchange processes with the lanthanum fluoride film play a role.

A key feature of gas-sensitive MOS sensors is its well-engineered technology, which allows for mass production at reasonable prices. Application examples are encountered in leakage control of oil pipelines as well as in the investigation of molecular processes when oxygen and hydrogen traces react in a high vacuum chamber.

In the above-mentioned gas sensors, the MOSFET is not arranged so as to act as an amplifier, but just as a diode. This is a special case which is meaningful only for the gas phase, not for liquids. In liquid solutions, however, the transistor function of the device can be utilized better. This is the basis of a successful class of chemical sensors known under the names ISFET or CHEMFET. Many other acronyms in connection with FET require explanation. A list is given in Table 3.1.

MOS sensors for liquid solutions, known as ISFET, ENFET or IMFET, are largely a close combination of a chemical receptor layer with the MOSFET acting as a voltage amplifier. Such a combination belongs to the field of electrochemical sensors, which are discussed in Chap. 7.

Table 3.1. Acronyms in connection with sensors on the basis of field effect transistors

Acronym	Meaning
FET	Field Effect Transistor Amplifying semiconductor component with inputs *gate*, *source* and *drain*. High input impedance at gate. Voltage amplifier.
MOSFET	Metal Oxide Semiconductor FET. Layer composition of MOSFETs, where input and output circuits are separated by an insulating oxide layer.
IGFET	Insulated-Gate FET. Generic term for FETs, where the gate is separated by an insulating layer from the the rest of the semiconductor body.
OSFET	Oxide Silicon FET. Specific case of ISFET and IGFET with SiO_2 as an insulating layer.
OGFET	Open Gate FET. Operation of an FET with gate input left open (operation as diode)
GASFET	Gas sensitive FET, e.g. with palladium as H_2 sensitive receptor layer
CHEMFET	CHEMically sensitive FET
ISFET	Ion Sensitive FET, ion sensitive FET
IMFET	Immunologic FET. Immunologic reaction (complex formation of antigen and antibody) is utilized
pH-FET	pH-sensitive FET
ENFET	Enzyme FET

3.1
References

Lundström I, Shivaraman, S, Svensson C, Lundkvist L (1975) Appl Phys Lett 26:55–57
Moseley PT (1991) New trends and future prospects of thick- and thin-film gas sensors
Sensors Actuators B 3:167–174

4 Mass-Sensitive Sensors

Mass-sensitive sensors in principle are balances reacting to small mass changes by the formation of a measurable electric signal. The best known sensors of this kind are based on the *piezoelectric effect* (Sect. 2.1.3). Such arrangements are turned into a chemical sensor by coating piezo crystals with a layer capable of selective absorption of certain substances. The resulting mass change can be measured electrically as a frequency shift.

The preferred material for piezo crystals is monocrystalline quartz, which is cut along certain crystallographic planes. If these planes are coated with metallic layers and the latter are provided with electric contacts, then an *oscillating crystal* results. Such crystals can be excited to oscillate by inserting them into a circuit with positive feedback, e.g. by combining them with a *negator circuit* (Fig. 4.1). The resulting oscillator circuit (Fig. 4.1, right) then provides an alternating voltage at its output. The frequency of this oscillation, the so-called *resonance frequency* f_0, is extremely stable. Dependence on external distortion sources like temperature changes is very low. The circuit in Fig. 4.1 can be used as a precise frequency reference; on the other hand its frequency can be measured with extremely high precision. This can be done simply by counting the periods of oscillation. This is the working principle of modern quartz watches, which attain their high accuracy in this way. In this way, modern quartz watches Even an average watch is a highly precise instrument, since a deviation of 1 min per week (worse than the average!) means a relative error of only 0.01%.

Oscillation of a piezo crystal is associated with sound propagation. Acoustic waves can propagate either through the crystal bulk or along its surface. To visualize the difference, one may imagine what happens if either an explosion

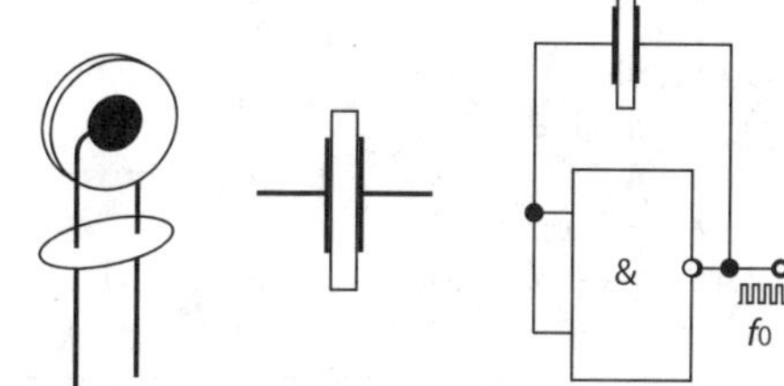

Figure 4.1. Oscillating quartz crystal (*left*) with graphical symbol (*centre*) and arrangement in an oscillator circuit (*right*)

occurs under water, or if a stone is thrown in a water basin, thereby causing surface waves.

Both propagation modes of acoustic waves may be utilized to construct mass-sensitive chemical sensors. Wave propagation through the bulk is the basis of so-called *bulk acoustic wave (BAW)* devices. Propagation along the surface is used in *surface acoustic wave (SAW) sensors*.

4.1
BAW Sensors

BAW sensors are largely classic oscillating quartz crystals with metallized planes for contacting and are covered by acceptor layers selectively interacting with the analyte. The analyte is absorbed and causes a mass change in the oscillating system, which gives rise to frequency shift. The *Sauerbrey equation* (Sect. 2.1.3) is the basis of quantitative evaluation.

Among the first applications of BAW sensors was the determination of hydrogen in gas mixtures. For that purpose, the crystal plans are metallized with palladium, which is characterized by a high capability to dissolve hydrogen. The hydrogen uptake is reversible, i.e. when the gas concentration decreases, then the mass of palladium layer also decreases. A sensor for mercury vapour works with a gold film at the crystal planes. Gold layers readily absorb mercury traces under alloy formation. After use, the sensor must be regenerated by heating the gold layer to ca. 150°C. More sensors designed to respond to different gas components are made by coating the metallizing layers by an additional acceptor layer, which can absorb the sample component. This acceptor layer should interact with the desired component as selectively as possible, without being changed by other parameters. Obviously, the extreme precision of the frequency measurement would be useless if the acceptor layer contained volatile parameters. Table 4.1 presents an overview of common compositions. Sensors, which need a foregoing chemical reaction, are not included in the

Table 4.1. Examples of adsorptive layers which can be used in mass sensitive chemical sensors

Analyte	Adsorptive layer
Hydrogen	Palladium; platinum
Mercury	Gold
Water vapor	Gelatin; lithium chloride; polyethylene glycol
Carbon dioxide	Dioctadecylamine
Ammonia	Pyridoxinhydrochloride; Ascorbic acid + Silver nitrate
Sulfur dioxide	Organic amines
Hydrogen sulfide	Smolder residue of chlorbenzoic acid, (acetonic extract); silver acetate
Hydrocarbons	Non polar gas chromatographic separation phases, e.g. carbowax 550

table. An example of this group is a carbon monoxide sensor which accepts the mercury vapour released when the analyte reacts with mercury(II)oxide at 210 °C.

There are adsorption layers which react with the sample irreversibly, e.g. ammonia with a mixture of ascorbic acid and silver nitrate or hydrogen sulphide with silver acetate.

Selectivities of BAW sensors differ according to their acceptor layer. Whereas palladium is highly selective for hydrogen, organic amines are only weakly selective for sulphur dioxide, and carbowaxes (well known as stationary phase in chromatography) interact with numerous non-polar gases, hence they are non-selective sensors for hydrocarbons. Response time also varies within a wide range.

In some cases, the mechanism of interaction with the analyte is unknown, e.g. with an adsorption layer obtained from an acetonic extract of the smouldering residue of chlorobenzoic acid. This layer is highly selective for hydrogen sulphide and can be used to determine traces down to 1 ppm in air.

Application of piezoelectric sensors in liquids is much more difficult than working with gases, since the surrounding liquid phase acts strongly, damping the oscillating crystal. Although the problems have been solved, real analytical applications are rarely encountered even today. The best investigated applications are antibody layers immobilized on quartz crystals which are used in immunoassays. Reaction with the corresponding antigen is highly selective, and the resulting mass change is measurable. Piezoelectric sensors in liquid phase do not obey the Sauerbrey equation.

In electrochemistry, the piezoelectric BAW sensor is known as an electrochemical *quartz crystal microbalance* (EQCM). This device plays an important role in electrochemical research. Metal deposition, corrosion and formation of passive layers have been studied successfully.

4.2
SAW Sensors

Piezoelectric sensors, where surface acoustic waves (SAW sensors) are used, came to the fore in recent years. The function principle is demonstrated in Fig. 4.2.

A common SAW arrangement consists of two interdigitated structures made of thin metallic films positioned some distance from each other at the surface of a piezoelectric crystal. One of the structures is the *transmitter*, the other one the *receiver* of the surface acoustic wave. In the course of transmission, the electric signal is converted into an acoustic wave and vice versa. Between the interdigitated transducers is a *retarding layer*. The latter is impregnated by a chemically sensitive film. Imposing an alternating voltage at the transmitter results in mechanic oscillations of the surface propagating along the surface as an acoustic wave. In turn, mechanical oscillations cause at the receiver

Figure 4.2. SAW arrangement (*top*) with simplified depiction of surface acoustic wave

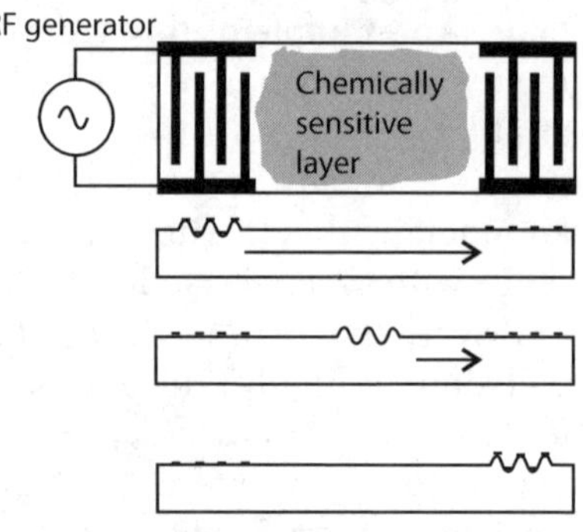

structure the formation of an AC voltage. The frequency of this voltage can be measured and depends on the distance between metallic structures. The resonance frequency of the piezoelectric crystal is influenced by this effect but also by the speed of the propagating wave. The sensor signal is formed by interaction of the sensitive layer with the acoustic wave along the retarding layer. Tiny mass changes of the latter have an effect on the run time of the wave and can be measured in terms of frequency shifts. Under common conditions, the frequency shift follows the following simplified equation:

$$\Delta f = (k_1 + k_2)f_0^2 \cdot h \cdot \varrho' ,\tag{4.1}$$

where k_1 and k_2 denote characteristic constants of the piezoelectric substrate, h the thickness of the coating, and ϱ' its density. For the commonly used quartz materials, k_1 and k_2 have values ranging from $-8.7 \cdot 10^{-8}$ to $-3.9 \cdot 10^{-8} \, \mathrm{m^2 \, s/kg}$. With exciting frequencies in the gigahertz range, a very high sensitivity vs. mass changes due to interaction with the analyte is achieved. Consequently, picogram amounts can be determined.

For SAW sensors, all the materials given in Table 4.1 except those with metallic conductivity can be used as chemically sensitive coatings.

SAW sensors can be miniaturized easily. The technology of microelectronics is applicable, so that mass production at reasonable prices is attainable.

A single piezoelectric substrate can carry several SAW sensor structures which can be processed independently of each other. The resulting sensor arrays are important for development of *'electronic noses'* and similar multi-component analysers (Chap. 10).

5 Conductivity Sensors and Capacitive Sensors

Commonly, information about the composition of samples can be obtained by measuring the electrical *conductivity* (or *resistivity*). In some cases, simply the resistance of the sample itself contains the desired information; in other cases the principles discussed for *resistive transducers* must be applied (Sect. 1.2.2). Normally this means that the interaction of sample components causes resistance changes in the receptor layer. Such changes can be measured and evaluated to extract analytical results.

Commonly, the resistance of a sensor layer is measured by means of an alternating current. The reasons for this will be given later. The result of such measurements is the AC resistance, the so-called *impedance*. The name of the corresponding method, *impedimetry*, can be retrieved in the term *impedimetric sensor*. Originally, impedimetry was a term used to describe highly sophisticated techniques which tried to obtain information about processes at polarized electrodes, i.e. electrodes participating in electrochemical processes (Sect. 2.2.6). Nowadays, the prevalent term for such research methods is *electrochemical impedance spectrometry* (EIS), whereas impedimetry is used for a large variety of the simplest *conductance sensors* applied in electrolyte solutions.

The AC resistance, a *complex* quantity (in mathematical terms), is composed of a *real* (*ohmic*) and an *imaginary* component. Of particular importance is the *capacitive resistance* as part of the overall AC resistance. If the capacitive contribution predominates, the device is a *capacitive sensor*. There is a smooth transition from capacitive to resistive sensor types.

The following cases are of particular importance for chemical sensors:

- The resistance of an electrolyte solution depends on the concentration of all ions in solution. Hence it contains analytical information a priori. In this case, the 'sensor' is a simple probe consisting of two metallic bodies (electrodes) dipping into the sample solution. The analytical result is obtained by conductance measurement.
- The receptor is represented by a layer of semiconductors, polymers or gels. The resistance, or dielectric constant, of this layer changes when interacting with the sample. The resistance change is measured and evaluated to draw conclusions about the sample composition.

5.1
Conductometric Sensors

The AC resistance (impedance) of electrodes is related in a highly complex manner to processes at the electrode surface, but also to the resistance of the homogeneous bulk solution. Conductometry might be considered an electrochemical method. For the sake of clarity, however, one should distinguish between methods connected with electrochemical processes at the electrodes and, on the other hand, methods dealing with properties of a homogeneous bulk solution. The latter is not a question of chemistry but of physical behaviour like ion mobility. In what follows, conductometric sensors are considered simple resistance probes.

One of the oldest instrumental methods for concentration determination is to measure electrolytic conductance. The measurement set-up is simple. It is sufficient to determine the resistance between two inert metallic electrodes. The well-known Wheatstone bridge can be utilized as with any other resistance measurement (Fig. 5.1). The bridge is balanced, i.e. the variable resistor R_v is adjusted to obtain zero at the instrument included in the circuit. Commonly this is an oscilloscope, since AC amplitudes must be detected. If the scope displays zero, the following condition is valid:

$$R_x = R_v \frac{R_1}{R_2} \,.$$

(5.1)

Strictly speaking, the measuring result contains not only ohmic quantities, but also some undesired capacitive contribution. Certain traditional methods can

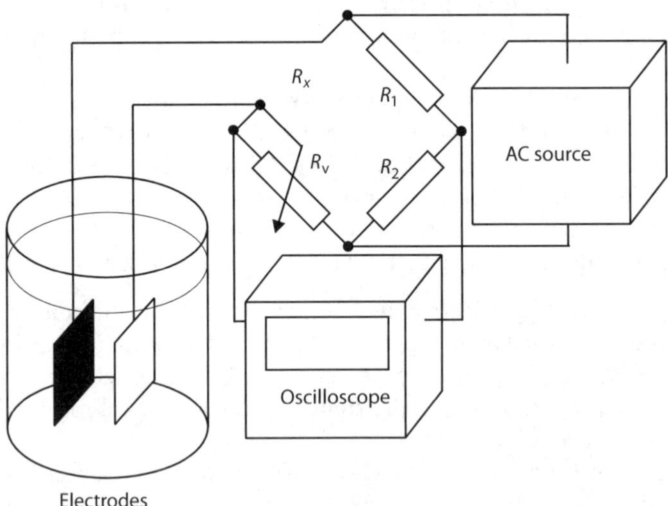

Figure 5.1. Conductance measurement in an electrolyte solution by means of Wheatstone bridge

keep this contribution negligible by special design of the electrodes. The AC current must be applied to avoid electrolytic decomposition and accumulation of reaction products at the electrode surfaces. Wheatstone bridges that automatically adjust are not very widespread for analytical conductance sensors. Instead, commonly the AC voltage between two additional electrodes located along a line between outer electrodes is measured (Fig. 5.2). *Four-point probes* designed in this way are widely used in miniature conductance sensors.

Conductance sensors in electrolytic solutions are only useful for determining the concentration of ions. The theoretical basis is *Kohlrausch's law of independent ion migration*. In the strict sense, this law is valid only for infinitely diluted solution. Each ion type has its individual characteristic *ion mobility u*. The concentration and mobility of an ion type are the quantities determining the ion's contribution to the overall conductivity. The result of the primary measurement is the electrolyte resistance R. The *specific resistance* of the solution, κ, is calculated taking into account the *cell constant c'*. The latter depends on the electrode area and the distance between electrodes. The cell constant is determined commonly by calibrating the probe using a solution with known κ. The relationship used is $\kappa = c'/R$. The overall conductance of an electrolytic solution is given by the following equation:

$$\kappa = F \cdot \sum u_i \cdot \alpha_i \cdot f_{\Lambda(i)} \cdot c_i , \tag{5.2}$$

where c_i denotes the concentration of individual ion types, u_i their mobility, and α_i the degree of dissociation. The constant λ_i is a special sort of activity coefficient, valid for moving ions in conductance measurement. F is the Faraday constant, corresponding to the electric charge of 1 mole of charge carriers with a unity charge number. The statement of Eq. (5.2) is that the specific

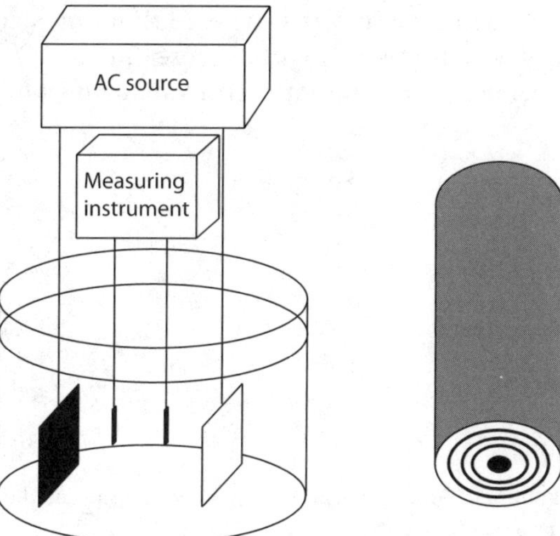

Figure 5.2. Conductance measurement by means of four-point-probe method. *Left*: Basic arrangement; *right*: cylindric probe with four electrodes

conductance κ can be interpreted as the sum of all ionic concentrations, each multiplied by its individual constants k_i':

$$\kappa = F \cdot \sum k_i' \cdot c_i \ . \tag{5.3}$$

Obviously, with conductance measurements only non-specific results can be obtained. Fortunately, the individual constants k_i' do not differ too much, except for H_3O^+ and OH^- in aqueous solution. It is allowed, therefore, to estimate the *total ionic concentration* or the *overall salinity of seawater* from the results of conductance measurements. Conductance sensors are thus encountered with all *seawater probes* commonly used in oceanography. In seawater, k_i' can be considered a universal constant since sodium chloride is present in huge excess compared to all other components.

Conductance sensors are widely used as detectors in an important special variant of liquid chromatography.

5.2
Resistive and Capacitive Gas Sensors

5.2.1
Gas Sensors Based on Polycrystalline Semiconductors

'*Chemoresistor*' and '*chemocapacitive sensor*' are terms sometimes used for gas sensor elements based on conductance measurements. One of the oldest and most commonly used chemical sensors is the so-called *Taguchi sensor* (Taguchi 1962, Seiyama et al. 1962). Taguchi sensors are ceramic devices made by pressing and sintering powdered solid materials. Common materials are semiconducting metal oxides like tin dioxide and zinc oxide (Fig. 5.3). Oxides used are always of the *n*-type. Millions of such sensors are in use worldwide to detect traces of *reducing gases* in the air, e.g. for leak detection in gas pipelines, indication of petrol vapour in filling stations, or for alcohol tests

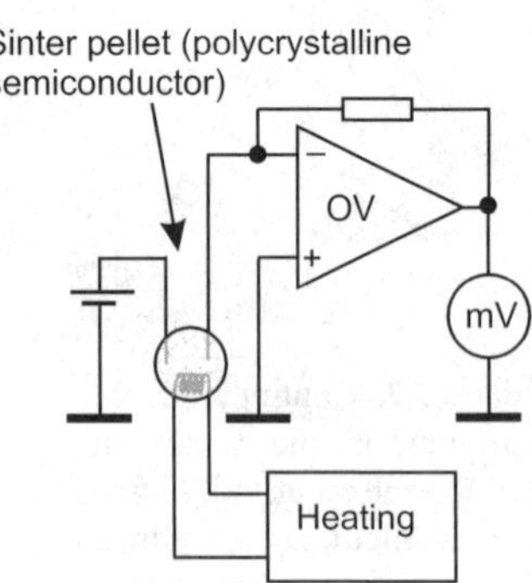

Figure 5.3. Polycrystalline gas sensor with associated measuring circuit

in exhaled air. Only recently has the Taguchi sensor been replaced by more reliable electrochemical sensor types.

The processes of signal formation in polycrystalline gas sensors are not completely clear; however, there are feasible explanations (Moseley 1991). The working temperature of the sensor is in the range 200 to 600 °C. The grains of the sintered body are covered by adsorbed oxygen. The latter withdraws electrons from the crystallites' bulk, causing the formation of oxygen ions at its surface (Fig. 5.4). Consequently, the charge carrier concentration in the grain volume decreases and a potential barrier is formed at the grain boundaries. In general, the electric conductance of the material is decreased by the extent of oxygen adsorption. The molecules of reducing gas, for their part, interact with the adsorbed oxygen, thus lowering the potential barrier and increasing the conductance of the sensor. This conductance change is reversible at the working temperature and can be measured by a simple set-up (Fig. 5.3). Polycrystalline sensors are robust and cheap, but their performance characteristics are changing slowly. Hence they are not very reliable. The results of alcohol testers with Taguchi sensors are not sufficient prosecutable evidence in a court of law.

In addition to SnO_2, also numerous other metallic oxides have been tested in polycrystalline gas sensors, among them Fe_2O_3, TiO_2 and mixed oxides like bismuth ferrite $BiFeO_3$. Numerous reducing gases like hydrogen, methane, carbon monoxide, ethanol vapour and hydrogen sulphide have been determined in air.

The conductance ($G = R^{-1}$) of the sensors depends on the concentration in a non-linear relationship following approximately the following equation:

$$G = \frac{1}{R} = k \cdot c_i^{n_i} \, , \tag{5.4}$$

where k and n_i are individual constants which must be determined empirically by calibration.

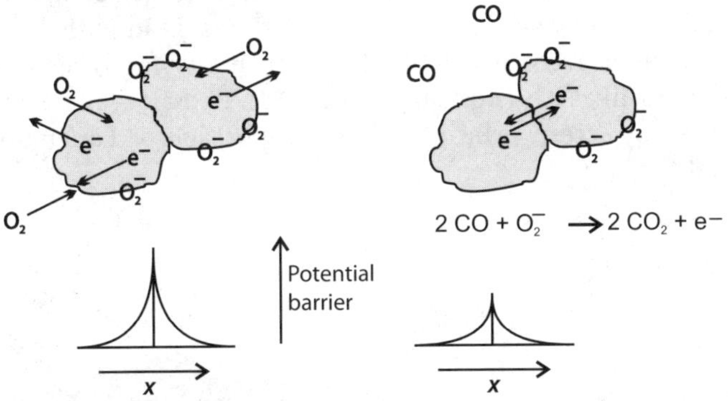

Figure 5.4. *Left:* Charge distribution following adsorption of oxygen at surface of *n*-type oxidic semiconductors and resulting potential distribution across grain boundary. *Right:* Effect of reducing gases on charge distribution and potential barrier

The functional characteristics of the Taguchi sensor can be applied to determine oxidizing gases if p-type semiconductors are used rather than n-type materials. The higher content of oxidizing gases, e.g. increasing oxygen partial pressure, tends to decrease the conductance, in contrast to the effect of reducing gases. Sensors of this type, however, do not play an important role in the oxidation of gases.

In the age of highly developed microsystems technology, sinter pellets seem outdated. Furthermore, it can be supposed that the response of such pellets is sluggish due to the relatively long distances which must be traversed by the gas molecules in order to reach equilibrium. Consequently, more and more sensors are manufactured in the form of planar structures following the scheme given in Fig. 5.5. On the surface of a ceramic alumina support are located metallic leads (commonly of noble metals) made by photolithography. The semiconducting oxide film is coated over this structure. On the opposite side of the ceramic support, a heating layer made of inert metal is attached. The sensitive oxide layer can be generated by means of established techniques like sputtering, vacuum deposition or chemical vapour deposition (CVD). There exists an alternative technology exclusively for SnO_2. Films of this material can be made easily by means of an aerosol of a tin(II)chloride solution in hydrogen chloride. The aerosol is sprayed on the hot substrate surface in an atmosphere of air. Films of this material are extremely robust and can even be used as resistive heaters.

As expected, homogeneous semiconductor sensors in planar technology are characterized by much lower response time compared to sensors in the form of sinter pellets. Modern thin-film techniques allow precise doping of the material.

Chemoresistors made of organic semiconductors have been proposed as an alternative to oxide ceramic sensors. Phthalocyanines (Fig. 5.6) have a behaviour similar to that of oxide semiconductors. Hydrogen atoms in the centre of the molecule can be substituted by metal atoms. Lead phthalocyanine has been studied intensively. It exhibits high sensitivity for gases which act as electron donors, like NO_2 (Bott and Jones 1984). Phthalocyanine layers can be manufactured by screen printing, vapour deposition or Langmuir–Blodgett film technology.

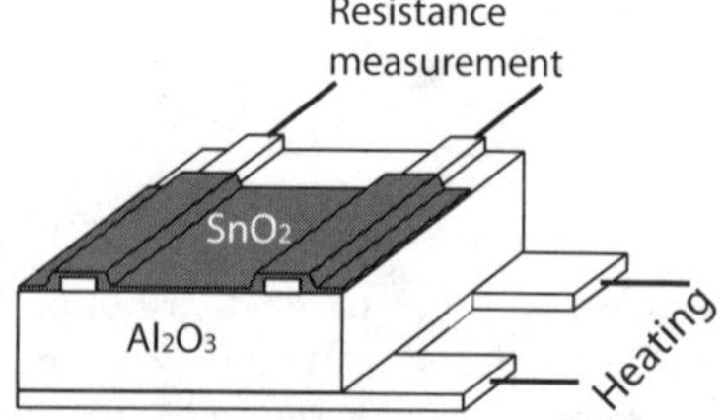

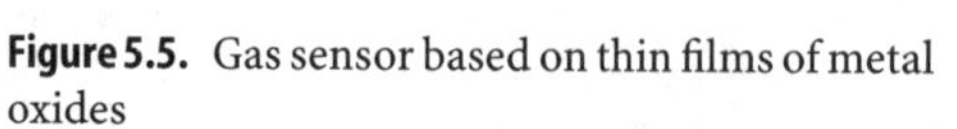
Figure 5.5. Gas sensor based on thin films of metal oxides

Figure 5.6. Structure of metal-free phthalocyanine

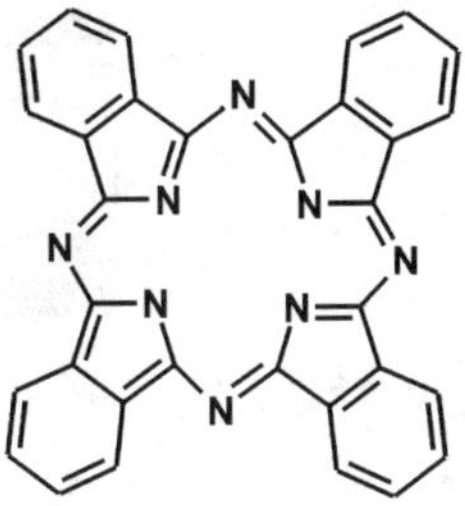

5.2.2
Gas Sensors Made of Polymers and Gels

A rather simple sensor design is based on a sensitive layer placed between two electrodes of metallic conductance (Fig. 5.7). The question whether the device is a resistive or a capacitive sensor cannot be answered a priori. First, the geometry must be considered. Resistive sensors have a larger ratio of electrode surface to receptor layer volume than capacitive sensors. Capacitive receptor layers may eventually have an infinite value of electric resistance. A measurable change in capacity is achieved by interaction of the dielectric with sample components. If the *dielectric coefficient* changes its value, then the capacity also changes.

Some types of chemoresistors and chemocapacitors for gaseous components are composed of electrically conducting polymers. For example, a polyphenylacetylene layer may interact with CO, CO_2, N_2 and CH_4. The result is a distinct variation of the AC resistance.

If solids are wetted, then their lattice components become hydrated, and mobile ions can form. As a result, their conductance increases. This behaviour is the basis for resistive humidity sensors. Commonly, a thin layer of the sensitive solid is placed between two electrodes. A useful material e.g. is phoshorous pentoxide.

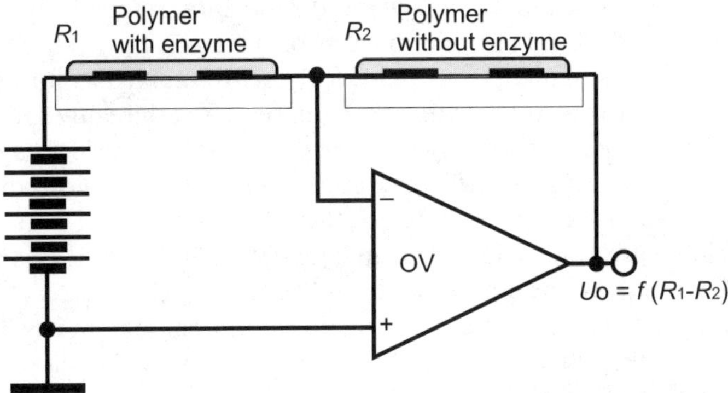

Figure 5.7. Basic design of chemoresistive (*left*) and chemocapacitive gas sensor (*right*)

Figure 5.8. Capacitive humidity sensor with porous β-alumina dielectric. *Right*: equivalent circuit

Capacitive humidity sensors commonly contain layers of hydrophilic inorganic oxides which act as a dielectric. Absorption of polar water molecules has a strong effect on the dielectric constant of the material. The magnitude of this effect increases with a large inner surface which can accept large amounts of water. An example of this type of dielectric is porous β-alumina. Colloidal ferric oxide, certain semiconductors, perowskites and certain polymers have also been used. β-alumina is characterized by ionic conductance. Materials of this type can be characterized by a complex resistance composed of real (ohmic) as well as capacitive terms. The behaviour of such solids can be symbolized by a model and an associated equivalent circuit as given in Fig. 5.8.

5.3
Resistive and Capacitive Sensors for Liquids

Chemoresistors for liquid phase (*impedimetric sensors*) have a design similar to that of gas sensors (Fig. 5.7). In contact with electrolytic solution, a specific electrochemical cell is established. With this cell, the measuring set-up cannot be arranged to respond to effects of a single electrode alone, as was possible with classical electrochemical impedance experiments (Sect. 2.2.6). Hence, with chemoresistors the equivalent circuit must consider both electrodes. For a sensitive layer with some intrinsic conductivity, for the low frequency range the conditions can be symbolized approximately by Fig. 5.9. C_f and R_f symbolize the film's capacity and resistance, respectively. C_i and R_i are the corresponding quantities of the sensor interface.

Impedance measurements are useful for resistive as well as for capacitive sensors. Analytes can affect the different components of the equivalence circuit in various ways. By *impedance spectroscopy*, i.e. by phase selective determination of the complex quantities, maximum sensitivity and selectivity can be achieved. Impedance in this case is not the preferred method in fundamental research; its greatest usefulness is in analytical applications. It is aimed at a calibration curve as linear as possible.

Among the examples of chemosensors for liquid phase is an interdigitated lead pattern coated with a hydrogel. When the pH is varied, the state of swelling of the gel changes. This might bring about a measurable variation in resistance. Enzymes can catalyse reactions accompanied by pH changes. If such enzymes

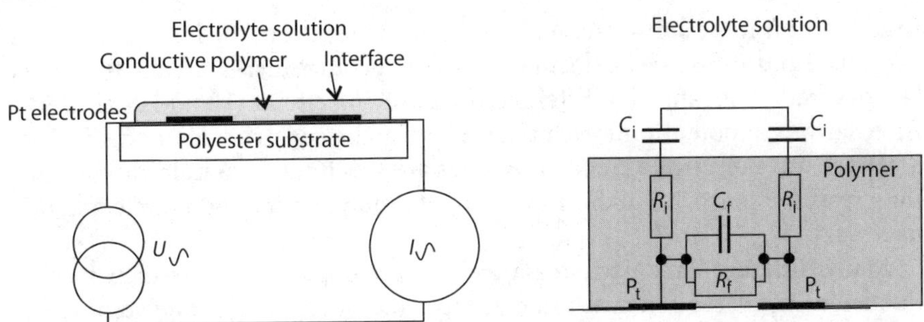

Figure 5.9. *Left*: Basic scheme of resistive and capacitive sensors with polymer or gel layers; *right*: simplified equivalence circuit for low-frequency range. C_f and R_f denote film's capacity and resistance, respectively. C_i and R_i denote corresponding quantities of sensor interface

are embedded in gel layers of the type mentioned, then a specific type of biosensor results. One example is a urea sensor (Sheppard et al. 1995).

Chemoresistors have proved useful for biosensors in particular. Many enzymes catalyse reactions which generate ions. Such enzymes can be immobilized easily in thin hydrogel films covering interdigitated structures. Increases in the ionic concentrations as a result of enzymatic reactions will strongly enhance the conductivity. The latter can be measured easily with little effort. Very useful is an electronic circuit measuring the difference in conductances of one sensor with an enzyme and another one without it (Fig. 5.10).

Nowadays, chemoresistive biosensors are very common. Most widespread are urea sensors, which are a good way to explain the function principle. The enzyme urease catalyses the following reaction:

$$CO(NH_2)_2 + H_2O \xrightarrow{\text{urease}} CO_3^{2-} + 2NH_4^+ \ .$$

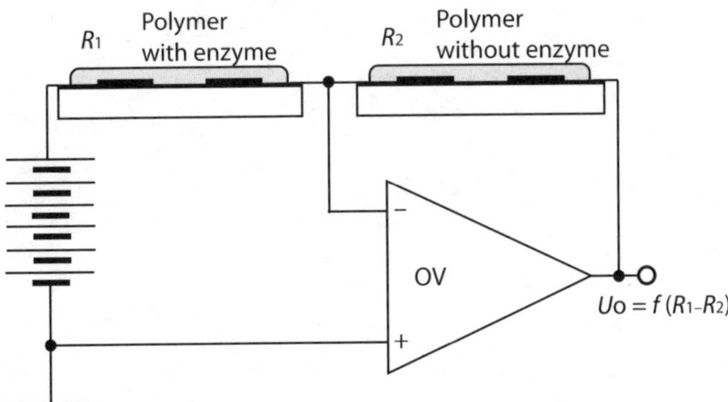

Figure 5.10. Circuit measuring difference in conductances of two resistive biosensors. R_1 is sensor containing enzyme, R_2 sensor without enzyme. The enzyme catalyses a reaction which generates ions

Urea is a non-ionic substrate. As a result of the reaction, the ionic substances carbonate and ammonia are formed. The conductance increases, and this effect is measured by means of a differential measuring circuit (Hendji et al. 1994). Alternatively, a potentiometric urea sensor can be constructed on the basis of a pH-sensitive electrode. Such electrodes are much more complex and fragile than resistive sensors. Furthermore, in potentiometry a reference electrode is necessary.

Many different enzymatic reactions are accompanied by changes in conductance. Amidases generate ionic groups, and phosphatases and sulphatases change the volume of charge carrying groups. All these reactions and others can be used in biosensors.

5.4
References

Bott B, Jones TA (1984) Sens Actuat 5:43–53

Fare TL, Silvia JC, Schwartz JL, Cabelli MD, Dahlin CDT, Dallas SM, Kichula CL, Narayan-swamy V, Thompson PH, Van Houten LJ (1996) Electrochemical sensors: capacitance. In: Taylor RF, Schultz JS (eds) Handbook of Chemical and Biological Sensors. Institute of Physics, Bristol, PA, pp 459–480

Hendji AMN, Jaffresic-Renault N, Martelet C, Shul'ga AA, Dzydevich SV, Sodatkin AP, Al'skaya AV (1994) Sens Actuat B21:123–129

Seiyama T, Kato A, Fujiishi K, Nagatani M (1962) Anal Chem 34:102

Sheppard Jr NF, Lesho MJ, McNally P, Francomacaro S (1995) Sens Actuat B28:95–102

Taguchi N (1962) Japan. Patent 45–3820

6 Thermometric and Calorimetric Sensors

6.1
Sensors with Thermistors and Pellistors

A thermometric sensor can be set up easily by coating the surface of a thermometer with a catalytic layer. If the catalysed reaction has a considerable heat effect, then the reaction heat is preferably released locally at the active surface. *Thermistors* are micro thermometers useful as a basis for thermometric chemical sensors. They mainly consist of a semiconductor body with a temperature-dependent conductivity. As an example, a hydrogen sensor is created by coating a thermistor with a thin layer of black platinum (Fig. 6.1, left). Hydrogen traces in the air burn in the catalytic area. The reaction heat causes the temperature to rise. A stationary temperature state is attained soon since the heat transport away from the sensor by heat conduction increases with temperature. The resulting temperature difference compared to ambient temperature can be measured in terms of resistance change. This difference depends on the hydrogen content in the air. Other combustible gases like hydrogen sulphide or carbon monoxide can be analysed by means of the same arrangement, but with different catalysts. It is essential to find a catalyst as selective as possible.

Thermistors are available in numerous sizes and shapes. Common forms are balls with diameters as small as 0.1 mm or thin films on a substrate. The tem-

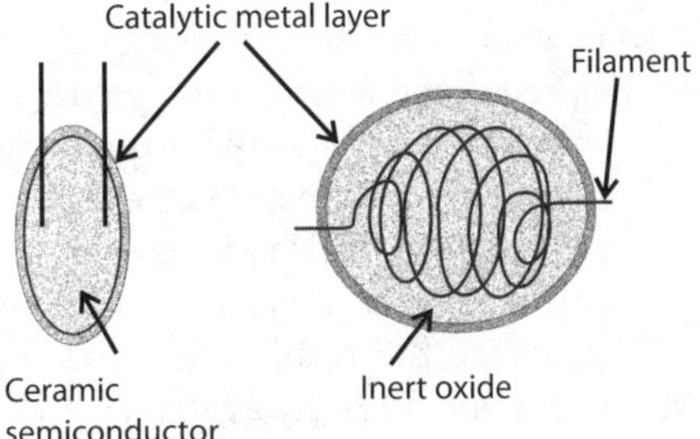

Figure 6.1. Thermometric (calorimetric) sensors. *Left*: thermistor, *right*: platinum resistance thermometer (pellistor)

perature coefficient of their conductance is very high. In a narrow temperature range, their resistance obeys the following equation:

$$R_T = R_{T_0} e^{\beta\left(\frac{1}{T} - \frac{1}{T_0}\right)} ,$$

(6.1)

where R_T and R_{T_0} are resistance values at temperatures T and T_0, respectively, and β is a constant specific to the actual sensor material. Its common value is ca. 5000 K. Consequently, the temperature sensitivity amounts to ca. 3 to 5% per Kelvin.

A further thermometer type used in sensors is the platinum resistance thermometer (Fig. 6.1, right). The device shown is known as a *pellistor*. A thin platinum coil is placed in an inert ceramic body. The temperature coefficient of the specific conductance of platinum is not very high. Hence pellistors are less temperature sensitive than thermistors. On the other hand, pellistors tolerate much higher temperatures than thermistors made of semiconducting oxides. The temperature coefficient of platinum resistance follows the relationship given in Eq. (6.2):

$$\varrho_t = \varrho_0(1 + 0.0039)t ,$$

(6.2)

where ϱ_0 denotes specific platinum resistance at $0\,^{\circ}$C and t the temperature in degrees centigrade ($^{\circ}$C). The specific resistance changes by ca. 0.4% per Kelvin, much less than the corresponding change with thermistors. Nevertheless, pellistors with metallic resistors are very common. The reason is, besides the high useful temperature range, their good long-term stability. Semiconductor thermistors are subject to ageing, i.e. their resistance changes slowly over time. The problem can be minimized to some extent by 'artificial ageing' (storage at increased temperature for a certain amount of time), but it cannot be avoided completely.

Commonly the temperature-dependent resistance change is measured by means of a Whetstone bridge. In most cases, it is sufficient to measure the bridge imbalance, without balancing the circuit. In an alternative measurement procedure, an electronic control circuit keeps the resistance (and consequently, the temperature) of the sensor constant. If reaction heat affects the sensor, a somewhat lower heat energy must be added continuously to keep the sensor at the predefined temperature value. This difference can be measured in terms of a power difference ΔP. This quantity is closely related to the heat effect of the reaction studied. Both quantities are *streams of energy*. For sensors of this type, the term *calorimetric sensor* is preferred. Operating the sensor with a constant current is advantageous since the curvature of the temperature probe does not influence the result. Also, the thermal convection is kept at a constant value. Overall, linearity as well as precision is improved compared to methods involving temperature measurement.

Thermometric and calorimetric sensors are discussed here only for use in gases. In the liquid phase, the conditions are much less advantageous, since the thermal conductivity of liquids is orders of magnitude higher than that of gases. Calorimetric measurements must be performed in closed vessels with thermal isolation, or at least in a flowing stream. Some application examples are encountered when thermometric sensors are used as detectors for flow-injection analysis.

6.2
Pyroelectric Sensors

A search for probes with higher temperature sensitivity resulted in expriments with *pyroelectric materials. Pyroelectricity* is related to *piezoelectricity*, as explained in Chap. 2, Sect. 2.1.3. Sometimes, both effects can be encountered in the same material. One could imagine that a volume increase of a piezo crystal at a higher temperature should result in effects similar to that caused by mechanical compression or expansion. A chemical sensor would emerge from a temperature change generated by a reaction that included the analyte. All the catalytic layers mentioned in the preceding chapter should be useful with pyroelectric sensors as well. To date, pyroelectric sensors have been used exclusively with gaseous samples.

Temperature change in a sensor does not depend alone on the heat effect of the utilized reaction, but also on environmental parameters like the convection or thermal conductivity of the medium. Improved reliability is achieved by means of a differential measuring circuit (Fig. 6.2). The temperature effect of the hydrogen sensor in Fig. 6.2 is a result of the heat effect of the absorption process of hydrogen at the active palladium layer.

A commonly used pyroelectric material is lithium tantalate $LiTaO_3$. For use in sensors, the material is doped with lanthanum traces. Polyvinylidene

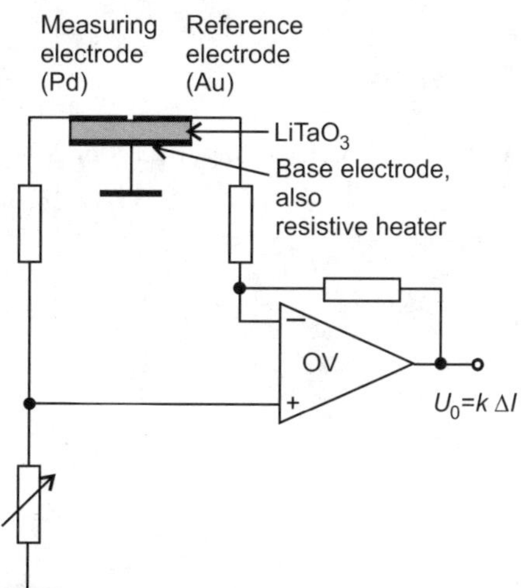

Figure 6.2. Pyroelectric sensor for hydrogen in differential measuring circuit

fluoride (PVDF) is a synthetic organic pyroelectric polymer. This interesting material consists of linear chain molecules where fluorine atoms alternate regularly with CH_2 groups, causing the molecule to attain a zigzag folding configuration. Foils made of this material can be manufactured in such a way that molecules are arranged in parallel. By subsequent application of a strong electric field, the molecules are rearranged with their fluorine atoms (their negative 'backbone') oriented to one side of the foil. Such foils have piezoelectric as well as pyroelectric properties. They can simulate the behaviour of human skin. There are, however, only rare examples of chemical sensors utilizing this material.

6.3
Sensors Based on Other Thermal Effects

In addition to the temperature-dependent properties discussed above, only a few other effects have been used for chemical sensors. An example worth mentioning is thermoelements, where the temperature coefficient of the contact potential between two different electric conductors is used. The most common thermoelements are combinations of platinum with its alloys as, for example, the platinum-rhodium alloy. The temperature coefficient of these thermoelements is not marked enough to be used in chemical sensors. Combinations including different semiconductor materials show a much greater effect. The temperature dependence of their contact potential is known as the *Seebeck effect*. When platinum was combined with semiconducting oxides like SnO_2, hydrogen traces could be detected (McAleer et al. 1985). It seems, however, that the high expectations did not pan out. Little research has been published on such sensors.

6.4
References

McAleer JF, Moseley PT, Bourke P, Norris JOW, Stephen R (1985) Sens Actuat 8:251–257

7 Electrochemical Sensors

Electrochemical sensors immediately generate electric signals. This is one of their main advantages and is one of the reasons for the close connection between the fields of chemical sensors on the one hand and electrochemistry, including the techniques of electrochemical experimentation, on the other. Electrochemists traditionally are skilled self-constructors of electronic instruments. Because of the nature of their work, they must understand how electronic instruments work. Such skills proved useful when sensors became an important application field of electrochemical science.

There are varying opinions about the question whether a specific device can be called an electrochemical sensor or not. A certain percentage of scientists would exclude conductance probes. The reason is that such probes are based not on running chemical reactions, but on the physical properties of ions which move in the inner volume of a solution far from where electrochemical processes take place. This problem can be solved by including the electrode interface in the field of conductance measurements. The processes at the electrode interface can be symbolized by defining equivalent resistances. All the partial resistances in an electrochemical cell can be combined to form a complex resistance called *impedance*. Contributions to this complex quantity come from the bulk solution as well as from the electrode interface. All the measurements in an electrochemical cell can be considered part of a unified *measurement of impedance*. Indeed, the terms 'conductance sensor' and 'impedimetric sensor' are often used as synonyms. Following these general considerations, electrochemical sensors can be classified in terms of the working principles listed in Table 7.1.

Table 7.1 does not contain 'capacitive electrochemical sensors'. They are based on measuring the *capacitive resistance* of a selective layer and are not really an independent group since basically the resistance is predetermined. In this specific case, all other contributions except the capacitive one are negligible.

The term 'coulometric sensor' is somewhat misleading. *Coulometry*, in its original sense, means to measure the amount of charge (the current-time product) for an electrochemical reaction and to calculate the corresponding amount of substance. Commonly, for reactions considered in this way, the total

Table 7.1. Transducer principles and measuring techniques of electrochemical sensors

Sensor	Transducer principle	Measured quantity
Potentiometric	Energy conversion	Voltage (high impedance)
Amperometric and "coulometric"	Limiting current	Current (low impedance)
Conductometric or impedimetric	Resistive	Resistance (= reciprocal of conductance)

conversion of the reacting species is assumed. This is the reason for using the term 'coulometric' for *amperometric sensors* where in the course of detection a high degree of conversion is achieved.

7.1
Potentiometric Sensors

Potentiometry has long been a significant part of instrumental chemical analysis. The experimental set-up is simple (Fig. 7.1). As reference electrodes, those *of the second kind* are useful. They are created by a combination of a metal/metal ion electrode and a stock of a sparingly soluble salt of the participating metal. Among the practical examples is the silver/silver chloride electrode with a salt bridge of potassium chloride solution, mentioned in Chap. 2, Sect. 2.2.6. The instrument for potential measurement must be charcterized by a very high input resistance. So-called *pH meters* are well suited for this purpose since they

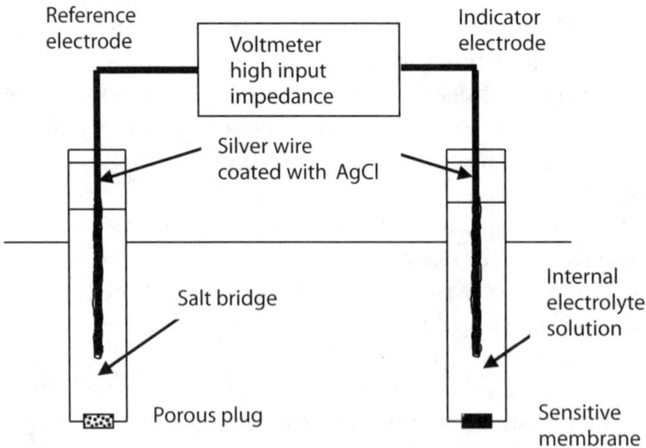

Figure 7.1. Experimental set-up for potentiometry

commonly have input resistances in the gigaohm range. Measuring electrodes used in classical potentiometry are often relatively large and expensive, since traditionally they are produced by small companies operating largely as in a cottage industry. With the advent of the sensor age, efforts were launched to miniaturize sensors and to mass produce them. Simultaneously, a tendency arose to apply the term sensor also for well-known traditional potentiometric probes. It seems better, however, to restrict the term sensor exclusively to devices characterized by miniature dimensions and by widespread availability. Nevertheless, it is useful to start with classical electrode forms when discussing potentiometric sensors. Classical potentiometric electrodes differ from potentiometric sensors only in their geometric dimensions.

Nernst's equation is the basis of all potentiometric measurements. The common form of this equation is that given in Eq. (7.1) for a working temperature of 25 °C:

$$E = E^{\theta'} \pm \frac{0.059}{z_i} \log a_i \, . \tag{7.1}$$

As expected, Eq. (7.1) describes activity rather than concentration. Potentiometric measurements always result in activity values. The number z_i is identical to the charge of the ion studied. The sign of the concentration-dependent term is positive for cations and negative for anions. A closer look at the equation shows that a concentration change by a factor of 10 should result in a potential variation of ca. 60 mV for single charged ions but 30 mV for double charged ions. Consequently, the concentration of single charged ions can be determined with higher precision. $E^{\theta'}$ is not identical to the standard potential E^{θ}. It is a constant which depends mainly on the reference electrode used and must be determined empirically for the actual set-up.

The thermodynamic deduction of the Nernst equation given in Chap. 2 ignores the fact that the value measured between working and reference electrodes is composed of numerous interface voltage values, the *Galvani potential differences* g_i. Figure 7.2 tries to demonstrate how a measurable voltage (EMK) could arise at a copper/copper ion working electrode and a hydrogen reference electrode. The ordinate distances drawn to symbolize g values are hypothetical,

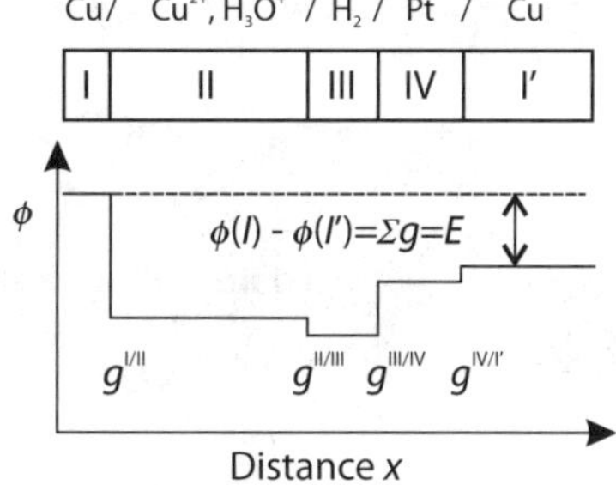

Figure 7.2. Potential difference measured between working and reference electrodes interpreted as sum of all Galvani potential values at each interfaces g_i, and alternatively as inner potential difference between outer phases $\phi(I)$ and $\phi(I')$

of course. Such interface potential differences are not measurable as a matter of principle. Alternatively, the emf of the cell can be interpreted as the difference between two *inner electric potentials* ϕ of two phases I and I' at the outer ends of the cell. Both these terminating phases consist of copper, since in every case there must be a connection from and to the voltage meter. Every phase existing in the universe can be characterized by a definite value of *inner electric potential (Galvani potential)*. These values are also of a hypothetical nature.

Among the galvani potential differences g_i contributing to the overall voltage (emf), one specific g is of analytical interest, namely that at the interface between sensor and sample solution. A potentiometric experiment should be designed in such a way that the result can be summed up in this single value. The precaution in meeting this requirement is to keep constant all other potential contributions except the one of interest. One can utilize the fact that each Galvani potential difference, although not measurable, will follow the Nernst equation. This fact follows from logical considerations, but alternatively it can be derived by means of a specifically defined quantity called the *electrochemical potential*. Use of this term is omitted here.

A difficult technical problem in the construction of miniature potentiometric sensors is the design of the reference electrode. All the well-established types (Fig. 7.3) require liquid-liquid junctions between an electrode of the second kind and the sample solution. Even with classical macroscopic constructions,

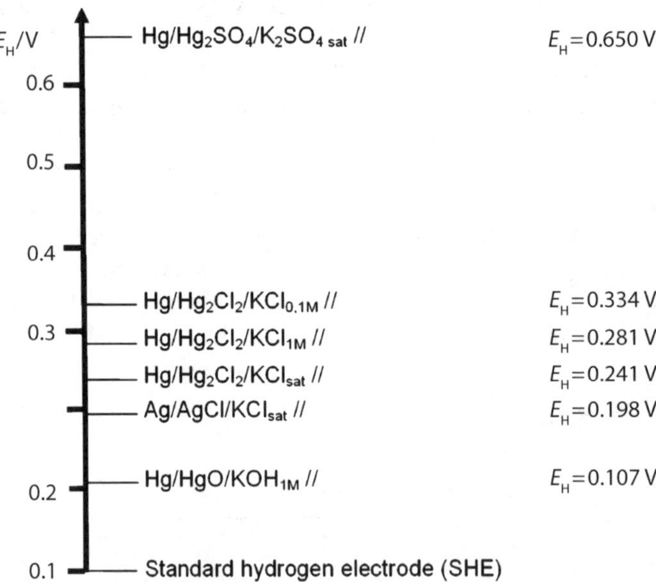

Figure 7.3. Standard potentials of common reference electrodes

this is the source of many problems. Porous plugs of ceramics and other materials have been proposed to stabilize the junction, as well as labyrinth seals, gel plugs or a thin solution film in the gap of a ground glass taper joint. Drying out such diaphragms causes massive problems. It is hardly possible to reactivate a reference electrode corrupted in this way. Even more difficult is constructing a miniature diaphragm for chemical sensors. Commonly, a *pseudo reference electrode* is preferred. It forms if an electrode of the second kind reaches a stable equilibrium with an ion in solution. It is simpler to coat some part of the sensor surface (e.g. a metallic silver area) with a layer of an active substance like AgCl. The sample solution must be spiked with a sufficient excess of chloride ions in this example. The combination of the three components mentioned (Ag and AgCl at the sensor surface, Ag^+ in the sample solution) results in the formation of a pseudo reference electrode. The chloride concentration, of course, should be kept constant during measurement.

7.1.1
Selectivity of Potentiometric Sensors

Selectivity is one of the most important advantages of potentiometric sensors. The ideal case would be the *specific* sensor which responds to only one single type of ion. This ideal cannot be reached in practice, but some types of sensor approximate it. Commonly, for every analyte a certain number of disturbing substances exist which cause *interferences*. The interfering ion 'simulates' the presence of a certain sample ion concentration. This distortion is not constant but affects lower sample concentrations, though to a lesser extent than it affects higher concentration values. To overcome this problem, different attempts have been described. The best known approach is the *Nikolskij equation* (*Nikolskij–Eisenman equation*), given in Eq. (7.2). This equation considers the *sample ion i*, which is in competition with the *interfering ion j*. The extent of interference is expressed in terms of the *selectivity coefficient* K_{ij}. The numbers z_i and z_j are the charge numbers of sample and interfering ions, respectively.

The selectivity coefficient expresses the ratio of sensitivities of interfering vs. sample ion. As an example, the value $K_{ij} = 10^{-2}$ means that the interferent must be present in a 100-fold excess compared to the sample to bring about an equal effect at the sensor:

$$E = E^{\theta'} \pm \frac{0.059}{z_i} \log \left(a_i + K_{ij} \cdot a_j^{\frac{z_i}{z_j}} \right) . \tag{7.2}$$

Equation (7.2) proves that the effect of interference decreases with increasing sample activity a_i. The selectivity coefficient can only give some indication of a possible distortion.

7.1.2
Ion-Selective Electrodes

As mentioned in a previous chapter, the art of manufacturing ion-selective electrodes (ISEs) consists in searching for the proper method to prepare the sensor surface in such a way that a Galvani potential difference results which should depend selectively on the activity of only one type of ion, if possible. The best option would be a *specific* electrode, but at the very least it should be *selective*.

IESs are classified into two main groups, depending on whether the interface in contact with the sample solution is a solid surface or a liquid surface. The resulting groups are *solid-membrane electrodes* and *liquid-membrane electrodes*. The term 'membrane' in this case does not always mean a thin film as usual, but can be a compact body. Among the IESs, the well-known glass electrode occupies a place of particular importance, more so than any other ISE, and it can be assigned to the group of solid-membrane electrodes as well as to that with a liquid membrane.

Galvanic cells can be set up with solid electrolytes rather than electrolytic solutions. Such a cell is the basis for a well-known potentiometric gas sensor, the *lambda probe*. The latter is designed to determine the oxygen content of combustion gases, e.g. in motor vehicles. The lambda probe can operate in two different modes, either potentiometrically or amperometrically.

Potentiometric Sensors with Solid Membranes

Different kinds of solid-membrane ISEs can be defined by considering how they make an electric connection between membrane and measuring instrument (Fig. 7.4). A most versatile embodiment is the connection via an internal reference electrode (Fig. 7.4, left). Such an electrode can hardly be realized in miniature form. An internal cavity is filled with a solution that develops

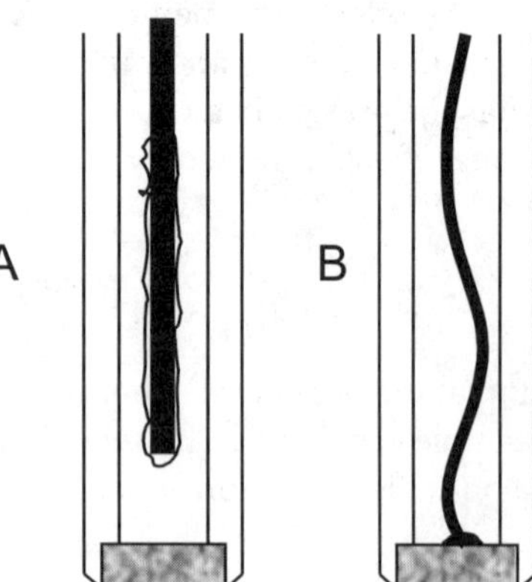

Figure 7.4. Ion-selective electrodes with solid membranes. **A** Contact via electrode of second kind. **B** Direct contact

a stable equilibrium with the electrode of the second kind impinging on this solution, e.g. a silver wire coated with silver chloride. The actual value of the equilibrium potential can be varied by changing the composition of the inner solution.

The direct contact shown in Fig. 7.4 (right) seems to be a simpler and less tedious construction. This is not the full truth, however, although indeed it is easy to attach a metallic wire to the remote side of a solid membrane. Some solids can be soldered, e.g. by means of indium metal. Often it is sufficient to press a wire into a somewhat undersized hole. Also, gluing with silver powder containing epoxy resin or similar conducting mixtures has proved useful. Unfortunately, the direct contact gives rise to a lot of problems. At the interface, a Galvani potential difference is formed which is strongly temperature dependent. The actual value of this potential difference depends on the properties of the measuring instrument. Furthermore, mechanical influences are known, e.g. the contact responds to mechanical attack like a microphone where the resistance changes in response to pressure variation. Such behaviour is characteristic for depletion layers forming at junctions like the p–n junction in semiconductors (Chap. 2, Sect. 2.1). The reason is the difference in conduction types of metallic lead and solid ISE membranes. The latter commonly are a solid electrolyte where ions are the charge carriers. Ions are mobile only in the membrane but not in the metallic contact. They are not able to cross the interface between both phases. When the measuring circuit is closed via a voltmeter, a depletion layer can be formed. The problems of the direct contact can be minimized in some cases by interposing a specific layer. An example is discussed later.

Examples of Solid-Membrane ISEs. In the 1960s, enthusiastic expectations subsided in connection with ISEs. Countless studies were devoted to new ISE types. Later, when a more realistic consideration arose, it proved that only a few ISE types were really useful in analytical practice. Some examples of well-tried sensors are given in Table 7.2, which lists examples in order of decreasing practicability.

The functional principle of solid-membrane ISEs can be explained best by considering them as a special sort of electrode of the second kind. In principle, every second-kind electrode can be used as a potentiometric sensor. Starting with a simple metal/metal ion electrode, e.g. the silver/silver ion electrode, the potential can be written as follows:

$$E = E^{\theta} + 0.059 \cdot \log a\left(\mathrm{Ag}^{+}\right) . \tag{7.3}$$

The activity $a(\mathrm{Ag}^{+})$ is related to the chloride activity in solution via a solubility equilibrium as soon as the sparingly soluble salt AgCl is present:

$$\mathrm{Ag}^{+} + \mathrm{Cl}^{-} \rightleftarrows \mathrm{AgCl_{s}} .$$

Table 7.2. Examples of solid-membrane ion-selective electrodes

Sample	Membrane	Interferences
F^-	LaF_3 monocrystal	OH^-
S^{2-}; Ag^+	Ag_2S	Hg^{2+}
Cl^-	$AgCl + Ag_2S$	Br^-; I^-; S^{2-}; CN^-; NH_3
Br^-	$AgBr + Ag_2S$	I^-; S^{2-}; CN^-; NH_3
I^-	$AgI + Ag_2S$	S^{2-}; CN^-
SCN^-	$AgSCN + Ag_2S$	Br^-; I^-; S^{2-}; CN^-; NH_3
Cd^{2+}	$CdS + Ag_2S$	Ag^+; Hg^{2+}; Cu^{2+}
Cu^{2+}	$CuS + Ag_2S$	Ag^+; Hg^{2+}
Pb^{2+}	$PbS + Ag_2S$	Ag^+; Hg^{2+}; Cu^{2+}

For salts with low solubility like AgCl, $a = c$ can be set. By insertion of the solubility product $K_L = c(Ag^+) \cdot c(Cl^-)$, E can be written as a function of chloride concentration: $c(Ag^+) = \frac{K_L}{c(Cl^-)}$. For E, we get

$$E = E^\theta + 0.059 \cdot \log K_L - 0.059 \cdot \log c(Cl^-) \tag{7.4}$$

or, simplified,

$$E = E^{\theta'} - 0.059 \cdot \log c(Cl^-) . \tag{7.5}$$

A chloride-coated silver wire can be used to determine silver-ion as well as chloride-ion concentrations. Indeed, simple silver/silver chloride electrodes are still utilized for analytical measurements, e.g. in seawater analysis to obtain a quick survey about local salt concentration changes. One problem with such simple sensors is the porosity of the sparingly soluble salt layer. Interaction of the homogeneous solution through the pores brings about some sensitivity, unlike dissolved oxygen, since a *redox electrode* is formed when metallic parts are in contact with the dissolved redox couple O_2/OH^-. It is therefore desirable to make the membrane as dense as possible to avoid penetration by dissolved redox couples. With dense membranes, however, new problems arise. Commonly, the pure deposit of a sparingly soluble salt is characterized by a rather low electric conductance. An extremely high resistance causes problems since the input resistance even of high-performance voltmeters has a finite value. Parasitic capacities at the input then may cause sluggish response times. The conductance of the solid membrane can be increased by the addition of conducting particles or by doping the solid.

The membrane should not only be dense but also as smooth as possible. A mirror-like finish without pores is desirable. Otherwise, sluggish response can be expected since all the concentration differences close to the interface should be in equilibrium. Concentration gradients must be balanced out by diffusion, and diffusion into and out of narrow pores is a slow process.

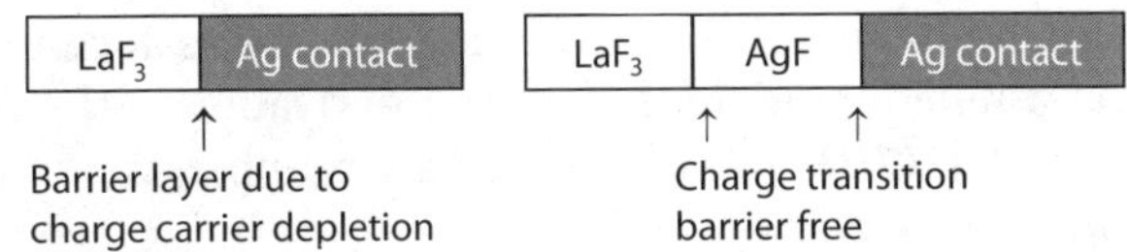

Figure 7.5. Barrier-free direct contact at lanthanum fluoride electrode

To date, the ISE with the best performance is the fluoride-sensitive electrode based on a lanthanum fluoride single crystal. The only meaningful interferent is the OH^- ion. A remarkable interference is encountered only at high pH, i.e. in a strongly alkaline medium. The conductivity of the single-crystal 'membrane' is increased by doping with europium. The direct-contact problem of this electrode has been solved in an elegant way. The procedure is useful also for other electrode types. It is based on the following considerations. In the bulk of the LaF_3 crystal, charge is carried mainly by mobile fluoride ions. Under certain conditions, at the interface with the metallic contact, depletion of charge carriers arises since ions cannot cross the boundary to the metal. The problem is solved by insertion of an intermediate layer where charge transfer towards the metal as well as towards the lanthanum fluoride is possible. The function is illustrated in Fig. 7.5.

Silver sulphide is a useful basis for a certain number of successful solid-membrane ISEs. Examples are listed in Table 7.2. Finely powdered silver sulphide, i.e. a washed precipitate, can be pressed by application of high pressure and increased temperature to form dense pellets which can be ground and polished. The conductivity of such pellets is relatively high, since anions are mobile in the crystal lattice. The pure Ag_2S itself is a reasonable sensor for sulphide ions. Mixtures of this substance with the sparingly soluble silver halides or pseudohalides retain the advantages of the silver sulphide body, but they are sensitive to halide or pseudohalide ions. The components in the membrane, e.g. Ag_2S and $AgCl$, take part in a well-defined equilibrium. It can be understood by realizing that the component $AgCl$ will dissolve according to its solubility product, generating a certain silver-ion concentration. The Ag_2S membrane will respond to this concentration. Eventually, the solid mixture in the membrane acts like a silver chloride electrode, but with a dense surface suited for polishing.

A similar mechanism is active in cation-sensitive electrodes on the basis of silver sulphide, like the electrodes sensitive to Cd^{2+}, Cu^{2+} and Pb^{2+}, mentioned in Table 7.2. These ions form sparingly soluble sulphides, which go in equilibrium with the silver sulphide base. It can be explained by assuming a mechanism where the salts generate a certain sulphide-ion concentration which gives rise to a response by the silver sulphide electrode. There exists a well-defined relationship between the measured potential E and the cation concentrations. The effect of the mentioned cations on the potential is only half that of single charged ions. The concentration dependence of such mixed membrane bodies is sometimes described as a *response mechanism of higher order'*. An alternative, more traditional classification is to call such electrodes *'electrodes of the third kind'*. For derivation of the relationships, one can start with Eq. (7.3). Taking Cd^{2+} as an example, the solubility equilibrium can be written as follows:

$$Cd^{2+} + S^{2-} \rightleftharpoons CdS \quad \text{mit} \quad K_L(CdS) = a(Cd^{2+}) \cdot a(S^{2-})$$

$$2Ag^+ + S^{2-} \rightleftharpoons Ag_2S \quad \text{mit} \quad K_L(Ag_2S) = a^2(Ag^+) \cdot a(S^{2-}) \ .$$

The silver-ion concentration can be expressed by $a(Ag^+) = \sqrt{\dfrac{K_L(Ag_2S) \cdot a(Cd^{2+})}{K_L(CdS)}}$.

On the basis of the above considerations, Eq. (7.6) can be derived. It is a specific form of the Nernst equation with concentration values rather than activities. The constant E' is a modified standard potential considering all effects of solubility equilibria:

$$E = E' + \frac{0.059}{2} \cdot \log c(Cd^{2+}) \ . \tag{7.6}$$

Miniature Shapes. Traditional forms of IESs are not really chemical sensors, since they lack the characteristic geometric and commercial attributes of sensors. There is considerable demand for small and cheap potentiometric sensors. Changing from traditional manufacturing to mass production is often accompanied by a loss of sensor performance. Working with miniature sensors often means estimating rather than measuring. On the other hand, miniature sensors offer more possibilities than traditional devices. Modern technology allows one to produce disposable sensors with performance sufficient for one single measuring process. Another new approach is the application of intelligent signal acquisition electronics to compensate for instabilities, drifting and other drawbacks.

Two problems must be solved preferably when miniaturization of potentiometric sensors is desired. It has been mentioned already that it is difficult to produce reference electrodes with miniature dimensions, and that the problem of the direct contact exhibits some problems. Obviously, the internal contact between the sensitive membrane and the voltmeter cannot be easily realized in the form of an electrode of the second kind with miniature sensors.

Two common forms of solid-membrane ISEs are the *coated-wire electrode* and structures made by *thick-film technology* (screen printing).

Coated-wire electrodes are metallic wires which have been dip-coated by a sensitive layer. Commonly several layers are coated consecutively. Often pastes of solid powders mixed with binders are used similarly to the well-known screen-printing inks. Like traditional set-ups, attempts have been made to overcome the problem of direct contact by inserting specific intermediate layers.

Thick-film patterns have been printed on different materials, among them e.g. impregnated paper. Ready-made screen-printing inks specifically designed for ion-selective structures are offered by different producers. Among them are mixtures for reference electrodes, e.g. silver powder mixed with silver chloride. The printed structures typically are formed by parallel stripes, one of them acting as receptor, the other one as pseudo reference electrode. A precisely measured concentration of halide ion must be added to sample solution to achieve a stable potential at the pseudo reference electrode. A large variety of potentiometric thick-film sensors is commercially available, many of them designed to be the basis for specific biosensors.

A much higher degree of miniaturization can be achieved by means of thin-film technology. However, complete integration of ion-sensitive structures on a semiconductor chip seems to be unusual. Instead, the combination of the receptor with an amplifying semiconductor component, the MOSFET (IGFET), proved more successful and seems to be better suited for application of microsystems technology. As a result, the ion-sensitive MOSFET (ISFET or CHEMFET) has been developed. This sensor type is discussed in a later chapter.

Potentiometric Sensors with Liquid Membranes

If the sample ion concentration must be determined, its interaction with the receptor surface should give rise to the formation of a definite local Galvani potential difference. A necessary precondition is that ions be able to cross the phase boundary. If in adjacent phases there exist charge carriers of different conductance types, then, as explained in Chap. 2 (Sects. 2.1.1 and 2.2.6), a 'partial charge separation' will occur, resulting in the formation of a double layer and in the generation of a contact voltage identical to the desired Galvani potential difference.

Charge transfer across a phase boundary can be provoked alternatively by an extraction equilibrium (Sect. 2.2.7). If a liquid immiscible with water is in contact with an aqueous sample solution and certain ions are soluble in both phases, then a concentration partition is established which can be interpreted by Nernst's law of partition, Eq. (2.44). The establishment of this equilibrium should be accompanied by voltage formation at the interface if the partitioning species is an ion.

Ions are polar particles. They are charged electrically and will always be surrounded by a hydrate shell, i.e. by a layer of water molecules. Normally, it cannot be expected that ions will spontaneously migrate towards a non-polar, hydrophobic environment. The solubility of 'normal' ions in a non-aqueous, organic solvent is very low. This situation can change drastically if ligands or other active substances present in the non-aqueous phaseare able to form stable complexes with the ions of interest present in the aqueous phase. As a result of interaction, the hydrate shell can be substituted by an environment less polar in nature. An outstanding example for such processes is connected with ligands possessing a polar molecular cavity and a non-polar external surface.

The ion studied is enclosed in the cavity, and the complex formed is slightly soluble in a non-polar solvent. In this way, the ion can be 'smuggled' through a phase boundary. Useful ligands often are chelate ligands, i.e. multidentate ligands which enclose the ion from different sides. Even more efficient are ligands with closed cavities which fit perfectly for the ion of interest and are equipped with specific groups binding it. A prominent group of such molecules is constituted by *neutral carriers*. A representative of this group is the natural antibiotic valinomycin, which exposes a specific interaction with potassium ion (Chap. 2, Sect. 2.2). The main effect of such active substances is to increase the equilibrium concentration of the sample ion in the non-aqueous phase or, in other words, a drastic change in Nernst's partition coefficient. As a result, the formation of a measurable Galvani potential difference at the phase boundary is markedly improved.

The transition of charge carriers through a phase boundary will not necessarily generate a homogeneous solution of sample ions in the bulk of the non-aqueous phase. To generate a measurable contact voltage, it would be sufficient if the charge transfer were restricted to a thin layer close to the phase boundary. Thus the phase boundary would act as an ion exchanger. Such conditions arise if the active substance interacting with the sample ion is not dissolved homogeneously in the non-aqueous phase but is interposed as an adsorptive layer between the phases. Amphiphilic molecules containing polar and non-polar terminating groups are usually able to accumulate at the phase boundaries of the discussed type, in particular if a long, hydrophobic alkyl chain is attached at the non-polar end. Molecules which can coordinate with sample ions at their non-polar side are called *liquid ion exchangers*. The phase boundary carrying a monolayer of such molecules acts like the surface of a solid ion exchanger. Commonly, a concentration-dependent Galvani potential difference forms at an interface of this kind.

The function of IESs with a liquid membrane can hence be ascribed to a variety of different chemical reactions and interaction equilibria. The extraction (partition) equilibrium is fundamental. Further contributions can come from complex formation, adsorption and ion exchange. Identical relationships are valid for voltage formation between metallic and electrolytic phases, as well as for phase boundaries discussed here. The potential which can be measured between the reference electrode and the terminal of an ISE with liquid membrane obeys the Nernst equation. A large variety of liquid-membrane compositions results from the large number of chemical phenomena participating in interface processes. Very important is the fact that ligands may be tailor-made to interact specifically with certain cations.

For practical use, it would be inconvenient to use liquid electrodes. It is useful to stabilize the non-aqueous phase. For the classical macroscopic design, different fixation methods are in use (Fig. 7.6). Very common are thin films of soft PVC. The 'softeners' which are traditional additives in PVC technology are highly viscous liquids with a high boiling point. They have proved to be

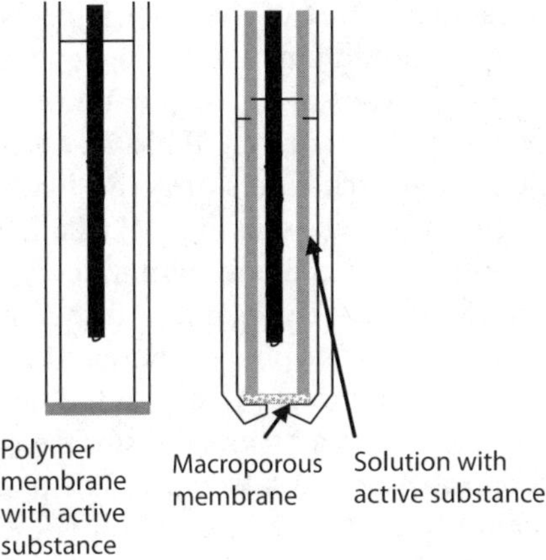

Figure 7.6. Ion-selective electrodes with liquid membrane

good solvents for active substances, e.g. for ion-selective ligands. By means of membranes containing softeners, well manageable electrodes have been constructed (Fig. 7.6). Such thin membranes are non-porous supports for a non-aqueous solution. Alternatively, macroporous polymer membranes are in use, e.g. filter membranes with a defined pore size of the Millipore type. The internal filling of such devices is a non-aqueous solution containing active

Table 7.3. Examples of ion-selective electrodes with liquid membrane

Active substance		Sample ion	Interferences
Liquid ion exchangers	$(RO)_2POO^-$	Ca^{2+}	$Na^+, Mg^{2+}, Ba^{2+}, Zn^{2+}$
	$RSCH_2COO^-$	Cu^{2+}	$Na^+, K^+, Ni^{2+},$ alkaline earths
	Crystal violet	NO_3^-	ClO_4^-
	Phenanthroline complexes of Ni^{2+} and Fe^{3+}	$NO_3^-; ClO_4^-$	$F^-, SO_4^{2-}, PO_4^{3-}, Cl^-,$ $HCO^{3-}, CN^-, NO^{2-}, Br^-$
Neutral carriers (ionophores); cryptands; cyclodextrines; calixarenes	Valinomycin	K^+	Na^+
	Monensin	Na^+	K^+, NH_4^+
	Dicyclohexyl-18-crown-6	K^+	Na^+, NH_4^+
	ETH 295	UO_2^{2+}	Numerous ions M^{2+}
	Cryptand-222	Zn^{2+}	Ca^{2+}, Cr^{3+}
	Per-O-actyl-α-cyclodextrin	Ephedrinium ion	
	Methyl-p-tert-butylcalix[4]-arylacetat	Na^+	Alkali ions

substances. Part of this solution fills the membrane pores, thus establishing a liquid-liquid junction to the aqueous sample solution (Fig. 7.6, right).

Some types of liquid-membrane ISEs of either practical or historical importance are listed in Table 7.3. The chemical formula of some meaningful liquid ion exchangers or active ligands are given in Figs. 7.7 and 7.8, respectively. Substance a in Fig. 7.7 (dialkylphosphate) illustrates the procedure for making a molecule amphiphilic by attachment of a long alkyl chain. At an interface between organic and aqueous phases, the alkyl chain will extend into the non-aqueous phase, whereas the ionic side will orient towards the aqueous phase and interact there with the sample ions. Thus, an ordered monolayer is formed which acts like the surface of a polyelectrolyte. At such an interface, ion exchange can take place. The phenanthroline complex of nickel ion (substance b in Fig. 7.7) is a hydrophobic cation which is soluble in non-aqueous liquids. It will accumulate at a phase boundary with an aqueous solution and can act there as an ion exchanger for single-charged anions like nitrate. It is the basis for a nitrate-selective ISE with liquid membrane. The dye crystal violet (substance c in Fig. 7.7) is a cation with a bulky organic side. A solution of this dye in the non-aqueous solvent nitrobenzene can be used to construct a nitrate-sensitive ISE.

Among the active substances useful for single-charged cations, in particular for the analytically important alkali metal ions, are naturally occurring antibiotics like the valinomycin mentioned already or monensin, which can be used to set up a sodium-sensitive electrode. Substance a in Fig. 7.8 (dicyclo-18-crown-6) belongs to the group of cyclic polyethers, the so-called crown

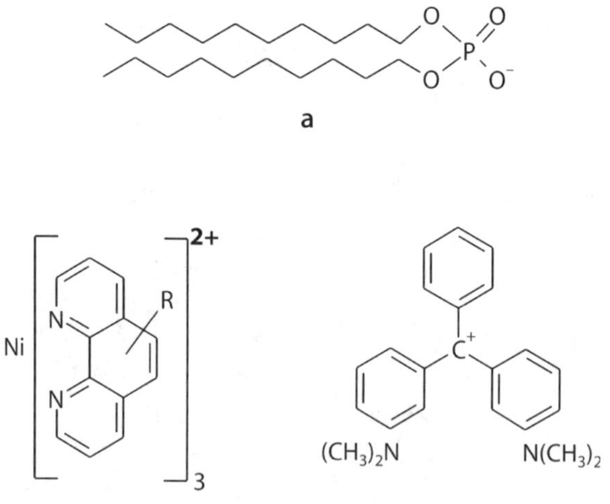

Figure 7.7. Ion exchanger substances used for ISEs with liquid membrane. **a** Dialkylphosphate. **b** Nickel phenanthroline complex. **c** Crystal violet

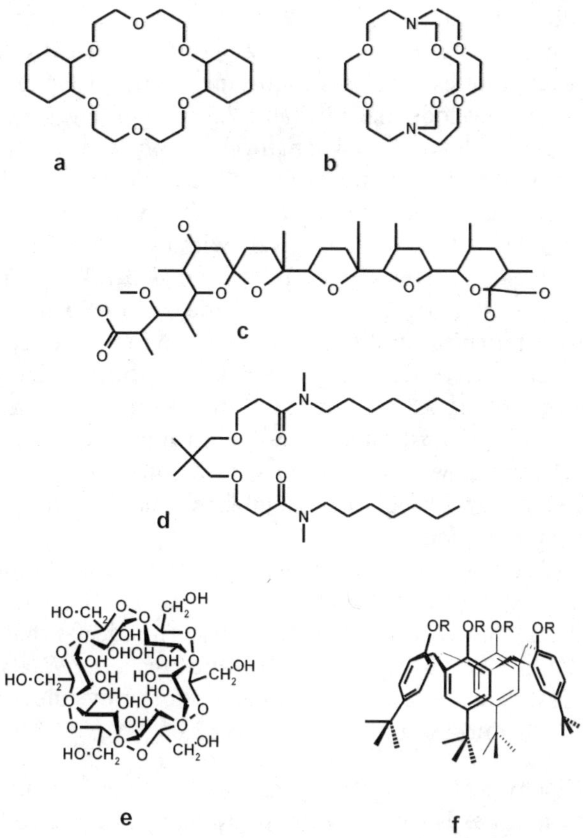

Figure 7.8. Ligands useful for ISEs with liquid membranes. **a** dycyclohexyl-18-crown-6. **b** Cryptand-222. **c** Monensin. **d** ETH 295. **e** α-cyclodextrin. **f** calix[4]arene

ethers. Substance a is a ligand which has been synthesized intentionally to react specifically with a specific cation. Potassium-sensitive electrodes have been built with substance a, like potassium electrodes containing valinomycin. There is considerable interest in the fast analytical determination of potassium ion, mainly in clinical chemistry, since this element is physiologically very active. Substance d in Fig. 7.8, the so-called ETH 295, has also been synthesized intentionally. It interacts selectively with the uranyl ion.

Some of the substances listed in Table 7.3 as examples of neutral carriers never became practically important. A large variety of such substances has been synthesized in the laboratory of the late W. Simon at ETH Zurich (e.g. the aforementioned ETH 295). Simon's group has compiled a list with specifications that characterize the performance of neutral carriers (Ammann et al. 1975).

In recent years, some recently synthesized substance classes have been tested with respect to their usability for selective recognition of ions in chemical sensors. Among them are the cryptands, which are ring systems closely related to crown ethers. Many studies have dealt with substance b in Fig. 7.8, known as

2,2,2-cryptand (4,7,13,16,21,24-Hexaoxa-1,10-diazabicyclo[8.8.8]hexacosan). It has been used to design a zinc-selective electrode (Srivastava et al. 1996). A very interesting class of substances is the group of *cyclodextrines* (e.g. per-O-octyl-α-cyclodextrine in Table 7.3 and e in Fig. 7.8). Cyclodextrines are cyclic compounds composed of natural carbohydrate components, which are formed by enzymatic decomposition of starch. The molecules form a toroid which is hydrophilic at its opening but internally acts like a hydrophobic bag. Certain sample ions interact selectively with cyclodextrines synthesized intentionally to fit closely. This steric selectivity may be used even for chiral recognition of optical isomers. As an example, the cyclodextrine mentioned above has been used for chiral studies of ephedrinium ion (Bates et al. 1994).

The *calixarenes* (e.g. methyl-*p*-tert-butylcalix[4]arylacetate in Table 7.3 and f in Fig. 7.8 are molecules that resemble a vase. They are synthesized by reaction of substituted phenols with aldehydes. The molecules have hydrophobic cavities which can hold sample ions. As an example, a sodium-sensitive electrode with polymer membrane has been designed as shown in Fig. 7.6, right (Diamond et al. 1988).

Active substances belonging to the groups of crown ethers, cryptands, cyclodextrines and calixarenes are members of a relatively new branch of science, the so-called *Host-guest chemistry*. The idea is that a small molecule or ion, the 'guest', is bound by electrostatic or van der Waals forces in the cavity of the 'host'. The steric constitution of the host molecule is crucial. This is a challenge for chemical synthesis.

Miniature Shapes. Miniaturization of liquid-membrane ISEs uses the same techniques as are used with solid-membrane electrodes. *Coated-wire electrodes* are manufactured by dip-coating a metallic wire with a polymer layer of e.g. PVC. The oldest example (Fig. 7.9, left) is a piece of platinum wire soldered to the internal lead of a coaxial cable. The end of the wire was formed into the shape of a ball by melting it in a hot flame. The polymer coating was applied by repeated dipping into the polymer solution (Freiser 1980).

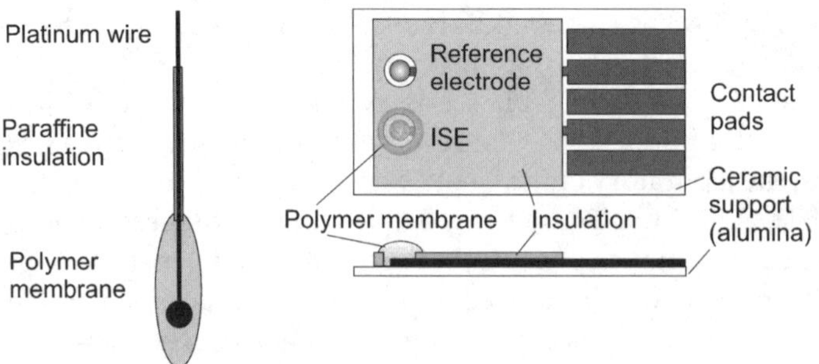

Figure 7.9. Shapes of miniature liquid-membrane and polymer-membrane ISEs. *Left*: coated-wire electrode, *right*: sensor in thick-film technology

Modern miniature potentiometric sensors are made by thick-film technology. An example is presented in Fig. 7.9 (right).

Solvents and ligands used to manufacture miniature sensors are identical to those used for electrodes with classical shapes. In miniature shapes, of course, the *direct contact* with the membrane seems to be mandatory. Problems resulting from the formation of depletion layers can be expected. An interesting approach, preferred for miniature sensors, consists in setting up an analogue of the connection via an electrode of the second kind (as preferred for classical ISEs). A screen-printed layer of silver and silver chloride particles, when covered by a polymer or gel layer containing chloride ions, functions much like a classic electrode of the second kind with an attached salt bridge. Unfortunately, the halide-containing layer is not very well protected against the affects of the sample solution. A certain improvement was achieved when contacts made of conducting polymers were used which had ionic as well as electronic conductivity. For example, a sodium-selective electrode with $NaBF_4$-doped polypyrrole (PPY) as contact was proposed by Cadogan et al. (1992). There are examples which seem to work reasonably well, although no precautions were taken against the typical problems associated with the direct contact. In such cases the redox couple oxygen/hydoxyl ion is apparently formed by penetration of oxygen into the polymer membrane. A redox electrode is formed which acts like an electrode of the second kind. Of course, such electrodes suffer from interference of oxygen dissolved in the sample solution.

For potentiometric miniature sensors, concessions must be made regarding linearity and stability. This can be tolerated since they are much cheaper than classical ISEs.

Introduction of the ISFET has brought real progress also for potentiometric liquid-membrane sensors.

The Glass Electrode

The glass electrode has been around since the early 20th century. It is the oldest and to this day the most important IES. It is designed to determine, quickly and reliably, the hydrogen ion activity or its common logarithm, the pH value. pH is one of the most important chemical parameters. Consequently, pH measurement is one of the most common analytical tasks. The relationship between activity and pH is logarithmic, and the measured electrode potential E is also a logarithmic function of activity. The voltage meter used can thus be calibrated easily to have a linear pH scale.

Efforts to miniaturize the glass electrode have not been successful. The classic design will no doubt survive into the future, although there have been proposals to replace it with small, more rugged sensors based on pH-sensitive liquid membranes. None of these proposals has enjoyed lasting success.

The exceptional position of importance of the glass electrode last but not least is based on the fact that it has properties of solid-membrane as well as

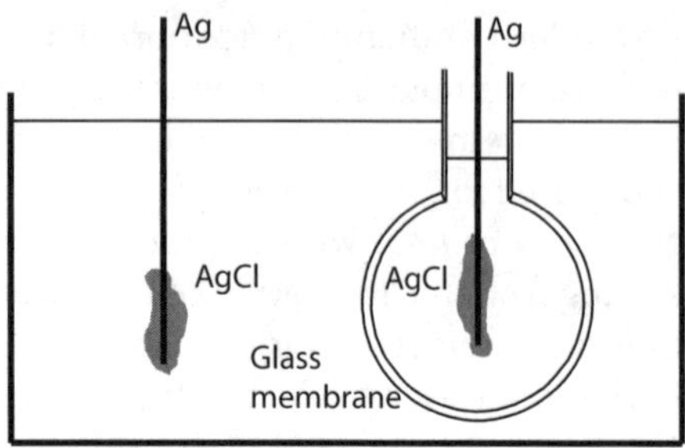

Figure 7.10. Galvanic cell for pH measurement with glass electrode (*right*) and reference electrode (*left*)

of liquid-membrane ISEs. The measurement set-up corresponds to that commonly used in potentiometry, as shown in Fig. 7.10, which depicts a *symmetric arrangement*. On either side of the glass membrane is located a reference electrode of equal composition. In the example chosen, the reference electrodes are silver/silver chloride electrodes. It is quite common to use an internal electrode of the second kind to establish an electric contact. Efforts have been made to apply a direct contact at the inner side of the membrane. The result was the so-called *enamel electrodes*. The problems with enamel electrodes are similar to those described in connection with ISEs as a whole. So far, enamel electrodes have been applied only in restricted fields.

An important constituent of the glass electrode is a thin membrane of sodium silicate glass (thickness ca. 0.4 mm), which is blown from a melt containing sodium oxide, quartz sand (SiO_2) and alumina (Al_2O_3). Variation of the alumina amount and addition of other substances (in particular uranium oxide) can be used to change the selectivity of the glass membrane. This is how sodium- and potassium-sensitive glass electrodes are designed.

Glass electrodes are being offered increasingly the in shape of *combination electrodes*, which contain an external reference electrode together with a salt bridge in a common unit (Fig. 7.11).

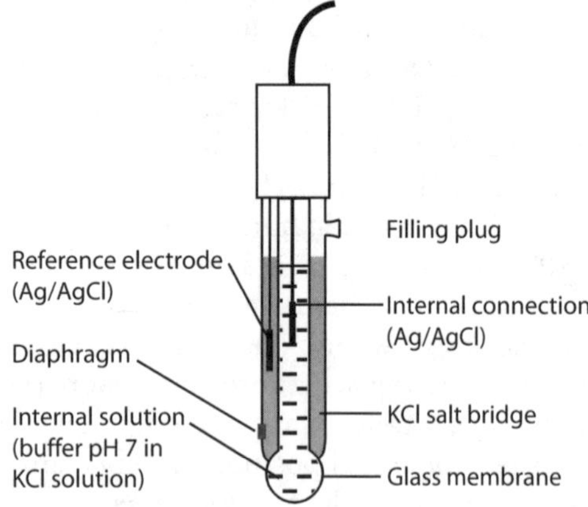

Figure 7.11. Combination glass electrode

The voltage determined for measurement is formed at the interface between the glass and the sample solution. Inside and outside the glass bulb, first the *hydrated gel layers* must be formed by soaking the electrode for 1 or 2h in slightly acidic water. The exposed silicate groups of the glass surface ($-SiO_4^{4-}$) react with water during the formation of the hydrogel containing the group SiO^-H^+. In equilibrium with aqueous solution, this group dissociates, releasing H^+, which is hydrated to give H_3O^+. This ion is attracted by electrostatic forces at the negative sites in the hydrated gel layer. As a result, an ion-exchange equilibrium with single-charged ions in solution is established:

$$-SiO^-Na^+ + H_3O^+ \rightleftarrows -SiO^-H_3O^+ + Na_{aq}^+ \; .$$

A commonly encountered potential difference at the phase boundary is formed when charge carriers have different mobilities in adjacent phases. According to the appropriate cell symbol, individual Galvani potential differences g_i can be attributed to the different phase boundaries:

Ag, AgCl /KCl$_{sat}$//sample solution/glass membrane/buffer solution /AgCl, Ag
$$\phi(I) \hspace{10cm} \phi'(I)$$
$$g_r \quad g_j \hspace{5cm} g_g \hspace{3cm} g_g' \hspace{2cm} g_r'$$

The measured electromotive force E between the metallic silver leads can be considered the difference between inner electric potentials $\phi'(I)$ and $\phi(I)$ of the silver phases. Alternatively, it would be correct to interpret the emf E as the sum of all the interface voltages g_i:

$$E = \phi'(I) - \phi(I) = g_r + g_j + g_g + g_g' + g_r' \; . \tag{7.7}$$

All the Galvani potential differences g_i except for g_g are assumed to be constant. The pH-dependent contribution g_g forms at the glass surface in contact with the sample solution. Consequently, the measured quantity E contains g_g as a constituent. The relationship is given by the corresponding form of the Nernst equation:

$$E = \text{const} + 0.059 \cdot \log a(H_3O^+) \; . \tag{7.8}$$

The *Nernst factor*, 0.059 V for room temperature, does not always reach its theoretical value. In experimental work, it is better to use the expression *Nernstian slope S* of the function $E = f(a)$. The slope must be determined empirically. The constant 'const' in Eq. (7.8) is a combination of all terms not dependent on concentration. In practical work, the name *asymmetry potential* (E_{as}) is preferred. This expression is derived from the expectation that the constant should be zero for a completely symmetric cell, i.e. if inner and outer solutions are of equal pH and if inner and outer reference electrodes are of identical types. In practice, E_{as} is not always zero but must be calibrated empirically by means of buffer solutions with known pH. By setting $-\log a(H_3O^+) = \text{pH}$, the common form of the Nernst equation for the glass electrode results:

$$E = E_{as} - S \cdot \mathrm{pH} \ . \tag{7.9}$$

Glass electrodes follow the above equation commonly in a pH range of 1 to 14. At higher pH, the signal caused by different single-charged ions becomes noticeable. This distortion is called the *alkali error* of the glass electrode. To make the best of it, this 'error' can be increased arbitrarily, e.g. by increasing the alumina content in the glass membrane. In this way, the surface of the hydrated gel layer is altered with the result that also alkali ions find sufficient space in the layer. As a result, glass electrodes selective for ions like Na^+, K^+ and Li^+ can be produced. Nevertheless, such electrodes remain sensitive to pH.

The Lambda Probe

The so-called Lambda probe to this day is the most successful and most widespread chemical sensor. It is used in all modern motor vehicles, where it serves to control the oxygen content in combustion gas. The combustion process in the motor is regulated to minimize the output of poisonous carbon monoxide.

The Lambda probe differs from all electrochemical sensors discussed so far in that it works with a solid electrolyte rather than an electrolyte solution. The sample is in the gaseous state, and the working temperature is ca. $500\,°C$.

At higher temperatures, the redox couple O_2/O^{2-} exists in the probe. It constitutes a redox electrode if a noble metal like platinum is present. Such a redox electrode, in contact with an electrolyte where the O^{2-} ion is mobile, can be used to measure potentiometrically the oxygen concentration in a gas. A useful solid electrolyte is zirconium dioxide ZrO_2. Mixing ZrO_2 with calcium oxide allows for the manufacture of ceramic objects. Such objects are characterized by a rather high electric conductance due to the ion migration in the crystal lattice. Ion mobility is facilitated by lattice defects (Sect. 2.1.1). Doping with yttrium oxide Y_2O_3 and other metal oxides generates positive holes, giving rise to anion mobility in the crystal. The potential at the redox electrode depends on the concentration of molecular oxygen. It can be measured as an emf if an equal electrode is set up to form the reference electrode. The latter must be exposed to a gas volume containing oxygen in known and constant partial pressure. Such a gas mixture is given by ambient air. Combining both electrodes results in a galvanic cell with a symmetric structure but with differing concentrations of the active constituent, a so-called *concentration cell*. The associated cell symbol can be written as

$$O_2 \ / \ Pt \ / \ ZrO_2 \ / \ Pt \ / \ O_2$$
$$p_r(O_2) \qquad\qquad\qquad p_m(O_2).$$

The concentration values of gaseous oxygen are commonly given in terms of partial pressure. The concentration in the reference compartment (ambient air) is given by $p_r(O_2)$, in the measuring compartment by $p_m(O_2)$. The solid electrolyte body is equipped with porous platinum electrodes which are gas permeable. An electrolysis current is carried by O^{2-} ions mobile in the solid electrolyte. The presence of a permanent current means the consumption of gaseous oxygen plus the formation of oxygen anions at the cathode, whereas at the anode oxygen anions are consumed and gaseous oxygen is generated. In this way, the electrolysis current will transport oxygen molecules through the solid electrolyte body. At either electrode, the equilibrium $O_2 + 4e^- \rightleftarrows 2O^{2-}$ is established. The emf between both electrodes obeys the Nernst equation:

$$E = E^{\theta} + \frac{0.059}{4} \log \frac{p_m(O_2)}{p_r(O_2)} .$$

(7.10)

With constant partial pressure $p_r(O_2)$ on the reference side, the Nernst equation can be written as follows:

$$E = E' + \frac{0.059}{4} \log p_m(O_2) .$$

(7.11)

A schematic view of a lambda probe is given in Fig. 7.12. A hollow part in the form of a finger made of zirconium dioxide ceramics protrudes into the studied gas compartment. The ceramic body is covered on either side by a gas-permeable platinum layer. Its inner volume is in contact with ambient air. The emf E is measured between the inner and outer platinum layers. The oxygen partial pressure in the compartment studied is calculated by means of Eq. (7.11).

Millions of exemplars of the lambda probe are in use with the Otto engine in road vehicles. One probe commonly is located in the exhaust pipe close to the engine. Exhaust gases have a temperature of ca. 900 °C. Petrol injection and related controlling tools are used to obtain a maximum degree of combustion in the engine. For quantitative characterization of this degree, the air-fuel ratio (AFR) is considered. An important quantity is the *lambda* (λ) number. The combustion process is symbolized by the following equation:

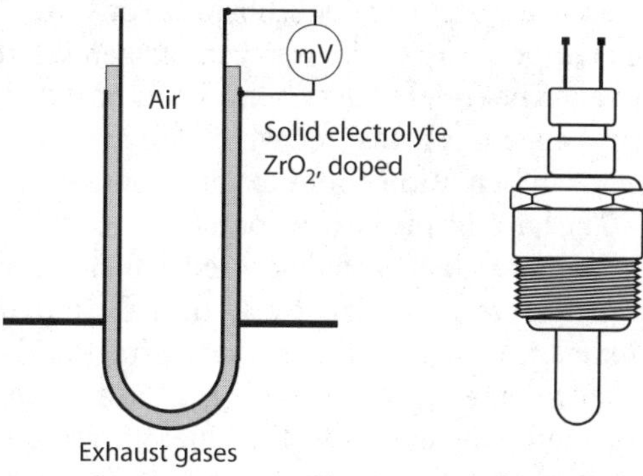

Figure 7.12. Lambda probe. *Left*: functional scheme, *right*: technical design

$$C_xH_y + (x + y/4)\,O_2 \rightarrow x\,CO_2 + y/4\,H_2O\,.$$

The AFR is *stoichiometric* if the added amount of oxygen is just sufficient to burn all the fuel components without any residue. The lambda number for the actual fuel-air mixture is defined by

$$\lambda = \frac{\left(\dfrac{m(\text{air})}{m(\text{fuel})}\right)_{\text{actual}}}{\left(\dfrac{m(\text{air})}{m(\text{fuel})}\right)_{\text{stoichiometric}}}\,. \tag{7.12}$$

The signal of the lambda probe in the exhaust pipe is fed into an electronic control circuit which keeps the actual lambda value close to 1, i.e. so that nearly complete combustion is achieved. The fuel-air composition under these conditions is a *rich* rather than a *lean* mixture. The exhaust gas then contains CO, NO and residues of hydrocarbons. These components are converted by means of the *three-way catalytic converter* into the less harmful gases N_2, H_2O and CO_2. If the lambda value is higher or lower than the optimum, then the poisonous gases canot be converted properly. The optimum AFR is ca. 14.7, i.e. 14.7 kg of air are needed to burn 1 kg of fuel.

The lambda probe is used to control many different combustion processes in addition to that in engines, e.g. in power plants. In a further application, it is included in automatic analysers designed for determining the *chemical oxygen demand* (COD) or the organic content in natural waters. After evaporating the sample, the dry residue is burnt in the presence of a catalyst. The oxygen demand is measured by means of a lambda probe and used to calculate the desired characteristic data. In such applications, the lambda probe is heated electrically. In metallurgical plants, specific lambda probes are useful for determining the oxygen content of a steel melt before casting. In this application, a high-temperature ISE sensitive to oxygen is set up. It contains a hollow zirconium dioxide probe with an inner rod of molybdenum. The space between is filled with a mixture of chromium and chromium dioxide. This mixture is used as a reference system which establishes the contact with the receptor. In the reference system, the equilibrium $4Cr + 3O_2 \rightleftarrows 2Cr_2O_3$ subsides. According to the temperature of the melt, an oxygen reference pressure $p(O_2)$ is generated. An external reference system is composed in the same way. Between both systems, an emf is measured which follows the Nernst equation, and consequently yields information about oxygen content in the melt.

The lambda probe can operate in two different ways. Until now, *potentiometric operation* has been discussed, which follows the transducer principle of *energy conversion*. Alternatively, the galvanic cell can be short closed. In this case, current is measured rather than voltage and the cell works as an *amperometric* or a *coulometric sensor*. This is a case of *current limiting transduction*. There are some advantages to this operating mode. Between the signal (a limiting current) and the oxygen concentration a linear relationship persists. Linearity is high over many concentration decades. Furthermore, amperometric lambda probes can be miniaturized more easily than potentiometric probes.

7.1.3
The Ion-Selective Field Effect Transistor (ISFET)

The metal-oxide field effect transistor (MOSFET) was described in Chap. 2. It is a nearly perfect voltage amplifier due to its inherent very high input resistance. The latter is a result of the fact that input and output circuits are isolated electrically from each other. Many pH meters have an input stage based on a MOSFET operational amplifier. It is well known that a serious distortion arises with large resistance voltage meters. This is caused by electric fields in the environment of the circuit. The lead from the voltage source to the meter is subject to induction of distortion signals; it picks up noise like an antenna. Such distortions must be overcome by elaborate screening precautions. A different approach proved to be more efficient. This approach is based on the idea that the line between voltage source and input terminal of the meter should be as short as possible. The MOSFET, as an amplifying component, can be located very close to an electrode as the voltage source. The electric lead from the source-amplifier combination is nearly free from distortions from the surroundings, since the signal is an amplified one and its impedance is low.

Arrangements like that in Fig. 7.13 have been proposed in different shapes. A hybrid sensor, e.g. consisting of a thick-film electrode with a reference electrode also screen printed, and a MOSFET attached by soldering (Afromowitz and Lee 1977). Such set-ups were only the first step towards an important invention of modern sensor technology. The next step was to reduce the connection between the ISE receptor and the gate of the MOSFET down to zero. This means that the gate electrode was coated directly by the sensitive layer. In this way was the ISFET (or CHEMFET) created. For mass production of this sensor type, nearly all the highly developed technological arsenal of microelectronics could be used. On the other hand, all the achievements of potentiometry with ISEs

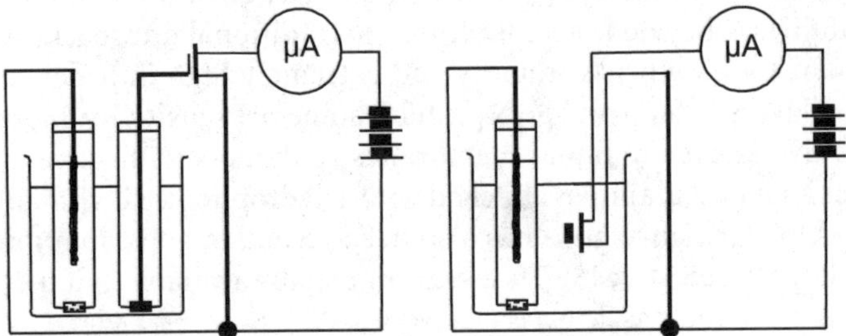

Figure 7.13. ISE with MOSFET and low impedance connection to measuring instrument (*left*). Measuring set-up with ISFET (*right*)

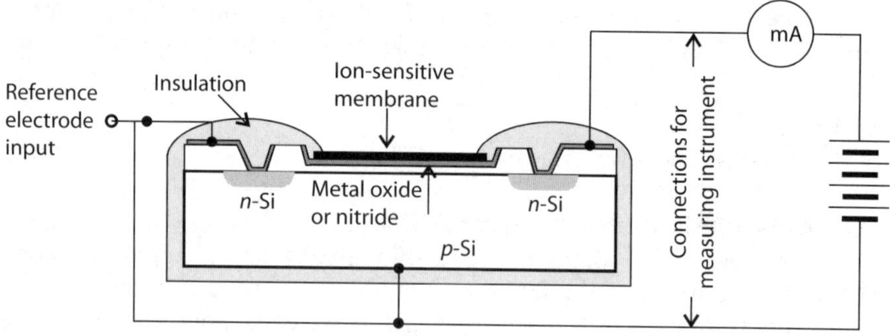

Figure 7.14. ISFET structure (simplified)

also became usable for ISFET sensors. This marked the formation of a bridge between two branches of science which had earlier been developed more or less independently of each other, namely electronics and analytical chemistry.

The ISFET (Fig. 7.14) structure is identical to that of the MOSFET. Commonly, two n-type zones are incorporated in a substrate of p-doped silicon. The substrate is covered by an insulating layer formed normally by silicon dioxide. Sometimes silicon nitride Si_3N_4 is used. A component made with such a composition would more accurately be called not MOSFET, but IGFET (isolated gate FET). Both oxides and nitrides are good insulators. The metallic gate layer is plated on top of the insulating layer. Finally, the ion-selective film is applied on top of the gate. The emf appearing between the reference electrode in solution and ion-selective membrane appears as a voltage between the *gate* and *source electrodes* of the MOSFET, as shown in Fig. 7.14. This gate-source voltage acts as an input voltage of the amplifier circuit and will bring about a linear current increase, which is measurable at the output of the circuit. Alternatively, an amplifier circuit with constant drain current can be used. The voltage between the reference and gate electrodes then appears with its amplitude unchanged, but amplified and with low impedance at the output.

All the ion-selective membrane materials mentioned in connection with solid- and liquid-membrane ISEs can be used with ISFETs. The most common are pH-sensitive ISFETs. They do not suffer from the traditional drawbacks of glass electrodes, namely high price, fragility and extremely high impedance. Even the insulation layers of SiO_2 and Si_3N_4 exhibit some pH sensitivity. More common pH-sensitive layers are liquid membranes (polymers with solvent) containing ion exchangers, e.g. amines attached with a hydrophobic alkyl chain at the opposite end. pH-sensitive glass has also been tested, in an adaptation of enamel electrodes. pH-sensitive ISFETs are commercially available, but they are still unable to replace the classic glass electrode.

Two problems have accompanied the development of the ISFET from the very beginning. The first one is the problem of *direct contact* for making contact with the sensitive membrane, which is inherent in ISEs. The second problem is encapsulation of the humidity-sensitive regions of the MOSFET substrate in such a way that only the gate (including the ion-selective layer) is exposed to solution. A permanent isolation of the vulnerable regions cannot be achieved by means of thin-film technology, since vapour-deposited thin oxide or nitride films are insufficient. To date such isolcation can only be achieved via encapsulation by polymer layers, which must be applied using procedures which are incompatible with microelectronics. Consequently, production of ISFET sensors is not really mass production, but a process containing elements of handcraft. Hence prices on ISFETs are rather high, and they are not as widespread as they were expected to be some years ago.

The technological progress achieved by the introduction of ISFETs is somewhat diminished by the fact that an external reference electrode is still necessary for measurement. Considerable effort has been devoted to miniaturizing the reference electrode. An acceptable, but not perfect, solution is the establishment of hybrid electrodes where the active substance (e.g. a mixture of silver and silver chloride) is screen printed. On top of such a structure, a salt-containing polymer layer can be screen printed. The latter will swell in contact with the sample solution forming a gel. Such devices are useful mainly for disposable sensors. An alternative solution, where silicon technology is used throughout, is depicted in Fig. 7.15 (Smith and Scott 1986). In this example, the basis is a pH-sensitive ISFET on an n-type silicon chip. A groove is formed by 'anisotropic etching'. The following technological steps (formation of p-doped zones by diffusion, generation of an isolating nitride film, metallizing with aluminium etc.) result in the formation of a MOS structure. Finally, by etching with hydrofluoric acid, a porous silicon layer is generated inside the groove. The latter acts as a diaphragm, which establishes the liquid–liquid junction to the AgCl pellet added later. The described process retains some advantages of microelectronics technology but suffers from some additional elaborate steps.

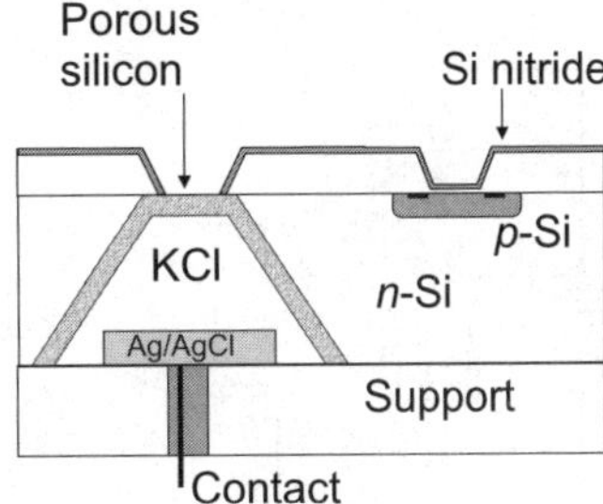

Figure 7.15. ISFET in silicon technology with integrated microreference electrode

7.1.4
Measurement with Potentiometric Sensors

Measuring Instruments

To ensure accurate measurement, instruments which work with ISEs must have a high input impedance. As a rule of thumb, their input resistance should be ca. 100 times higher than the internal resistance of the cell containing the ISE. Since several ISEs (e.g. the glass electrode) have resistances in the range of megaohms, a good pH meter should have at least an input resistance of gigaohms. Even low-cost pocket pH meters fulfil this requirement today.

The lead from the working electrode to the meter must be shielded carefully to avoid distortion by external electric fields of the surroundings.

With ISFETs, measurement is much easier than with ISEs. There are two ways to design the measuring circuit (Fig. 7.16). In the circuit with a *constant gate voltage*, the drain current I_D of the ISFET is a function of the gate voltage U_G according to $I_D = k \cdot U_G$. The current follower shown in Fig. 7.16 (Chap. 2, Sect. 2.4.1) sets the potential of the source terminal virtually to zero (mass potential) and outputs a voltage proportional to the gate voltage. In the circuit with a constant *drain current*, the voltage value between the reference electrode

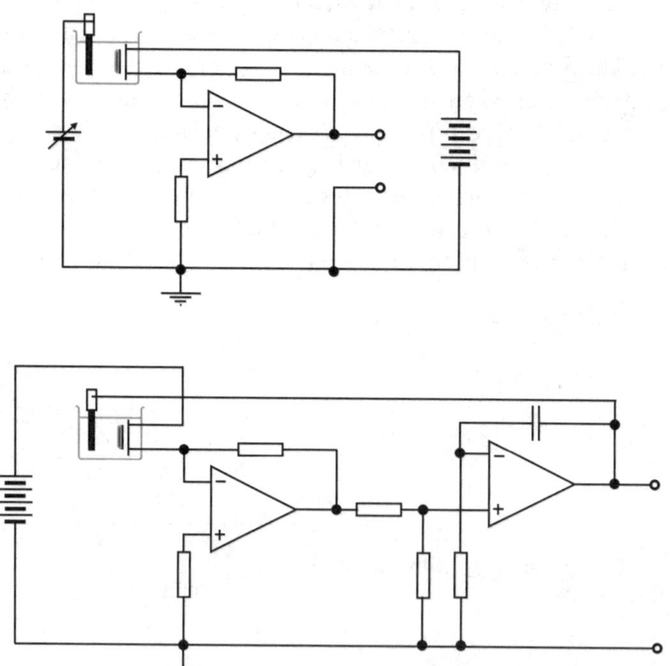

Figure 7.16. Connection of ISFETs with measuring instruments. *Top*: constant gate voltage, *bottom*: constant drain current

and the ion-selective membrane appears unchanged at the output, but with low impedance. This circuit is most common with ISFET applications. In this circuit, all the details of a sensor signal are followed and may be manipulated by microcomputers.

Experimental Conditions

Results of potentiometric measurements are always activity values. For chemical analysis, however, concentration values are desirable. In order to obtain a clear relationship between the concentration and the measured signal, the activity coefficient should be kept constant. This requirement can be fulfilled by establishing a high value of *ionic strength I*. The actual value of ionic strength determines activity coefficients of all ion types present in solution [Eq. (2.2)]. The higher it is, the more stable it is against variation by addition of foreign components. Independent of ionic strength, the pH should be kept constant during measurement. Also, the temperature should not vary too much. The temperature dependence of the Nernst factor will cause an error of 2% per Kelvin for single-charged ions.

A common practice is to mix a metered volume of the sample solution with a known amount of a buffer solution which also sets the ionic strength at a high value. Such buffer solutions are commercially available under the name TISAB (total ionic strength adjustment buffer). Commonly they also contain some antioxidant (e.g. ascorbic acid) designed to exclude negative effects of atmospheric oxygen. Useful compositions can be found in application notes of electrode suppliers. Here, as an example, a TISAB suitable for fluoride determination by means of a fluoride-sensitive electrode is used. The buffer is prepared starting with a solution of 1 M sodium chloride, 1 M acetic acid and 1 mM citric acid. By adding solid sodium hydroxide, the pH is adjusted to a value of 5.5. This pH is stabilized due to the function of the acetic acid/acetate buffer system. The value is optimum for fluoride determination. A higher pH, according to strong alkaline reaction, would cause a too strong interference of OH^- ions.

Calibration

The relation between measured value E and the logarithm of concentration c is established by calibration. This procedure is necessary since every sensor has its own characteristics that must be considered in order to get precise analytical results. Calibration can be done either by plotting a *calibration graph* or alternatively by the *standard addition* technique.

Calibration graphs are produced by performing a second experiment following the literal measurement. In this second experiment, a series of measurements is done with known concentrations of the substance to be analysed.

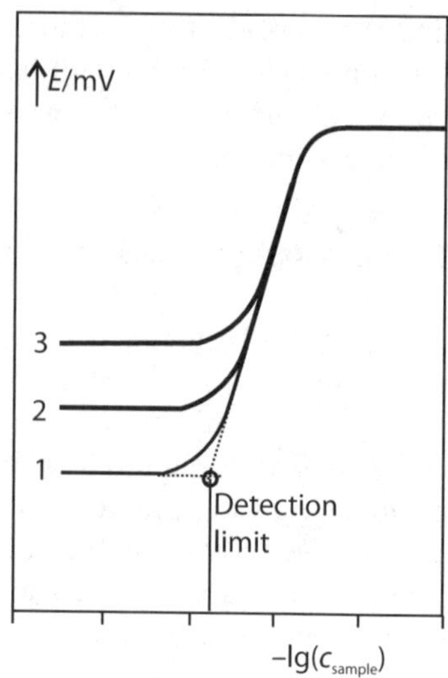

Figure 7.17. Typical calibration graphs for ion-selective ions. *1* without interfering ion; *2* and *3* with interferent

All conditions should be as close as possible to the sample analysis. The graph $E = f(\log c)$ is utilized as a reference to determine the concentration c_x which belongs to the measured value E_M. If potentiometric calibration graphs are not linear, then a *non-Nernstian behaviour* is perceptible, which means that the electrode may produce inaccurate results.

Figure 7.17 depicts the typical shape of calibration graphs for potentiometric sensors. At the upper limit, i.e. at very high analyte concentrations, the electrode does not obey the Nernst law. The lower limit of solid-membrane electrodes is marked commonly by their own solubility. The electrode itself generates the concentration indicated. This threshold concentration is increased if interfering ions are present. According to the *Nikolskij equation*, depending on the *selectivity coefficient K_{ij}*, above a specific value of interferent concentration c_j only the interfering ion generates a signal.

Calibration graphs suffer from the disadvantage that they must be recorded in a separate experiment, outside the sample solution. In many cases, unknown components in the sample solution will affect the result. Even if efforts are made to make the reference solution similar to the sample solution, this can be successful only if all its constituents are known. Otherwise, some error will result.

The *standard addition method* avoids this kind of error. In this case, all the measurements take place in the original sample solution. The latter is spiked by one or some consecutive additions of a standard, i.e. known amounts of the substance to be analysed. If possible, these additions are done such that the concentration of all the other constituents is kept constant. Commonly, a large volume of the sample (conditioned by TISAB addition etc.) is spiked with a small volume of standard solution. For calculation of the result c_x, the standard concentration c_S and the Nernstian slope S [as defined in Eq. (7.9)] are combined:

$$E_1 = \text{const} + S \cdot \log c_x \,, \tag{7.13}$$

$$E_1 = \text{const} + S \cdot \log \frac{c_x v_x + c_S v_S}{v_x + v_S} \,. \tag{7.14}$$

By subtraction, one gets

$$0 = S \cdot \log \left(c_x \frac{c_x v_x + c_S v_S}{v_x + v_S} \right) \,. \tag{7.15}$$

By rearrangement, an expression for the desired concentration c_x results:

$$c_x = \frac{c_S}{10^{\frac{E}{S}} \left(1 + \frac{v_x}{v_S}\right) - \frac{v_x}{v_S}} \,. \tag{7.16}$$

Errors with the single standard addition may result if a non-linear relationship exists between E and c in the interval considered. This error can be recognized and compensated if a series of additions is made rather than a single addition. Such a procedure is similar to a *titration*. Fortunately, the results can be extracted by means of a linearized function as proposed by Gran. A typical *Gran plot* is shown in Fig. 7.18. Assuming that each value c_S corresponds to the concentration increase caused by the appropriate addition, the following formula can be written:

$$E = \text{const} + S \cdot \log(c_x + c_S) \,, \tag{7.17}$$

$$\frac{E}{S} = \frac{\text{const}}{S} \cdot \log(c_x + c_S) \,, \tag{7.18}$$

$$10^{\frac{E}{S}} = \text{const}' \cdot (c_x + c_S) \,. \tag{7.19}$$

In the last equation was set $\text{const}' = 10^{\text{const}/S}$. A graphic representation of the function is depicted in Fig. 7.18.

Determining Selectivity Coefficient

The selectivity coefficient can be determined by means of a simple method which has been recommended by IUPAC. First, a solution of the interfering ion j is prepared. The concentration is chosen to induce a marked signal. Next, consecutive additions of the sample ion are made in such a way that the ionic strength is not changed. It seems that the intererent solution is 'titrated' with the sample ion. The resulting curve is depicted schematically in Fig. 7.19. At the intercept of both extrapolated straight lines, the corresponding signal $E(intercept)$ might be caused by the interferent activity $a_j(intercept)$ as well as

Figure 7.18. *Gran plot* for linearizing a series of multiple standard additions

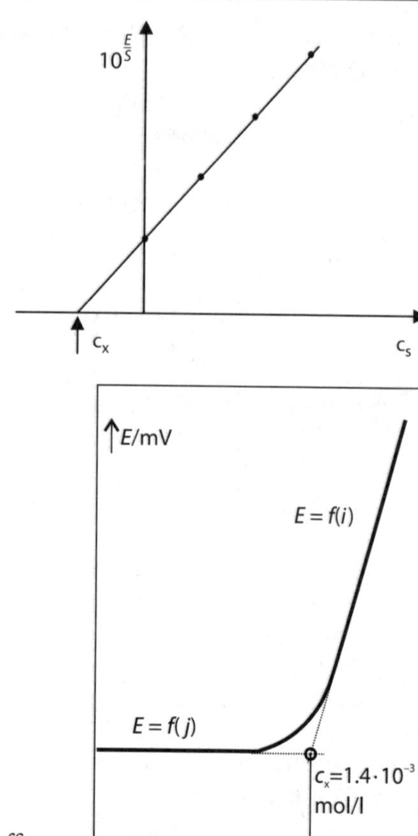

Figure 7.19. Determination of selectivity coefficient by consecutive additions of sample ion to solution of interferent ion

by the sample ion activity $a_i(intercept)$. According to the Nikolskij equation, the following condition is valid:

$$a_i(intercept) = K_{ij} \cdot a_j^{\frac{z_i}{z_j}} (intercept) .$$

(7.20)

Activity values are predetermined and well known, hence K_{ij} can be calculated.

7.2
Amperometric Sensors

With amperometric sensors, the electrode is polarized by a predetermined potential, and the resulting electrolysis current is measured. Under certain conditions, saturation is achieved, and the limiting current is proportional to the actual concentration. Instead of the transducer principle of *energy conversion* (with *potentiometry*), we now have the *current-limiting transduction*.

Under these conditions, an electric current is an expression of the reaction rate. Now the electrode is not necessarily in thermodynamic equilibrium with the analyte. It is not necessary to wait until the equilibrium has been established. Consequently, commonly the response time of amperometric sensors is magnitudes shorter than that of potentiometric sensors (Morf and de Rooij 1995). The strict concentration proportionality of the measured quantity (the limiting current) is another appreciated property of this sensor group.

Amperometric sensors can be miniaturized more easily than potentiometric ones. Mass production is usual and yields low prices. Hence there is a tendency in favour of amperometry, in particular with biosensors.

7.2.1
Selectivity of Amperometric Sensors

An 'amperometric sensitivity coefficient' K_{ij}^{amp} has been proposed (Wang 1994). It has not found general acceptance, however. The reason is that amperometric sensors do not have an inherent selectivity, but selectivity must be generated by additional efforts. Without such efforts, amperometric sensors are hardly selective, although an individual value of *half-wave potential* $E_{1/2}$ (Chap. 2, Sect. 2.2.6) is assigned to the electrode reaction of every electrochemically active substance. To distinguish between two substances, their half-wave potentials should differ by at least 200 mV. However, the useful potential range of a common electrode is restricted to ca. 1.5 V. Hence, in a single voltammogram, only a few substances can be determined simultaneously. At an electrode operating in amperometric mode, generally more than one process is running at a time. In most cases, several substances are simultaneously reduced or oxidized. There exist only a few exceptions to this rule. Among them is the lambda probe mentioned in a previous chapter. Its inherent selectivity is based on the fact that between the electrodes in a solid electrolyte only one specific ion is mobile (the O^{2-} ion).

In amperometric operations, analyte molecules must come into contact with the electrode surface continuously to allow electrode reaction. Filtering the ion flow before a reaction is one way to increase selectivity. Useful filter layers are permeable only for specific molecules. Such layers can be formed by *permselective membranes* or by monolayers containing *intramolecular* or *intermolecular channels*, respectively. Alternatively, a layer of selective catalyst molecules can be located in front of the electrode surface. With catalysts, one of the possible electrode reactions is preferred to such an extent that the electrolysis current (as an expression of the reaction rate) depends nearly exclusively on the analyte concentration. Oxides of copper, nickel, cobalt and ruthenium have been used as catalytic additives for carbon paste electrodes (Chen et al. 1993). For carbohydrates, a good selectivity was achieved with Cu_2O, for amines with NiO. Some examples of selectivity enhancement of amperometric sensors are listed in Table 7.4.

Table 7.4. Selectivity enhancement of amperometric sensors

Membrane	Example	Selectivity enhanced for
Permselective membranes		
Dialysis membranes	Cellulose acetate	Small molecules
Lipophilic membranes	Phospholipids	Non-polar molecules
Gas-permeable membranes	Macroporous PTFE foil	Dissolved gases (O_2; Cl_2)
	Silicone rubber	Oxygen in water
Ion exchanger membranes	NAFION (cation exch.)	H^+, diverse cations
	Polyvinylpyridine (PVP)	Anions
Membranes with steric recognition		
Layers with intermolecular channels	Calixarenes; crown ethers	Alkali metal ions
Layers with intramolecular channels	Cyclodextrines with electroactive markers	Cycloheyanol, etc.
Modified electrodes and selectively catalytic layers		
Inorganic	Pd/IrO_2 in PVP on platinum	Sulfite and SO_2
	Ni-oxide	Amines
	Cu-oxide	Carbohydrates
Natural substances	Enzymes	Biologically active substances
Electrolytic cells with selective conductivity of the electrolyte	ZrO_2-Y_2O_3-solid electrolyte cells in amperometric mode	Oxygen in gases

Permselective membranes exclude specific molecules from transport due to their size or other properties. With dialysis membranes, the molecular size is critical. An important application is the exclusion of protein molecules which may contaminate the electrode surface. Lipophilic membranes block the transport of polar molecules, and in ion-exchanger membranes, only ions with a defined charge sign are mobile. Best known are Nafion membranes (Du Pont), which act as cation exchangers in wet state. Anions cannot penetrate the membrane.

Membranes with *steric recognition* of the analyte have properties similar to that of permselective membranes. As mentioned in connection with potentiometric sensors, some molecules (ligands) can coordinate with specific analyte molecules due to their size. As a result, the bound analyte molecules may cross a phase boundary and will become mobile in the subsequent non-aqueous phase. With potentiometric sensors, partition equilibria were established in this way with subsequent formation of a potential difference across the interface. With amperometric sensors, charges must be transferred over a distance in order to establish a permanent electrolysis current. To support this process, the particles bound by selective ligands should be mobile in the non-aqueous layer, or alternatively the analyte ions should be able to jump easily inside a matrix from one ligand to the next one and so forth. These requirements are not fulfilled by all the ligands which have proved useful for potentiometry. On the other hand, some receptor ligands are able to coordinate with analyte molecules selectively, but their chemical bond is so strong that their mobility is decreased nearly to zero. Layers containing such ligands are useful neither for potentiometric nor for amperometric sensors. They can be utilized, however, in conductometric, optical or mass-sensitive sensors. Properties of this kind are characteristic for paracyclophanes as well as for polymer layers generated by the novel method of *molecular imprinting*.

The different ways to enhance selectivity are not strictly separate from one another. For example, we may consider many enzymes. They catalyse a certain reaction like other selective catalysts; on the other hand they resemble selective ligands in their capacity for steric (molecular) recognition.

7.2.2
Electrode Design and Examples

Amperometric sensors commonly have the shape of microelectrodes. As discussed in Chap. 2, Sect. 2.2, at microelectrodes a constant diffusion-limited current is being established without the need to generate convection externally. Such electrodes are well suited for obtaining concentrations proportional signals in situ, e.g. when a biological sample is punctured by the electrode. Most common shapes of microelectrodes are microdisks, interdigitated lead structures, and narrow bands on a support. Also widely used are macroscopic shapes with a diffusion barrier or with a protecting layer covering the active membrane which contains immobilized molecules.

Amperometric sensors should have a high electric conductance. Membranes with intrinsic conductivity, e.g. conducting polymers, are a good basis for immobilizing active molecules. Alternative methods are arranging the active molecules to form monolayers close to the electrode surface or imbedding them in carbon pastes. Many new techniques have opened up in connection with the development of biosensors. Among them are layers of the so-called *redox polymers*. In the matrix of such layers, specific molecules or groups are fixed which do not only increase the conductivity but also catalyse redox reactions. A further highly important development is immobilization of active molecules via formation of *self-assembled monolayers* (SAMs). These novel techniques will be discussed in more detail in Sect. 7.4 of this chapter dealing with amperometric biosensors.

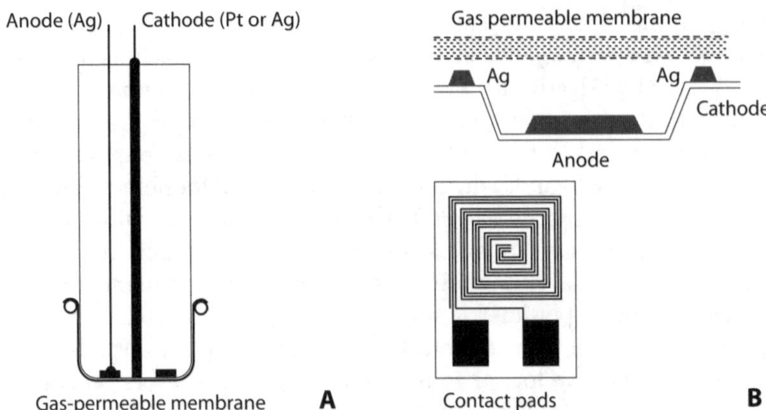

Figure 7.20. Clark sensor. (A) classical design, (B) miniature shape

A very important amperometric sensor is the Clark sensor for determination of dissolved oxygen. The classical macroscopic design is used widely, but miniature forms are also common (Fig. 7.20).

The Clark sensor is equipped with a gas-permeable membrane located in direct contact with the working electrode surface (Fig. 7.20). The membrane separates the sensor from the sample solution. In this way the electrode surface is protected. Selectivity is increased since only gaseous components can reach the active area. Membranes of macroporous polytetrafluorethylene (PTFE) are used preferably. The material is strongly hydrophobic. Water in liquid state is unable to permeate the membrane pores. Unlike liquid water, gases easily diffuse through the pores. Membranes of silicon rubber have also been used with the Clark sensor. They are permeable for oxygen since it is dissolved homogeneously in the matrix.

Oxygen dissolved in the sample solution diffuses through the membrane. The working electrode is located very close to the membrane, separated only by a thin electrolyte film. Commonly the active electrode area is a small, flat surface of platinum. The polarization potential imposed is high enough to reduce each oxygen molecule arriving at the electrode. As a result, the local oxygen concentration behind the membrane is close to zero. Hence the concentration profile between the sample solution and the electrode surface depends only on the oxygen content in the sample. The thickness of the diffusion layer in this case is identical to the thickness of the gas-permeable membrane. This is clearly a constant thickness, and consequently the current-voltage curve is of sigmoidal shape, assuming the scan rate is not too fast. An amperometric sensor is established by imposing a constant polarization potential and measuring the resulting limited current (Fig. 7.21). Of course, the imposed potential value should be in a rather negative region where the current is diffusion limited.

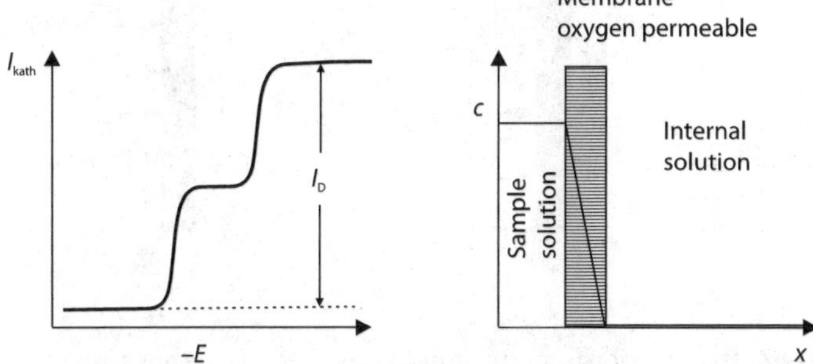

Figure 7.21. Clark sensor: effect of diffusion barrier in front of active electrode area. *Left*: current-potential diagram; *right*: concentration profile for working potential in limiting current region

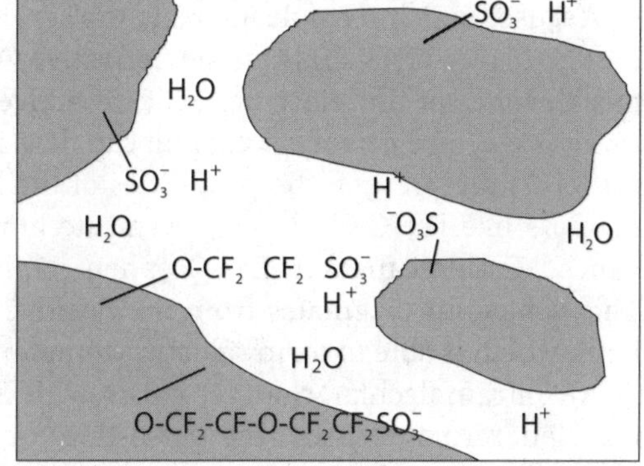

Figure 7.22. Structure of a Nafion cation exchanger membrane. Hydrophilic clusters are surrounded partially by hydrophobic PTFE regions. Cations like H^+ can move from one hydrophilic cluster to the next etc.

The miniature form of the Clark sensor depicted in Fig. 7.20 contains an anode and a cathode as well as a small electrolyte stock (chloride solution containing gel-forming agents) behind the membrane. The technology for manufacturing such sensors differs from standard microelectronic techniques. It had to be developed especially for sensor applications (Suzuki 2000).

Among cation-exchanger membranes, the cation exchager NAFION (Du Pont) plays an outstanding role. Its basic matrix consists of PTFE, an extremely hydrophobic polymer insoluble in nearly all known solvents. In NAFION membranes, at the surface of a PTFE foil, fluorine containing side chains are attached which carry sulfonic groups. This way, hydrophilic regions arise which in wet, swollen state can bind cations by electrostatic forces. Cations are mobile in the molecular cavities of the polymer (Fig. 7.22). A common application of NAFION membranes is exclusion of anion interferences. The surface of the working electrode is coated by the ion exchanger. For that purpose, a NAFION solution is spread on the surface and allowed to dry.

Ligands belonging to the groups of *calixarenes* and *cyclodextrines* were mentioned in connection with potentiometric sensors (Fig. 7.8). They can be

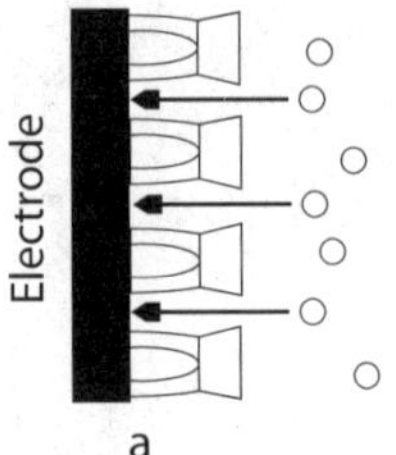
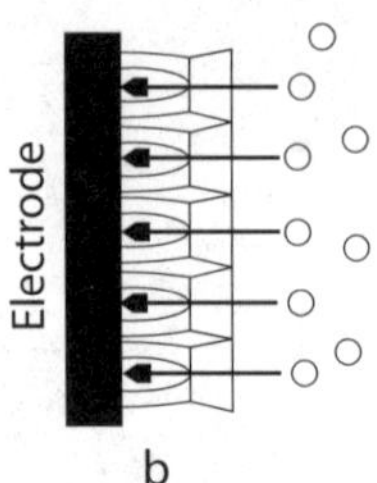

Figure 7.23. Two sorts of molecular channels formed in monolayers of calixarenes or cyclodextrines. **a** Intermolecular channel. **b** Intramolecular channel. Permeability for small molecules is controlled by the analyte

used to construct selective amperometric sensors as well. For that purpose, the molecules are arranged in the form of ordered monolayers, either generated by means of the Langmuir–Blodgett technology or in the form of SAMs. Such layers can set up molecular channels of either the intramolecular or intermolecular type (Fig. 7.23).

A sensor with intermolecular channels is an electrode covered by a Langmuir-Blodgett layer of a calix[4]arene derivative (Yagi et al. 1996). The layer is permeable for a specific electrochemically active substance, the so-called *marker*. In this state, an electrolysis current can flow. If alkali metal ions (Cs^+, Rb^+, K^+, Na^+, Li^+) are present, the properties of the layer change drastically. Anionic markers like $Fe[(CN)_6]^{3-}$ can permeate more easily, and their signal is enhanced. Cationic markers like $Co[phen_3]^{2+}$, however, show a decreased signal. This behaviour originates from the positive surface charge of the calixarene layer, which is able to form selective complexes with the cations mentioned.

An intramolecular channel exists with SAMs formed by aggregation of a cyclodextrine derivate, the per-6-thio-β-cyclodextrine at a gold electrode (D'Annibale et al. 1999). Ferrocenylcarboxylate or hydroquinone are used as electrochemically active markers. Their signal decreases if certain neutral molecules like trimethylcetylammonium are present in the sample solution. The strongly selective complex formation with 'cones' of cyclodextrines somehow 'clogs' the internal channels.

The above examples demonstrate that electrochemically inactive substances can be determined by amperometric sensors if membranes of controlled permeability are utilized.

Numerous catalytic layers made by different immobilizing techniques have been used to enhance the selectivity of amperometric sensors. One example chosen from a large variety was a platinum microelectrode coated with polyvinylpyridine carrying a layer of anodically deposited palladium/iridium oxides (Shi et al. 2001). Polyvinylpyridine is an electrically conducting polymer. The sensor has been used to determine sulphur dioxide in gaseous phase as well as sulphite in solution.

The lambda probe, normally known to be a potentiometric sensor, can operate in amperometric mode as well. The sensor described in what follows was developed especially for this mode (Fig. 7.25). As with other variants of the

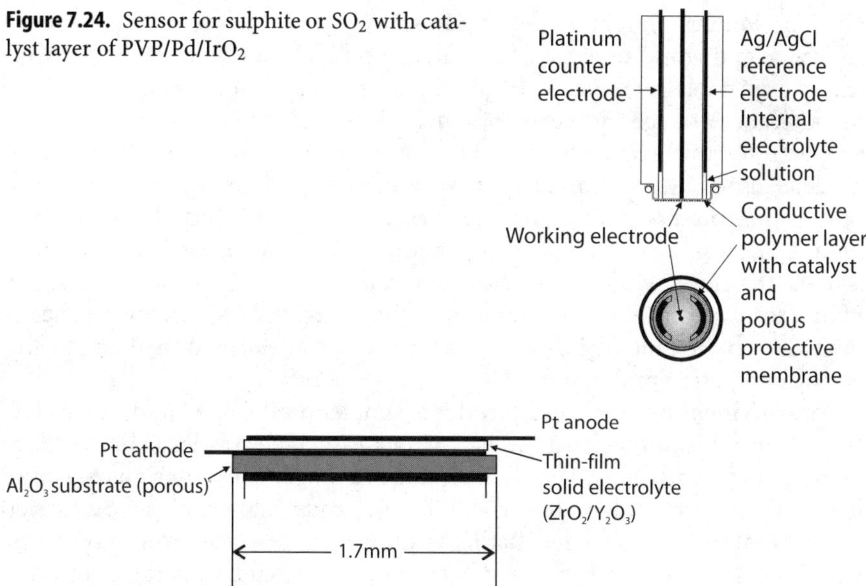

Figure 7.24. Sensor for sulphite or SO_2 with catalyst layer of PVP/Pd/IrO$_2$

Figure 7.25. Miniature lambda probe in thin-film technology. For differential measurement, two exemplars are necessary

sensor, a solid electrolyte carries two porous noble-metal electrodes on either side. One side is in contact with the sample gas, the other side with a reference gas volume. If the electrodes are short closed, a permanent electrolysis current arises, since in the solid electrolyte lattice O^{2-} ions are mobile. Electrolysis takes place, i.e. on one side O_2 is reduced to O^{2-}, whereas on the other side the opposite process occurs. Below the line, oxygen is transported from one side to the other, where the original concentration is lower. The electrolysis current measured depends on the concentration ratio persisting on either side, and it is limited by the diffusion of the analyte. Working in this mode, the sensor follows the principle of *current-limiting transduction*. Its miniature design is very practicable, and mass production is possible. Figure 7.25 depicts an amperometric lambda probe in miniature design.

7.2.3
Measurement with Amperometric Sensors

Amperometric with macroelectrodes is done using approved technical equipment. There are numerous commercially available devices with equal design. The electrochemical cell works with three electrodes (Chap. 2, Sect. 2.4), i.e. *working, auxiliary* and *reference* electrodes. They are driven by a *potentiostat* designed as an analogue electronic circuit. Digital potential control instead of analogue control circuitry did not stand the test of time, since it does not function fast enough. The potentiostat provides for the voltage between reference and working electrodes which should always be equal to an arbitrarily adjustable reference voltage. The actual state of the art is to generate the reference voltage by means of a digital-to-analogue converter (DAC) on an electronic card controlled by a PC. The actual value of reference voltage is predetermined by a computer program and may be either a permanent value (amperometric mode) or it may be varied linearly with time (voltammetric mode). The response of the electrochemical cell is an electrolysis current which is fed in the aforementioned PC card and transduced there in a series of digital numbers by means of an analogue-to-digital converter (ADC). Such series of numbers can be plotted easily on the screen or transformed mathematically, e.g. by integration or derivation.

Measurement with microelectrodes is simpler on the one hand, since electronic control is not necessary. The current at such electrodes is in the range of nano- to picoamperes. The IR drop at the homogeneous solution resistance is negligible. Current load of the counter electrode is low and can be carried by usual reference electrodes. Rather than using a three-electrode circuit, the variable voltage is simply imposed between reference and working electrodes. Measurement of the current response is a technical problem, however. Very low current is distorted by external effects. When low-noise operational amplifiers became available, opportunities to work with microelectrodes was enhanced. Such amplifiers are inserted into the potentiostat commonly in the form of a current follower between working electrode terminal and ground. The current follower keeps the working electrode potential equal to zero (ground potential), and it generates an output voltage proportional to the actual electrolysis current. With low-current circuits, damping and filtering units are necessary and must be adjusted carefully.

In commercially available potentiostats, specific additional electronic modules are provided for picoampere measurements. For classic investigations,

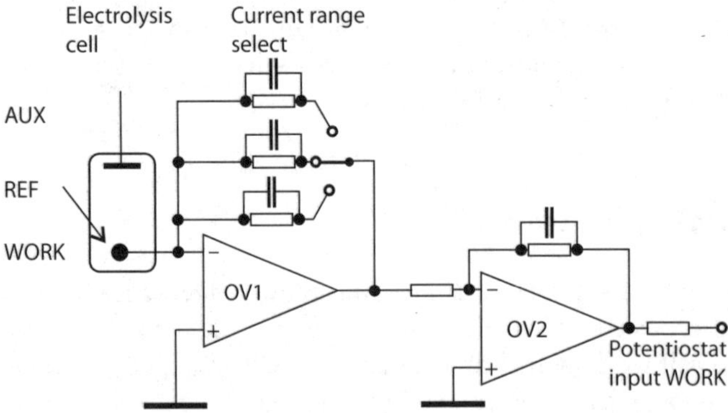

Figure 7.26. Preamplifier for picoampere measurements with microelectrodes

such modules would be too sluggish. Also, additional preamplifier circuits based on special low-noise amplifiers have been proposed which would be used prior to the input of a common potentiostat (Fig. 7.26). In the input stage (OA1 in the figure), specific low-noise operational amplifiers with high-input impedance are used. The frequency response of RC elements for different measurement ranges must be adjusted very carefully. Considerable efforts has been devoted to blocking distortions from the power supply and from other sources.

Even with the latest low-noise amplifiers, a distortion-free function of picoampere amplifiers is only possible if the complete set-up is enclosed in a Faraday cage (made of metal sheets or grids) which must be grounded.

7.3
Sensors Based on Other Electrochemical Methods

Impedance measurement can be considered a third way to evaluate electrochemical sensors besides potentiometry and amperometry. Electrochemical impedance studies in a narrower sense deal with phenomena at the electrode surface. The overall impedance of a chemosensor also includes effects of charge carrier properties far from the electrode. This was visualized by equivalence circuits presented in Chaps. 2 and 5. By individual experimental design, the study can be focused more on processes at the electrode surface or otherwise on ion properties in homogeneous solution. Even the variation of the dielectric constant in a layer will affect the overall impedance. If impedimetry is designed only to acquire data corresponding to ionic properties or value of the dielectric constant, it is not really an electrochemical method, in a strict sense.

Among the class of impedimetric sensors are real electrochemical sensors where a chemical reaction is the source of information. Of special importance are biosensors with SAMs, which are considered in more detail in the following chapter (Knichel et al. 1995, Rickert et al. 1996).

7.4
Electrochemical Biosensors

7.4.1
Fundamentals

Biological Recognition as a Principle of Selectivity

Following the actual definition (Chap. 1, Sect. 1.2), biosensors are characterized by a receptor function which is implemented by biologically active substances that are able to recognize selectively other substances by a biological mechanism. This *biological recognition* is based on the *lock-and-key principle* in the majority of cases. It means that molecules are identified by their size.

The selectivity of biosensors is astonishing. Bioactive substances sometimes may identify reliably one specific substance among a matrix of millions of others.

Biosensors either work as *biocatalytic* or as *bioaffinity* sensors. In biocatalytic sensors, mostly *enzymes* are immobilized at an electrode surface to act as selective catalysts. Enzymes catalyse slow reactions. The reaction rate, under appropriate conditions, is a quantitative measure for the substrate concentration. If the sample is one of the reactants, i.e. if the sample itself is the substrate, then chemical sensors can be created on this basis. The reaction rate can be measured in terms of an electrolysis current if amperometric biosensors are applied. An alternative approach is to indicate the reaction product of the catalysed reaction selectively, as is done with potentiometric biosensors. In bioaffinity sensors, as the second group, commonly very stable complexes with sample molecules are formed and bound strongly to the sensor surface. The extent of this complex formation is a quantitative measure of the sample concentration. It can be measured indirectly, since many properties of the electrode are changed by complex formation. In bioaffinity sensors, mostly the antibody-antigen reaction is utilized.

Nucleic acid sensors can be classified as another group of biosensors.

Immobilization of Biologically Active Substances

Bioactive substances designed to act as receptors must be immobilized at an electrode surface. There are some differences depending on the question whether biocatalytic or bioaffinity sensors must be realized. Potentiometric and amperometric sensors also are characterized by different requirements. In principle, with potentiometric sensors a higher impedance can be tolerated, whereas amperometric sensors require a good electric conductivity. Nevertheless, immobilization techniques are similar for both groups.

Adsorption is the simplest way to fix bioactive molecules at an electrode surface. Commonly, adsorbed molecules are bound by weak forces, resulting in less durable sensors. Adsorption is useful to immobilize enzymes, antibodies and nucleic acids.

Frequently a simple contact between an electrode surface and a solution is sufficient to generate reasonable adsorptive layers without adding further reagents. This is valid in particular for carbon-containing surfaces. In some cases, however, an adsorptive bond with a carbon surface may affect the enzyme function to such a degree that it is deactivated. Sometimes even denaturation occurs. The majority of molecules is bound only weakly, and a part of them is later lost by progressive desorption. Electrodes with adsorptive layers preferably are used for tentative experiments. They are useful mainly for fast tests.

Enzymes, antibodies, nucleic acids and other bioactive substances alternatively can be bound by covalent chemical bond at solid surfaces. This kind of immobilization commonly brings about monolayers that are located directly at the electrode surfaces, similar to many adsorptive layers. In contrast to adsorption, a very stable bond is achieved. Hence, electrodes with chemically bound molecules are robust and durable. Immobilization commonly is performed by a two-step procedure. First, the surface is prepared to generate tether groups. Normally, nucleophilic groups are formed, such as carboxylic, amino-acid, hydroxy, thiol, and phenolic groups. Following this chemical activation, active molecules are attached to the surface groups. Metallic surfaces require preparation steps which are different from those for surface groups and useful for carbon surfaces. The arsenal of synthetic methods to bind enzymes covalently is limited. Due to the time-consuming nature of this process, there are few commercial applications available.

A highly efficient way to bind enzymes covalently has been derived from SAMs, which were introduced in Chap. 2, Sect. 2.3. Originally, gold electrodes were functionalized in a complex series of reaction (Willner et al. 1993). An alkyl chain was fixed via an SH group at the gold surface. At the opposite end, the chain contained an amino group that was linked to the enzyme by means of a coupling reagent. On top of the resulting layer, multiple layers of other enzyme molecules were formed, which could even be linked finally with the mediator ferrocene by covalent bonding.

The first step in SAM formation is often given as adsorption, although in fact it consists in the formation of a very stable covalent bond. Enzyme electrodes on the basis of SAMs are robust in general. Useful sensors are manufactured on the basis of gold films made by thin-film techniques.

Imbedding in *carbon pastes* or in *conducting organic salts* is a useful immobilization method for many bioactive substances. Most of them, like enzymes and antibodies, are proteins. Complete biological cells and even microorganisms have also been embedded. Such substances and objects are sufficiently hydrophobic to be compatible with the ingredients of carbon pastes. Carbon pastes are prepared by mixing carbon powders (graphite, spectral carbon or glassy carbon particles) with an organic, water-insoluble binder like liquid paraffin or silicon oil. The paste is pressed into a tube equipped with a metallic contact at the remote end. Carbon pastes are useful for a large variety of applications. Even small *organs* of animals can be fixed in this way. The electrochemical function of carbon pastes is similar to that of compact carbon electrodes covered by an adsorptive film. Particles in the paste also normally exhibit at their surface an adsorptive bond with the active substance. By metallizing the particles, a more selective function can be achieved (Wang et al. 1995).

Alternatively to carbon pastes, conducting organic salts have been used to imbed bioactive components, in particular to immobilize enzymes. The redox mediator tetrathiofulvalene acts as electron donor and forms a solid, conducting salt with the electron acceptor tetracyanoquinodimethane. This salt has a low melting point and can be mixed with proteins to give a conducting paste. This 'binder' fulfils a second function, i.e. that of a mediator (Bartlett 1990).

Inclusion in polymers or hydrogels frequently is used to fix biologically active substances or microorganisms. Polymer layers are easily coated on solid surfaces. They are solvents for active substances, but they can also include large molecules, cells or microorganisms which are 'glued' in this way to the surface. A disadvantage is their low electric conductivity. To overcome this drawback, conducting particles can be added. Polymer layers commonly have only a low capacity to absorb water. This is a disadvantage with enzymes, which need water to exhibit their specific functionalities.

Hydrogels are well suited to immobilize enzymes. They can contain water up to 98 percent. Inclusion in hydrogels is one of the oldest methods of immobilization. It was applied first with potentiometric enzyme electrodes. The gelating agents used were gelatine and algines, the latter frequently attached with calcium-containing side chains. Synthetic gels like polyacrylamide and polyvinylalcohol have also been applied frequently. With amperometric sensors, besides the low conductance, slow diffusion of the sample inside the gel matrix presents additional problems. It has a negative influence on response time. A further problem is the continuous washout of enzyme. Such loss can be minimized by crosslinking enzyme and matrix. The resulting aggregates are less soluble and are kept in the matrix without lowering their efficiency. A widely used crosslinking agent is glutaraldehyde whose function follows the scheme given below:

$$\boxed{E}-NH_2 \quad + \quad OCH-(CH_2)_3-CHO \quad + \quad H_2N-\boxed{PVA}$$
$$\downarrow$$
$$\boxed{E}-N{=}CH-(CH_2)_3-CH{=}N-\boxed{PVA}$$

It is possible to generate a layer of crosslinked enzyme molecules alone, without binder, at an electrode surface.

To avoid washout of enzyme, permselective membranes can be clamped across the gel layer. Dialysis membranes (commonly foils of cellulose acetate) are well suited for that purpose. They can be penetrated easily by small molecules, but bulky protein molecules (among them enzymes) are retained.

It is not surprising that *electropolymerization* has been used to make functional layers for biosensors. A fruitful development of functional layers started with this application. The method was used preferably to immobilize enzymes, either by embedding in polymer layers or by linking with polymer surfaces which had been functionalized before by attachment of amino groups. The stability of embedded molecules has been further improved by crosslinking with glutaraldehyde.

An interesting kind of covalent fixing is the *avidin-biotine reaction*. It is used exclusively for biosensors. In this reaction, a small molecule is somehow 'enveloped' by a large molecule.

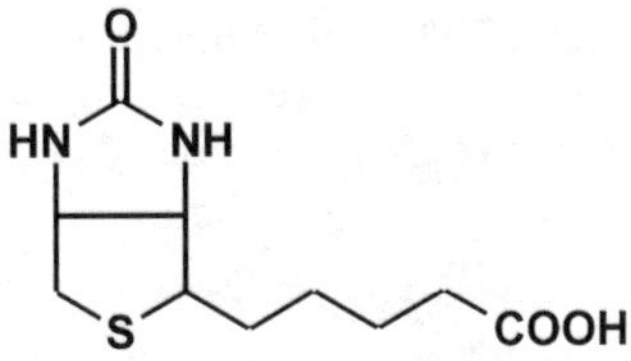

Figure 7.27. Biotin

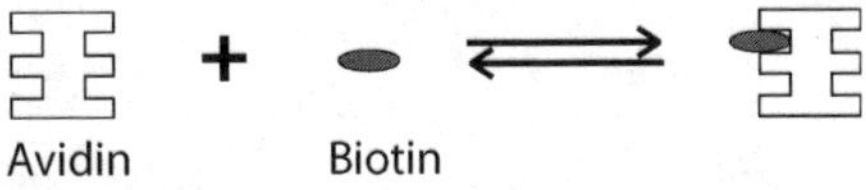

Figure 7.28. Scheme of avidin-biotin reaction

Avidin is a high-molecular protein found in egg white (albumen). Alternatively, *streptavidin* extracted from *steptomyces* can be used in place of avidin. Avidin has a strong affinity to *biotin*, which is found in egg yolk. Biotin, also known as vitamin H, is a low-molecular, water-soluble B-complex vitamin (Fig. 7.27). One avidin molecule can bind up to four biotin molecules. The bond is very strong and stable against extreme pH. Since avidin only requires the bicyclic system of biotin, the carboxylic group of the latter can be used for further linkages. For sensor application, commonly the transducer surface is modified by avidin. The probe molecule is *biotinylated* and linked to the transducer surface. Nucleic acids and proteins, among them enzymes, can be biotinylated easily.

The avidin-biotin reaction can be used to link several molecular monolayers which lie on top of each other with avidin and biotin molecules arranged in alternating mode. A large variety of configurations results from different combinations of the simple reaction depicted schematically in Fig. 7.28. The stability of the avidin-biotin complex is extremely high.

7.4.2
Classes of Electrochemical Biosensors

Enzyme Sensors

Potentiometric enzyme electrodes were the first biosensors. In such sensors, an IES was coated by an enzyme layer acting as a biocatalyst for reaction of a specific substance. The reaction product subsequently was detected by the ISE. This feature had been transferred soon to ISFETs. A special term, *ENFET*, has even been informally proposed for enzyme-modified ISFETs. Attempts were also made to utilize further biological interactions for recognition of analytes and construction of potentiometric biosensors. *Immunologic sensors* on the basis of *antigen-antibody reaction* may be called *IMFETs* if they are built on top of a MOSFET. Immunologic reactions, however, may be used much more efficiently in combination with other transducers besides potentiometric

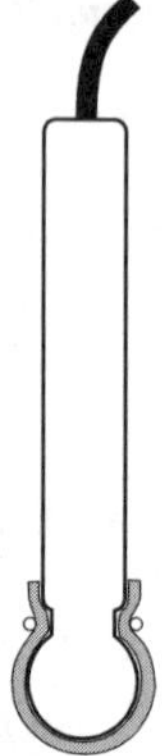

Figure 7.29. Potentiometric urease sensor with enzyme entrapped in a gel. The gel layer, supported by a nylon tissue, covers the surface of an ammonium sensitive glass electrode

ones. To this day, potentiometric biosensors remain unique to a few enzymatic reactions.

Among the oldest biosensors is a urea sensor (Guilbault and Montalvo 1969). As depicted in Fig. 7.29, the enzyme urease is fixed in a hydrogel layer of polyacrylamide which is coated on the surface of a glass electrode with nylon gauze as support. Urea diffuses into the gel, where a catalysed hydrolysis takes place:

$$CO\left(NH_2\right)_2 + H_2O \xrightarrow{\text{urease}} CO_3^{2-} + 2\,NH_4^+ \,.$$

The reaction products carbonate and ammonia change the pH, which is detected by the glass electrode, thus allowing one to estimate the urea content. A better function can be achieved by utilization of a special ammonium-sensitive glass electrode, as proposed by the inventors.

Actual potentiometric biosensors are commonly designed on the basis of ISFETs, i.e. they are ENFETs. Inclusion in hydrogels is no longer done. Covalent bonding, preferably at carbon surfaces, is preferred. Alternatively, embedding in polymer layers is performed, today preferably in polymers generated electrochemically. Also widespread are PVC coatings with specific softeners (as solvents for active substances) as well as layers of silicon rubber and polyurethane.

Table 7.5 presents some examples of practically useful potentiometric enzyme sensors. It is no coincidence that a large variety of glucose sensors has been developed. The demand for glucose sensors in medicine is high, since easy-to-handle analytic procedures for blood sugar control are required to help people with diabetes. One of the sensors in Table 7.5 utilizes an enzyme pair which finally releases fluoride as a reaction product. Fluoride is detected efficiently by a fluoride-sensitive lanthanum single-crystal electrode. Other prominent products of enzymatic reactions are H_3O^+, which can be detected with a pH-sensitive glass electrode, and H_2O_2, detectable with a redox electrode containing noble metal or carbon particles.

Figure 7.30. Concentration profile in stationary state at potentiometric enzyme sensor

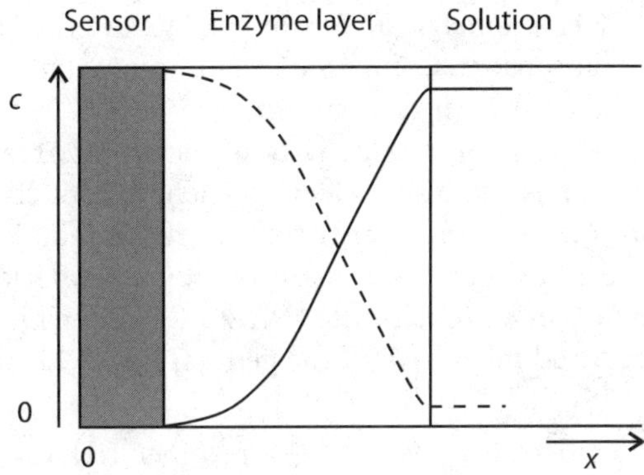

The formation of a stationary concentration of reaction product at the electrode surface is a prerequisite for proper functioning of potentiometric enzyme sensors. Stable conditions are established after a certain settling time where reaction speed is equal to the diffusional transport of the analyte to the electrode (Fig. 7.30). This time is represented by the response time of the sensor. In stationary state, the maximum concentration of reaction product is located at the electrode surface (Vadgama 1990). The enzyme layer should be designed to ensure that this optimum state is reached again following a change in sample composition. Response times of potentiometric biosensors are different. Commonly they assume values of some minutes.

Table 7.5. Potentiometric enzyme biosensors

Sample	Enzyme	Reaction	Product detected
Urea	Urease	$CO(NH_2)_2 + H_2O \xrightarrow{\text{urease}} CO_3^{2-} + 2NH_4^+$	NH_4^+
Glucose	Glucose-oxidase	$glucose + O_2 \xrightarrow{\text{glucoseoxidase}} H_2O_2 + gluconolactone$	H_2O_2
Glucose	Glucose-oxidase; peroxidase	$glucose + O_2 \xrightarrow{\text{glucoseoxidase}} H_2O_2 + gluconolactone$ $H_2O_2 + 4\text{-fluraniline} \xrightarrow{\text{peroxidase}} F^- + polymer\ products$	F^-
Neutral lipides	Lipase	$lipide + H_2O \xrightarrow{\text{lipase}} glycerin + fatty\ acids + H^+$	H_3O^+
Lactate	Lact-oxidase	$lactate + O_2 \xrightarrow{\text{lactoxidase}} H_2O_2 + pyruvate$	H_2O_2

Amperometric biosensors to date are the largest and most important group of biosensors. Compared to potentiometric sensors, they are characterized by a much shorter response time.

The extreme selectivity of enzymatic reactions can be utilized best with enzymes catalysing electron exchange. Such enzymes are *oxidases* and *dehydrogenases*. The former catalyse redox reactions including oxygen [Eq. (7.21)], the latter such reactions with the participation of the coenzyme *nicotinamide adenosine dinucleotide* (NAD). Considering the oxidized form NAD^+, or the reduced form NADH, the general Eq. (7.22) is obtained:

$$\text{substrate} + O_2 \xrightarrow{\text{oxidase}} \text{product} + H_2O_2 \,, \tag{7.21}$$

$$\text{substrate} + NAD^+ \xrightarrow{\text{dehydrogenase}} \text{product} + NADH \,. \tag{7.22}$$

The substrate (i.e. the analyte) and its reaction products normally are not electrochemically active. The course of reaction is followed by evaluating the consumption or generation of species belonging to the redox couples O_2/H_2O_2 or $NAD^+/NADH$, respectively. The applied potential is adjusted such that the electrolysis current reflects either the consumption of the oxidizing substance or an increase of byproduct. Modified Clark electrodes are widely used to determine the oxygen loss caused by an enzymatic reaction. Also in use are redox electrodes which detect any H_2O_2 or NADH which formed during the reaction. Such electrodes work in a medium potential range. The redox potential of $NAD^+/NADH$ amounts to ca. 0.8 V vs. saturated silver/silver chloride electrode. Hence sensors containing this coenzyme can operate in air-saturated solution.

All the enzymes mentioned in connection with potentiometric sensors can be used also with amperometric sensors. However, for the amperometric mode a much greater variety of useful reactions is available. Some examples are given in Table 7.6.

The function of simple amperometric enzyme sensors (first-generation sensors) includes the following steps (simplified): diffusion of substrate towards the electrode, reaction with immobilized enzyme, and regeneration of enzyme to give the original form by oxygen or $NADH^+$. The linear dependence of an electrolysis current on substrate concentration can be achieved only under the condition that diffusion is made the slowest step in the series of processes, since the slowest step determines the overall reaction rate. Diffusion control would not prevail if oxygen were depleted in the near-electrode region with the result that the enzyme could not be regenerated.

The selectivity of enzymes is based on the lock-and-key principle, i.e. the substrate molecules are included in a perfectly fitting cavity of the enzyme molecule. The redox-active centre of the molecule is positioned deep inside the molecule and barred from direct contact with the electrode surface. Regeneration cannot be executed by direct electron transfer, even if the enzyme

Table 7.6. Amperometric enzyme sensors

Sample	Enzyme	Reaction	Product detected
Polyphenol	Polyphenol-oxidase	polyphenol + O_2 $\xrightarrow{\text{PPO}}$ o-quinone	o-quinon
Cholesterol	Cholesterol-oxidase	cholesterol + O_2 $\xrightarrow[\text{ferrocene}]{\text{chOx}}$ cholestenone + H_2O_2	H_2O_2 /Fc
Ethanol	Alcohol-dehydrogenase	C_2H_5OH + NAD^+ $\xrightarrow{\text{EDH}}$ CH_3CHO + $NADH^+$ + H^+	$NADH^+$
Lactate	Lactatemono-oxygenase	lactate + O_2 $\xrightarrow{\text{LMOx}}$ acetic acid + CO_2 + H_2O_2	H_2O_2
Pesticides[a]	Acetylcholin-esterase	acetylcholine + H_2O $\xrightarrow{\text{ACE}}$ choline + acetic acid cholin + $2O_2$ + H_2O $\rightarrow$ betain + H_2O_2	H_2O_2

[a] by inhibition of acetylcholine esterase activity

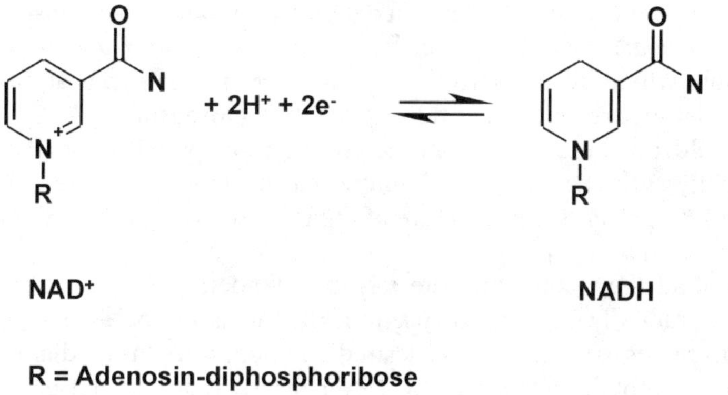

R = Adenosin-diphosphoriboce

NAD⁺ + 2H⁺ + 2e⁻ ⇌ NADH

R = Adenosin-diphosphoriboce

Figure 7.31. Redox equilibrium of nicotinamide adenine dinucleotide (NAD)

molecule is in touch with the electrode surface. This problem has been solved mainly by the introduction of *redox mediators.*

Second-generation amperometric enzyme sensors are characterized by reversible, dissolved redox active substances (*mediators*) which are incorporated into the sensor matrix. Redox couples to be used as mediators should be reducible or oxidizable at moderate potentials, and their reactions should be fast. Mediators are mobile inside the sensor matrix and can react with spent enzyme molecules under regeneration of their original form. Mediators are 'electron shuttles' with a function following the simplified scheme given in Fig. 7.32. It remains unclear which processes occur inside the enzyme molecule during the course of electron transfer. According to one theory, inside the molecule there might be two contact sites, one reducing and one oxidizing, which are inter-

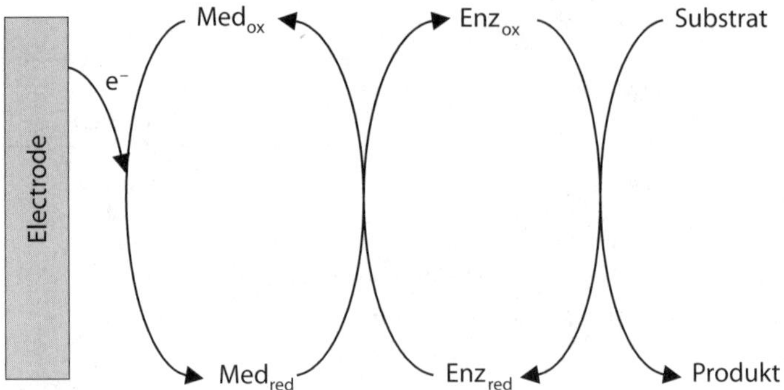

Figure 7.32. Reactions occurring in an enzyme electrode with mediator. Med_{ox} and Med_{red} are the oxidized and reduced forms of the mediator, respectively

connected by a conducting pathway. This would explain the fact that enzymes can take part in redox reactions without uptake or release of charges.

Some authors prefer to speak about a *third generation* of biosensors. This term is not used consistently. In some cases, it denotes a combination of sensor and its electronic evaluation unit, a so-called *biochip*. Also, the *reagentless biosensors* (sensors where all the active compounds are immobilized at an electrode) are sometimes referred to as third-generation biosensors.

Mediators should not be merely reversible redox couples, but they also should react rapidly with enzymes. Commonly used substances are listed in Table 7.7. Figure 7.33 presents the structure of the most important mediator, the ferrocene/ferrocenium couple.

The problem of all the soluble mediators is their tendency to be washed out of the matrix gradually. It would be useful if the mediator could also be immobilized. This seems to contradict the desired function, since the mediator molecules should be mobile independently. Nevertheless, it seems that both

Table 7.7. Common mediator substances and their redox potentials. E^* potential of an equimolar mixture of the couple at pH 7 vs. standard hydrogen eletrode

Mediator	E^*/V
$Os(bpy)_3^{3+}/Os(bpy)_3^{2+}$	0.84
Ferricinium/ferrocene	0.44
$Fe(CN)_6^{3-}/Fe(CN)_6^{4-}$	0.36
quinon/Hydroquinon	0.28
Methylene blue	0.01
Methyl viologen	− 0.44
Tetracyanoquinodimethane (TCNQ)	0.252
Tetrahiafulvalene (TTF)	0.216

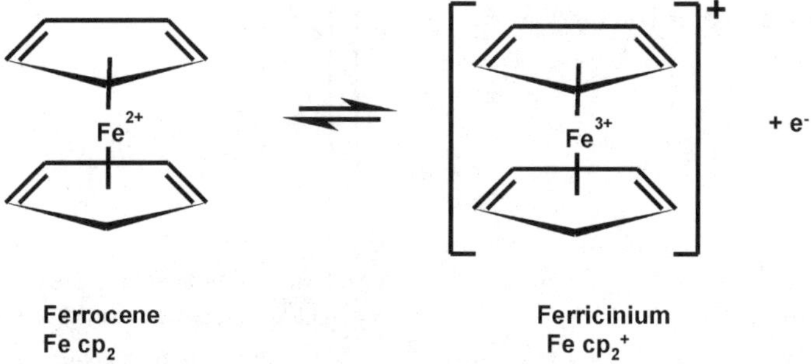

Figure 7.33. Redox equilibrium of ferrocene (dicyclopentadienyl-$Fe^{2+/3+}$)

requirements (immobilization, but mobility inside the matrix) are fulfilled by *redox polymers*. Most successful are electrodes where a mediator substance, the osmium bipyridyl complex, has been attached to a polymer structure forming a three-dimensional redox network (Fig. 7.34). Enzyme molecules are fixed at the surface of the polymer body (Heller 1990). The development status of enzyme sensors with redox polymers is very high. Up to four layers with different functionality have been attached to one sensor (Kenausis et al. 1997).

At present, amperometric glucose sensors are the most important and the most common biosensors. Somewhat less common, but also important, are lactate sensors. There is a large variety of shapes, but one can state the general ambitions for miniaturization and mass production. Sensors where the concentration of enzymatically generated oxygen is analysed are based on miniature forms of the Clark probe (Sect. 7.2.2). Frequently, thin sensors are encountered which can be stuck into a sample (Fig. 7.35). In this example, a small platinum anode is surrounded by a cylindrical silver body acting as cathode. Both are covered by a thin foil of cellulose acetate containing the enzyme. An outer layer of collodium or a dialysis membrane is used to protect the sensor (Wilson and Thévenot 1990). Different variations of this set-up are commercially available and widely used, although they belong to the group of *first-generation biosensors*.

Structures made by screen printing on ceramic or polymer board make up a large part of the commercially available enzyme sensors. Very often the

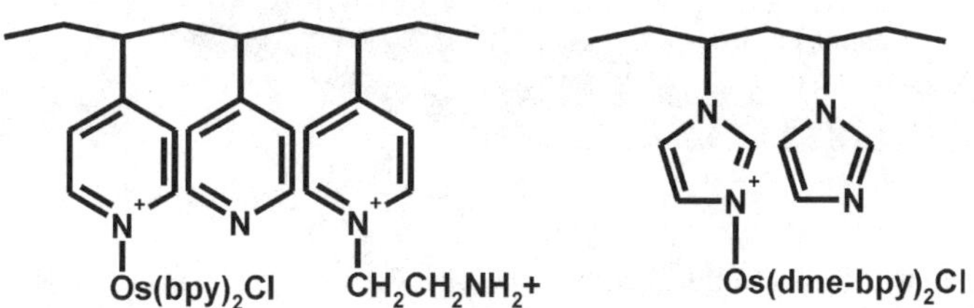

Figure 7.34. Structure of redox polymers with osmium species

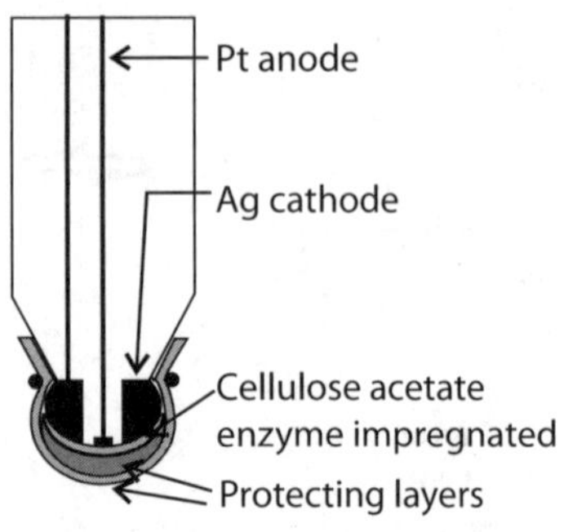

Figure 7.35. Enzyme sensor based on Clark probe

scheme given in Fig. 7.36 is encountered. A layer containing silver and silver chloride particles acts as reference electrode. In equilibrium with chloride ions in the membrane, this layer provides a constant reference potential. Other screen-printed layers act as working and counter electrodes. They contain particles of noble metals or glassy carbon. At present, ruthenium dioxide is also used as a material for the working electrode. This substance has been used widely in printed electronic circuits to form resistances. RuO_2 turned out to be useful for redox electrodes since it is catalytically active for many redox reactions. A solution of the enzyme glucose oxidase can be applied directly onto the printed electrode structures. The structures are stabilized then by crosslinking with glutaraldehyde. Finally, the surface is coated by a protecting layer, commonly a polymer membrane permeable for diffusion.

Enzyme sensors in thick-film technology, according to the design in Fig. 7.36, are commercially available. Most common are glucose sensors of this type. The protecting layer allows one to use the sensors for glucose determination in blood samples without pretreatment.

Enzyme electrodes where the active molecules are chemically bound to a solid surface are not very widespread yet. One example is an amperomet-

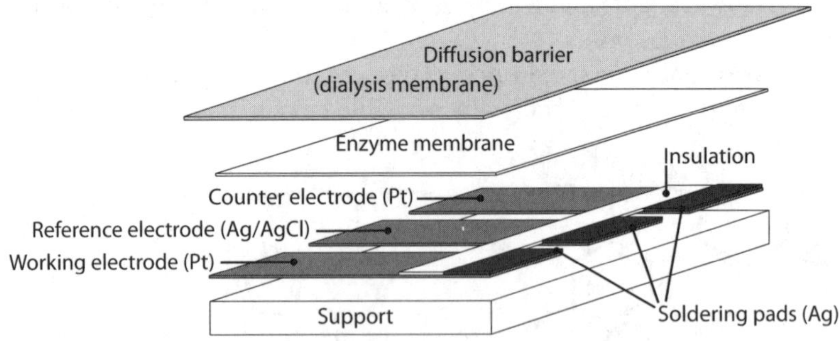

Figure 7.36. Amperometric sensor in thick-film technology. For working and counter electrodes, also layers containing RuO_2 are used

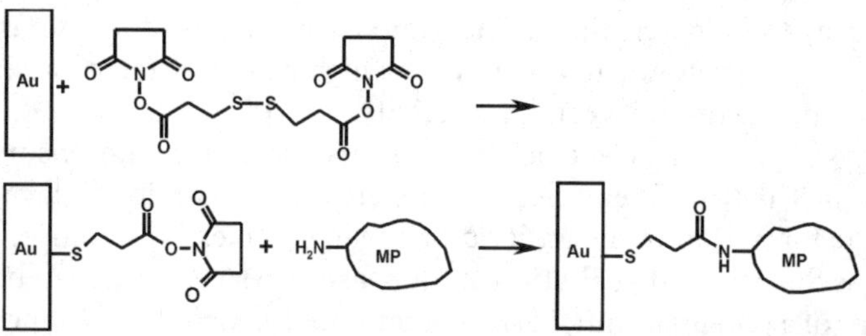

Figure 7.37. Adsorption of DTSP (dithiobis[*N*-succinimidylpropionate]) at a gold electrode and covalent bonding of horseradish peroxidase (HP) at active ester groups of SAM formed

ric sensor with horseradish peroxidase bound to a graphite surface that was functionalized by cyanur chloride (CC) (Cardosi 1994).

SAMs are used increasingly to immobilize enzymes. A simple example is sketched in Fig. 7.37 (Darder et al. 1999).

A specific problem related to enzyme sensors is their restricted shelf life, since enzymes are not very stable. In dry state, they can be stored for some time. Disposable sensors are activated when they are humidified but lose their activity soon after use. Efforts are being made to improve the stability of enzyme sensors by chemical treatment. It has been proposed that the active substance in polyelectrolytes be immobilized with polysugars (Gibson and Hulbert 1993). The enzyme-polyelectrolyte complex formed seems to be stabilized electrostatically by some kind of Faraday cage. In this way, the enzyme activity remains preserved somewhat longer.

Immunosensors Immunosensors are an alternative way to make use of the extreme selectivity of biological receptors. Living organisms can generate antibodies to dispose of nearly every existing substance. Consequently, nearly every substance can play the role of antigen. In immunosensors, antibodies are considered to be reagents which selectively form stable complexes with the antigen that constitutes the sample. After isolating and cleaning, the antigen is immobilized at an electrode surface. In this way, a probe is prepared which should be able to bind a single sample among thousands or even millions of different substances. Although this reaction is disturbed somewhat by parallel non-selective adsorption processes, immunosensors are characterized by a considerable selectivity. The next question is how to generate an electrochemical signal.

Immunosensors are bioaffinity sensors, i.e. antigen and antibody form a stable complex, but there is no decomposition of reactants or formation of further reaction products. Consequently, analytical information in amperometric immunosensors cannot be obtained directly by evaluation of an electrolysis current reflecting the reaction rate, as in biocatalytical sensors. The common way to obtain amperometric signals is chemical modification (*labeling*) of the sample (the antigen) to be determined. The sample can be linked e.g. to an electrochemically active group or to an enzyme which can be detected after complex formation by means of its catalytic activity.

Alternative ways of obtaining electrochemical signals with immunosensors are based on the fact that different electric properties of an antibody layer are changed by the formation of the antibody-antigen complex. The complex formation changes the charge distribution at the sensor surface, which can be measured in terms of a potential shift, i.e. by potentiometry. Also, conductance variation can be evaluated by impedimetry.

For immobilizing the isolated and cleaned antibody, methods similar to those described in Chap. 2, Sect. 2.3 can be used as. There is, however, a peculiarity. Since the most common type of antibody molecules is Y-shaped (Chap. 2, Sect. 2.2.8), the molecules must assume a specific orientation at the sensor surface. The antigen-binding sites should be oriented towards the sample solution (Fig. 7.38). This orientation is best achieved if the molecule is linked covalently at its group opposite to the antigen-binding site. For that purpose, reactive groups are bound to carbohydrate groups at the region near the molecular 'hinge'. Alternatively, the antibody is modified by the *protein G* which binds specifically at the 'foot' of different antibodies. This protein can be immobilized easily at an electrode surface in advance. Very useful is binding of antibody molecules in the form of SAMs. It must be born in mind that antibodies are voluminous molecules with considerable space requirements. A common practice is to form first a monolayer composed only partially of bifunctional molecules (molecules carrying a group for surface binding at one end and for linking with the antibody protein at the other end). Together with the bifunctional molecules, shorter, monofunctional molecules are fixed at the surface. The latter 'dilute' the bifunctional layer.

A somewhat different method is modification of the antibody molecule itself by a 'linker' carrying a thiol group at its free end. Thiol groups readily bind to gold surfaces. Regardless of the advantages of SAMs, immobilization by simple adsorption is also used.

Potentiometric immunosensors have been assembled on the basis of a thin semiconducting titanium dioxide film. The latter was covered by an activated polymer membrane containing covalently bound antibodies. Complex forma-

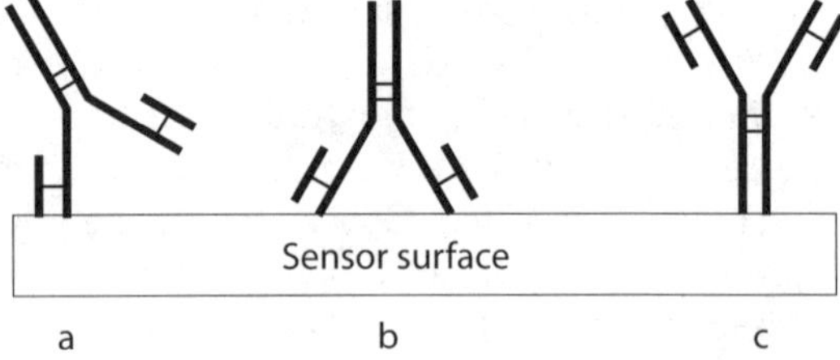

Figure 7.38. Orientations of antibody molecules at solid surface. Only position c is useful for sensor application

tion with the antigen caused a potential shift, which is a measure of the antigen content in solution (Yamamoto et al. 1983).

Such electrochemical immunosensors proved meaningful only where the signal was obtained by impedance measurements. A first example was a sensor for determination of methamphetamine in urine (Yagiuda et al. 1996). The antibody *anti-methamphetamine* was immobilized at two adjacent platinum electrodes and crosslinked by treatment with glutaraldehyde. In contact with the sample solution, conductivity of the antibody layer between the electrodes decreased as a result of antigen binding. This sensor is a member of the group of chemoresistors.

Layers of synthetic peptides (molar mass ca. 3000 Da) which represent a specific site of an antigen have been bound via 'linkers' or 'spacers' at a gold surface. As a result, a SAM-based impedimetric biosensor has been realized (Rickert et al. 1996). The signal was obtained by evaluation of the impedance spectra recorded in the presence of the reversible redox system ferrocyanide/ferricyanide. The redox reaction of this 'indicator' is blocked when the antigen binds with the antibody. For this sensor, the term *impedimetric sensor* is appropriate. It has been applied to detect the foot-and-mouth disease virus. Such application justifies the considerable efforts expended on the preparation of synthetic peptides.

Sensors with Whole Cells, Microorganisms and Organs Living organisms have always been used to observe changes in the environment, i.e. as *biomonitors*. Biosensors result from combining organisms with transducers in such a way that measurable signals are obtained. Commonly, enzyme action is also the determining source of information in sensors based on whole organisms, cells or organs. Often it is useful to leave enzymes in their natural environment. In this way, a better stability of the biocatalysts is achieved and costs are minimized. Disadvantages are longer response time, lower selectivity and worse reproducibility.

The most elementary biosensors are fruit pulps or slices which have been combined with amperometric electrodes. A well-known example is the 'bananatrode' (Wang and Lin 1988). This sensor, most useful for demonstration experiments, contains a paste mix of banana pulp, nujol and carbon powder which has been pressed into a glass tube with an electric contact (Fig. 7.39). The mass contains the enzyme polyphenolase, which catalyses the oxidation of polyphenols, among them important biological messengers like dopamine. The sensor can be tested by means of simple compounds like catechol, which can be detected in beer. As a result of air oxidation, *o*-quinone is formed. The latter is an electrochemically active compound which can be detected e.g. by differential-pulse voltammetry.

With fruit pulps of eggplant, apple and potato, sensors similar to that with banana pulp have been assembled. In all these cases, the enzyme polyphenoloxidase is the active agent. Products of oxidation are also *o*-quinones.

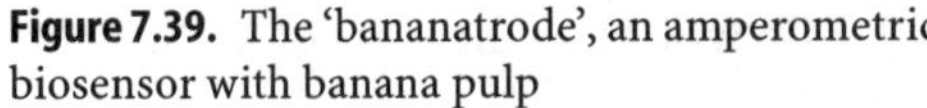

Figure 7.39. The 'bananatrode', an amperometric biosensor with banana pulp

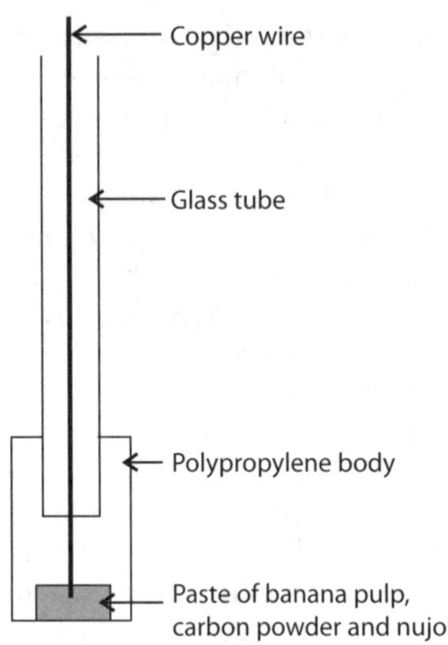

Biosensors have also been set up with plant tissue slices rather than pulps. As an example, a potato slice spiked with glucose oxidase has been used to detect phosphate and fluoride (Schubert et al. 1984).

Isolated biological cells or cell fragments (e.g. membrane particles) have been used as receptor layers in biosensors. Such compositions allow interesting experiments, but they are not suited as a basis for commercial sensors.

Biosensors with living organisms can be realized with bacteria, algae, fungi and protozoae. Commonly, the function is based on the enzymes contained in the organisms. Often it is simpler and cheaper to grow a culture of bacteria or fungi and immobilize the complete culture than to isolate the enzymes, a tedious process. Here, the microbe plays the role of enzyme container.

Microorganisms can be immobilized in the same way as biologically active substances. Common ways are imbedding in gels or polymers as well as inclusion behind a membrane. Modified Clark electrodes are widely used (Sect. 7.2.2). The biological activity of many microorganisms is connected with consumption or formation of oxygen. Bacteria consume molecular oxygen by respiration, whereas some microalgae generate oxygen by photosynthesis. Pollutants in the sample will affect these activities and can be detected by a changed sensor signal. A typical arrangement is shown in Fig. 7.40. The microbes are immobilized on a membrane positioned between a dialysis membrane and the oxygen-permeable membrane of the Clark sensor, mostly of macroporous PTFE.

Examples of microbial biosensors on the basis of Clark electrode are sensors containing the microalga *Chlorella vulgaris*, which was covered by an alumina membrane (Pandard et al. 1993). Also bacteria cultures of *Bacillus subtilis* and

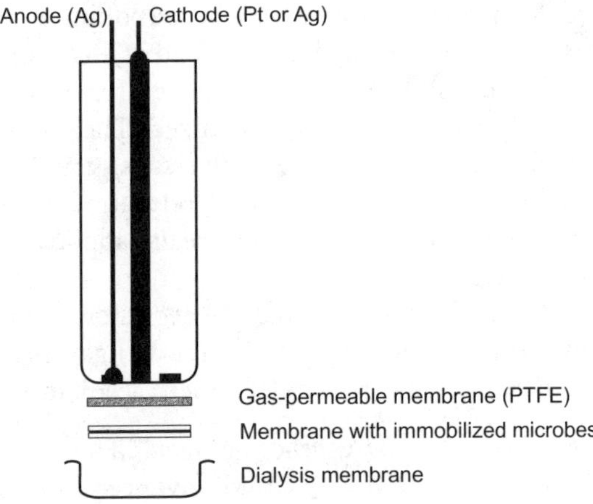

Figure 7.40. Microbial biosensor on basis of Clark probe

Bacillus licheniformis entrapped by a polycarbonate membrane have been used (Li and Tan 1994).

Biosensors with microorganisms are good monitors for toxic substances in natural waters. Also the *biological oxygen demand* (BOD) can be determined by means of biosensors.

Probes with organelles or with whole organs clearly perform like biological recognition mechanisms. Among the examples described was a sensor containing the immobilized antennae of a potato beetle (*Leptinotarsa decemlineata* Say) (Schroth et al. 2001). It was connected via an electrolyte bridge with the gate terminal of a field effect transistor. Traces of guajacol in air caused a detectable change in gate potential. Traces of guajacol and related compounds are released when foils of the potato plant are injured. The potato beetle is able to recognize such traces over a distance of many kilometres. A 'bio-FET' has even been assembled by entrapping an intact living beetle. A similar object has been built up with the antennae of small crabs. Such sensors respond to traces of trimethylaminoxide.

Nucleic Acid Sensors Molecules of nucleic acids exhibit extraordinary potentials. They are electrochemically active due to oxidation of the base guanine, and they can act as ligands which are able to bind many foreign substances. Due to these properties, they should be interesting receptors for chemical sensors. On the other hand, it would be useful to have analytical sensors for nucleic acid determination. The most important sensors in connection with nucleic acids are *hybridization sensors*. These allow one to search in solution for a specific DNA type and to identify in this way an individual biological species, a group of individuals or even a single individual. Hybridization sensors implement a specific form of the *genetic fingerprint*.

Sensors for DNA and with DNA. Traditional methods for electrochemical determination of nucleic acids, preferably of t-RNA (transfer ribonucleic acid) and DNA (deoxyribonucleic acid), are based commonly on voltammetric stripping analysis. The nucleic acid is adsorbed at a carbon electrode from stirred solution with a constant potential imposed for a definite time period. Subsequently, the deposited material is oxidized. The current flowing during the oxidation step is the source of the analytical signal. Such procedures are not selective. They require tedious sample pretreatment steps and numerous reagents. The term 'chemical sensor' is not really applicable for the electrodes used in such processes.

Adsorptive bonding of DNA at carbon surfaces is relatively strong. Molecules in the resulting monolayers act as ligands for heavy metals and other substances, which can be determined by voltammetry after adsorptive deposition.

DNA Diagnostics and 'Genetic Fingerprint'. The general interest in DNA diagnostics is growing due to the rapid advances in knowledge of the human genome. Progress in microtechnologies has facilitated broader application of nucleic-acid-manipulation techniques. At present, the most important applications of DNA diagnostics are methods for recognition of polymorphism and genetic mutation. It is necessary to detect reliably only one base-pair mismatch in the double helix. Similar methods can be used with the *genetic fingerprint*, i.e. to identify one specific individual based on its DNA.

Commonly, the diagnostic process starts with the generation of a sufficient stock of nucleic acid material, since the original DNA sample commonly is present only in trace amounts. By means of the polymerase chain reaction (PCR) a small amount of DNA can be amplified exponentially. This reaction allows enzymatic replication of nucleic acid molecules outside a living organism. In the material produced, the genetic sequence of interest must be present. Indeed, only a tiny section of the molecule is necessary to detect sufficiently the defect site or to identify the individual of interest.

Actual versions of DNA diagnostics are:

- *Electrophoresis* on agarose gel. DNA is cut enzymatically into short sections which are separated by electrophoresis. As a result, a characteristic image is generated. It is used for identification.
- *Hybridization.* An *oligonuclotide* is immobilized at a surface and used as a *probe*. Oligonucleotides ('oligos') are modules of DNA strands. The oligonucleotide chosen must reflect a characteristic section of the large, single-stranded molecule. The sample solution contains the DNA sample which has been resolved into single strands (the DNA double string has been 'molten') by thermal treatment in advance. In contact with the probe, the complementary strand forms a 'duplex' (the double strand) if the exactly matching counterpart of the immobilized oligos is present in the sample solution. This process is called *hybridization*. It is associated with property changes that are the source of a useful signal. Frequently, signals are obtained optically by measurement of *fluorescence* or *chemiluminescence*. Optical techniques are laborious, since the sample DNA must be marked

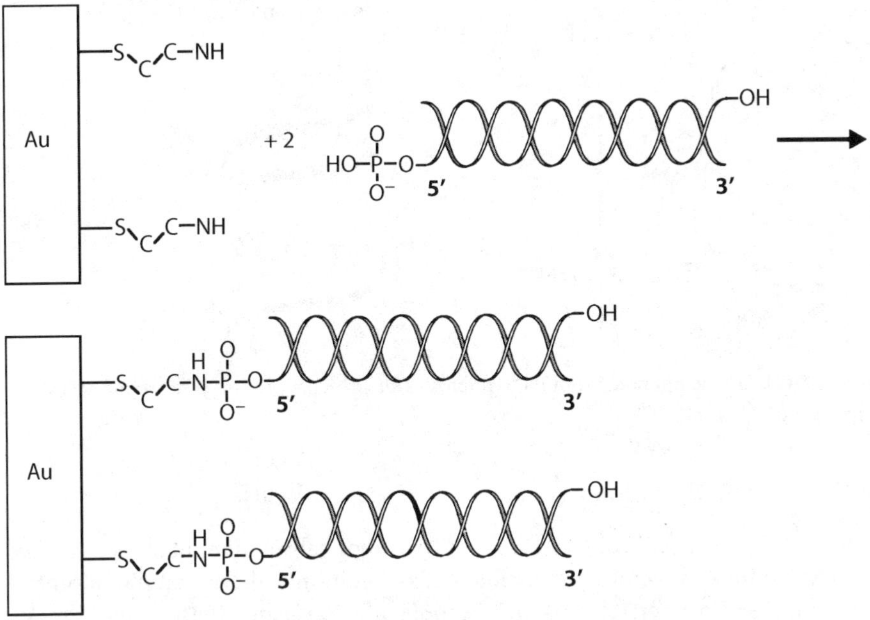

Figure 7.41. Covalent attachment of dsDNA at gold surfaces via SAMs

prior to analysis, e.g. by attachment of a fluorescing group. In contrast, electrochemical techniques allow one to detect hybridization without prior chemical modification of the sample.

Immobilization of DNA. Nucleic acids can be immobilized by means of methods equal to that used for protein molecules, i.e. adsorption, crosslinking, inclusion in gels or polymers, covalent chemical bonding with formation of SAMs, and finally, also the avidine-biotin complex formation (Sect. 7.4.1). SAMs with nucleic acids have also been used on silicon, where patterns have been generated by photolithography. Immobilization of DNA or oligonucleotides on glass or nylon surfaces is the basis of so-called DNA chips which combine numerous single probes. They are useful for simultaneous determinations (Ramsay 1998).

A characteristic procedure for immobilization of dsDNA or corresponding oligonucleotides is shown schematically in Fig. 7.41 (Zhao et al. 1999). The SAM on a gold surface was prepared in advance. It is allowed to react with DNA in the presence of the reagent 1-ethyl-3-(3-dimethylaminopropyl)carbodiimide-hydrochloride (EDAC).

Electrochemical Hybridization Sensors. Hybridization detection means recognizing the degree to which an immobilized single strand has formed hybrids (duplexes) with the studied complementary counterpart. The following electrochemical methods for detection of hybridization are currently in use:

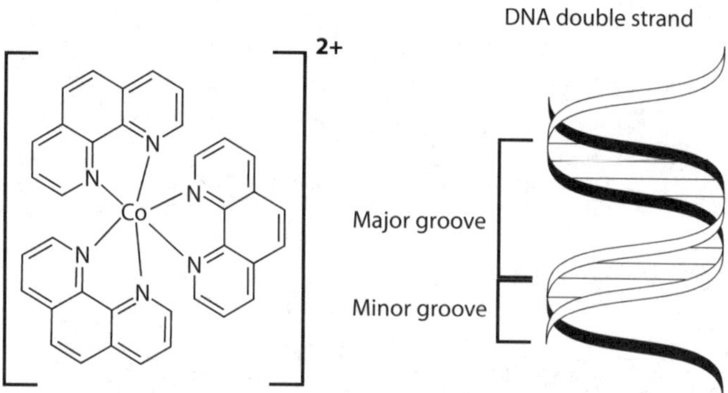

Figure 7.42. Cobalt-phenanthroline Co(phen)$_3^{2+}$ (*left*). Incorporation into *minor groove* of DNA (*right*)

- *Hybridization detection by means of reporter molecules*

Reporter molecules (indicators) are electrochemically active substances which are oxidizable or reversibly reducible. Most commonly, the cobalt phenanthroline complex Co(phen)$_3^{2+}$ is used. The molecule is attached in the *minor groove* of the DNA molecule (Fig. 7.42). There exist more redox systems which are incorporated in this way (intercalated). They must carry a positive charge (Millan et al. 1992, 1994, Millan and Mikkelsen 1993). Polynucleotides have been immobilized to form probes by means of covalent chemical bonding at the surface of glassy carbon and carbon paste electrodes. Cyclic voltammograms have been recorded before and after exposure to sample solution.

Assuming that at first the electrode surface is covered by a dense layer of ssDNA molecules, the transport of reporter molecules towards the surface is strongly inhibited, and as a result, the electrochemical signal is low. After hybridization, the signal increases since the double strand in general carries a larger negative surface charge. Consequently, the indicator cation displays a stronger interaction with DNA. It cumulative and can reach the electrode surface more readily. The difference in both signals provides information about the degree of hybridization. Figure 7.43 depicts the process.

- *Indicatorless and catalytic detection of hybridization by means of synthetic polynucleotides*

The inherent electrochemical activity of DNA is utilized in processes without reporter molecules. It is well known that the guanine base of DNA can be oxidized at moderate potentials. This property can be utilized if the molecules constituting the probe are not oxidizable themselves. For that purpose, oligonucleotides are synthesized which contain inosine instead of guanine base molecules. Inosine is practically inert at potentials where guanine is oxidized. If the target molecule combines with the synthetic probe during

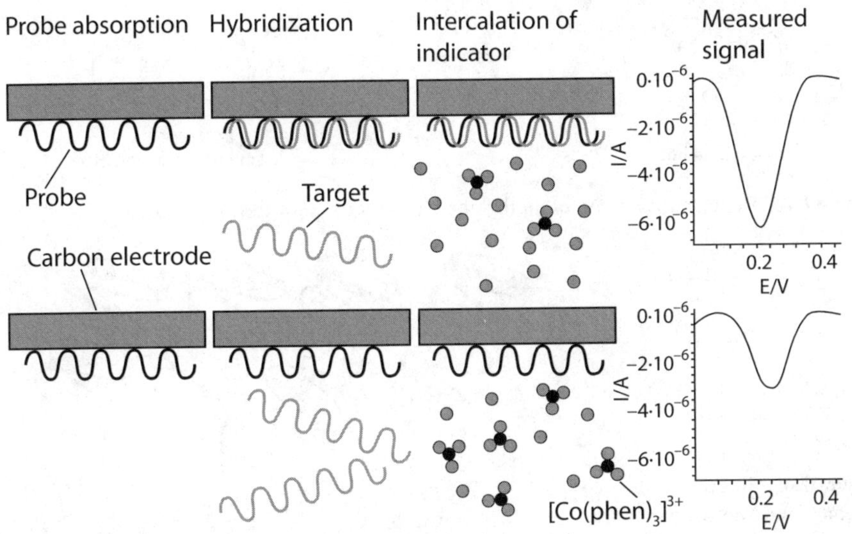

Figure 7.43. Hybridization detection by means of reporter molecule cobalt phenanthroline

hybridization, due to the guanine content of sample molecules, an oxidation current can be measured (Wang et al. 1998). The method has been advanced by utilization of catalytically active metal complexes like the bipyridyle complex of ruthenium(II/III). In this case, the soluble complex transfers electrons from the guanine of dsDNA towards the electrode surface (Thorp 1998).

- *Hybridization detection by charge transfer along dsDNA axis*

Some results suggest that electrons can move along the axis of an intact dsDNA molecule. Electronic conductivity of single strands is much lower. Charge transfer is inhibited if the hybridization is imperfect, i.e. if a double strand with base mismatches is formed. Only one mismatch in a molecule can disturb the charge transport severely. On this base, a detection method for hybridization can be established (Kelley et al. 1999).

An electrochemical hybridization sensor can be designed using special intercalator substances. Useful intercalators are electrochemically active reversible redox couples which are able to be inserted (intercalated) into the so-called π-stack of the dsDNA molecule. They are flat molecules of planar size such as e.g. methylene blue. The redox reaction of methylene blue is reversible. It can be reduced to give its colourless form (leukomethylene blue), which is readily reoxidized (Fig. 7.44).

If one intends to utilize the conductivity of the dsDNA for hybridization detection, it is important to generate a well-ordered SAM of DNA double strands. This is achieved best with oligonucleotides of moderate length, which correspond to a molecular section of native nucleic acids. Most commonly an oligonucleotide of 15 base pairs is used with an alkyl chain containing an SH

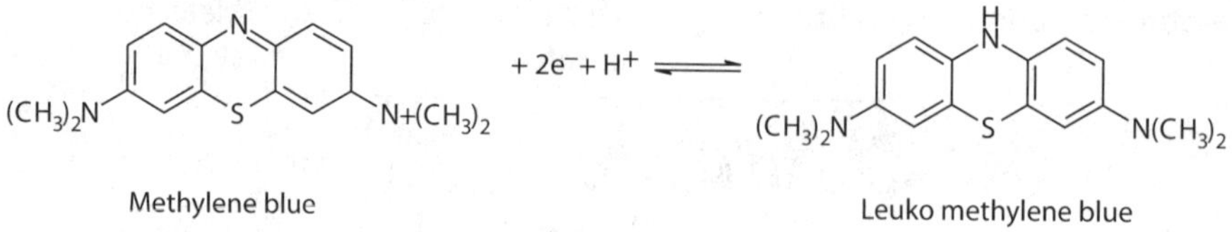

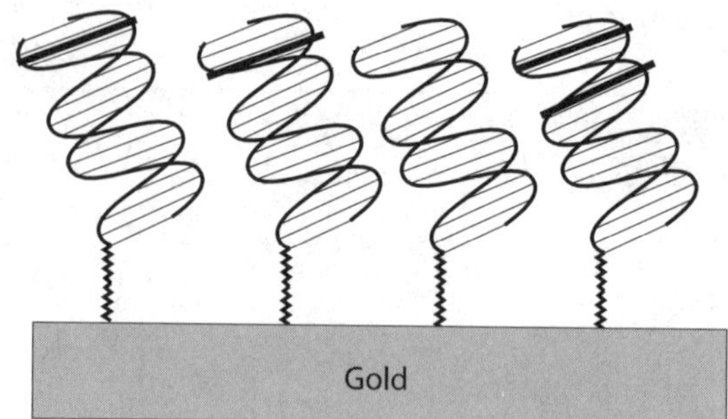

Figure 7.44. Redox equilibrium of methylene blue and its colourless leuko form

Figure 7.45. Intercalation of planar redox active molecules into double strand of DNA

group attached at the 5′ end as linker. SH groups readily bind covalently at a treated gold surface. As a result, a SAM of perpendicular oligonucleotides is formed (Kelley et al. 1997). The intercalator is inserted always in an upper position, far from the electrode surface (Fig. 7.45).

Electrons must cover the entire distance between electrode surface and intercalator molecule in order to undergo a redox reaction. A single base mismatch can interrupt the electron transport, hence the electrochemical signal is reduced drastically. The preferred electrochemical detection methods have been cyclic voltammetry and chronocoulometry. Intercalated molecules of methylene blue are reduced. Mismatches in the duplex formed after hybridization result in much lower signal intensity compared to perfect double strands.

Non-specific adsorption is a problem with all electrochemical hybridization detection methods. Molecules of nucleic acids are adsorbed at free sites of the electrode surface, i.e. at 'holes' in the DNA monolayer, the so-called pinholes. Such molecules contribute to the electrochemical signal, i.e. they simulate hybridization events.

7.5
References

Afromowitz MA, Yee SS (1977) J Bioeng 1:55

Ammann D, Bissig R, Güggi M, Pretsch E, Simon W, Borowitz IJ, Weiss L (1975) Helv Chim Acta 58:1535

Bartlett PN (1990) In: Cass AEG (ed) Biosensors: A Practical Approach. IRL, Oxford, pp 47–97

Bates PS, Kataky R, Parker D (1994) J Chem Soc Perkin Trans 2:669

Cadogan A, Gao Z, Lewenstam A, Ivaska A, Diamond D (1992) Anal Chem 64:2496–2501

Cardosi MF (1994) Electroanalysis 6:89–96

Chen Q, Wang J, Rayson G, Tian B, Lin Y (1993) Anal Chem 65:251–254

D'Annibale A, Regoli R, Sangiorgio P, Ferri T (1999) Electroanalysis 11:505

Darder M, Takada K, Pariente F, Lorenzo E, Abruña HD (1999) Anal Chem 71:5530–5537

Diamond D, Svehla G, Seward EM, McKervey MA (1988) Anal Chim Acta 204:223–231

Freiser H (1980) Coated wire ion-selective electrodes. In: Freiser H (ed.) Ion-Selective Electrodes in Analytical Chemistry, vol. 2. Plenum, New York, pp 85–105

Gibson TD, Hulbert JN (1993) Anal Chim Acta 279:185–192

Guilbault GG, Montalvo JG (1969) Anal Lett 2:283–293

Heller A (1990) Accounts Chem Res 23:128–134

Kelley SO, Boon EM, Barton JK, Jackson NM, Hill MG (1999) Nucl Acids Res 27:4830–4837

Kelley SO, Jackson NM, Barton JK, Hill MG (1997) Bioconjugate Chem 8:31–37

Kenausis G, Chen Q, Heller A (1997) Anal Chem 69:1054–1060

Knichel M, Heiduschka P, Beck W, Jung G, Göpel W (1995) Sens Actuat B28:85–94

Li F, Tan TC (1994) Biosens Bioelectron 9:445–455

Wang J, Rivas G, Fernandez JR, Lopez Paz JL, Jiang M, Waymire R (1998) Anal Chim Acta 375:197–203

Millan KM, Mikkelsen SR (1993) Anal Chem 65:2317–2323

Millan KM, Saraullo A, Mikkelsen SR (1994) Anal Chem 66:2943–2948

Millan KM, Spurmanis AJ, Mikkelsen SR (1992) Electroanalysis 4:929–932

Morf WE, de Rooij NF (1995) Sens Actuat A51:89–95

Pandard P, Vasseur P, Rawson DM (1993) Water Res 27:427–431

Ramsay G (1998) Nat Biotechnol 16:40–44

Rickert J, Göpel W, Beck W, Jung G, Heiduschka P (1996) Biosens Bioelectron 11:757–768

Schroth P, Schöning MJ, Lüth H, Weissbecker B, Hummel HE, Schütz S (2001) Sens Actuat B78:1–5

Schubert F, Renneberg R, Scheller FW, Kirstein L (1984) Anal Chem 56:1677–1682

Shi G, Luo M, Xue J, Xian Y, Jin L, Jin J-Y (2001) Talanta 55:241–247

Smith RL, Scott DC (1986) IEEE Trans BME 33:83–90

Srivastava SK, Gupta VK, Jain S (1996) Anal Chem 68:1272–1275

Suzuki H (2000) Electroanalysis 12:703

Thorp H (1998) Trends Biotechnol. 16:117–121

Vadgama P (1990) J Membran Sci 50:141–152

Wang J (1994) Talanta 41:857–863

Wang J, Lin MS (1988) Anal Chem 60:1545

Wang J, Lu F, Angnes L, Liu J, Sakslund H, Chen Q, Pedrero M, Chen L, Hammerich O (1995) Anal Chim Acta 305:3–7

Willner I, Riklin A, Shoham B, Rivenzon D, Katz E (1993) Adv Mater 5:912–915

Wilson GS, Thevenot R (1990) Unmediated amperometric enzyme electrodes. In: Cass AEG (ed) Biosensors: A Practical Approach. IRL, Oxford, pp 1–17

Yagi K, Khoo SB, Sugawara M, Sakaki T, Shinkai S, Odashima K, Umezawa Y (1996) J Electroanal Chem 401:65–79

Yagiuda K, Hemmi A, Ito S, Asano Y (1996) Biosens Bioelectron 8:703–707

Yamamoto N, Nagaoka S, Tanaka T, Shiro T, Honma K, Tsubomura H (1983) Analytical Chemistry Symposium Series, No. 17 (Chemical Sensors), pp 699–704

Zhao Y-D, Pang D-W, Hu S, Wang Z-L, Cheng J-K, Dai H-P (1999) Talanta 49:751–756

8 Optical Sensors

8.1
Optical Fibres as a Basis for Optical Sensors

The rich analytical facilities of optics and spectroscopy were discussed in Chap. 2, Sect. 2.4.3. Traditional optical instruments are typical laboratory instruments. Commonly, they require abundant space, a vibration-free working area and regular maintenance. The sample solution is normally placed in a *cuvette* with two optical plates in parallel at its extremes. Going from this technology to chemical sensors would not be simply a matter of miniaturizing the traditional set-up. For application of optical and spectroscopic techniques in optical sensors, nearly exclusively the phenomenon of *light conduction* is utilized. Light conduction means that incident light in an optical dense medium can stay 'enclosed' as a result of multiple total reflections, so that it can be conducted over large distances, even if the conducting path is curved. In contrast to daily experience, light conductors can guide light 'round the corner'. Optical guides can be of cylindrical or planar shape (Fig. 8.1).

The most common devices in sensor technology are thin cylindric waveguides, the so-called *optical fibres*. They consist of transparent materials. The first experiments with such fibres or fibre bundles date back to the beginning of the sensor age. Probes assembled with optical fibre bundles could be dipped smoothly into liquids suggesting a certain similarity to electrodes. For such probes, the terms *optrode* or *optode* became common.

Optical guides without participation of chemical reactions can extract their signal only by evaluating given properties of the media studied, i.e. light absorp-

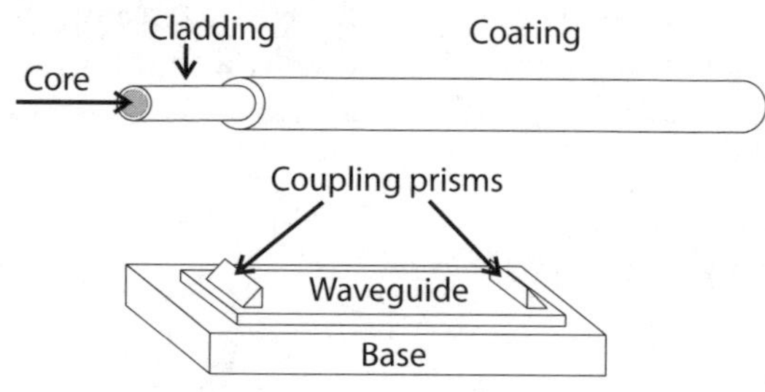

Figure 8.1. Basic set-up of light guides. *Top:* cylindric, *bottom:* planar shape

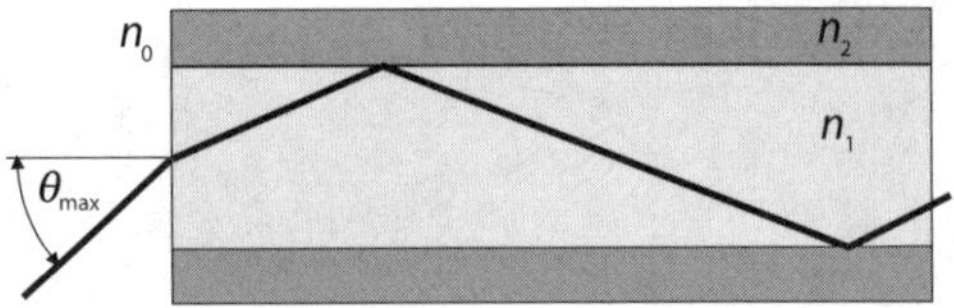

Figure 8.2. Light conduction by multiple total reflection in a single optical fibre. Incident light is accepted only inside a cone with the maximum incident angle θ_{max}

tion, light reflection, luminescence or the refractive index. Optical properties of liquids commonly depend on composition. With *extrinsic* optical sensors, the light guide is a medium for light 'transport' only but is not involved in signal formation. With *intrinsic* optical sensors, however, properties of the guide itself are affected by the analyte composition, possibly by means of an auxiliary medium, the *mediator*. The light guide in such cases actively contributes to signal generation.

Generally, a cylindrical light guide consists of the fibre itself as the *core* surrounded by a *cladding* layer and a light-impermeable *jacket*. The refractive index of the core (n_1) is always higher than that of the cladding (n_2). As a result of this arrangement, the light inside the guide is conducted like electricity inside a cable. Light beams are reflected towards the inside of the guide by internal total reflection. They can leave the cable only if the cable is bowed strongly. Incident light is conducted in nearly non-dissipative state if it reaches the fibre inside a cone which is narrower than the maximum incident angle $\theta_{max} = \sin^{-1}(n_2/n_1)$, as shown in Fig. 8.2.

Preferred materials for light guides are fused silica, glasses or specific polymers. The latter are cheap and can be manufactured easily. If sensor use is restricted to visible light, polymer fibres are sufficient commonly. Fused silica is characterized by maximum spectral transparency and thus useful for light conduction over long distances.

The popular image of total internal reflection is based on the *corpuscular* model of radiation. The alternative *wave model* explains different properties. A result of this model is the theory of light propagation by clearly distinguished methods, the *modes* (Fig. 8.3).

Optical fibres of the *single-mode type* have a core diameter of some micrometers at maximum, whereas multimode fibres can have a diameter of up to 1 mm (Fig. 8.4). The terms are a consequence of an optical theory which considers permitted field distributions along the fibres.

The *numerical aperture* (NA) is an important parameter for the familiar light beam approach. It provides a connection between the maximum angle of incidence θ_{max} and the refractive indices of corresponding materials:

$$NA = n_0 \cdot \sin\left(\theta_{max}\right) = \frac{\sqrt{n_1^2 - n_2^2}}{n_0} . \tag{8.1}$$

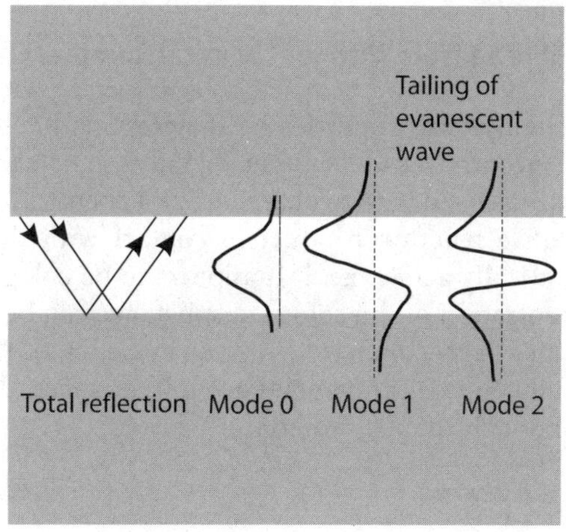

Figure 8.3. Light propagation in thin fibres. Corpuscular approach (*left*): Internal total reflection. Interpretation of light as wave (*right*): propagation in form of *modes*. The *evanescent wave* penetrates the surrounding medium

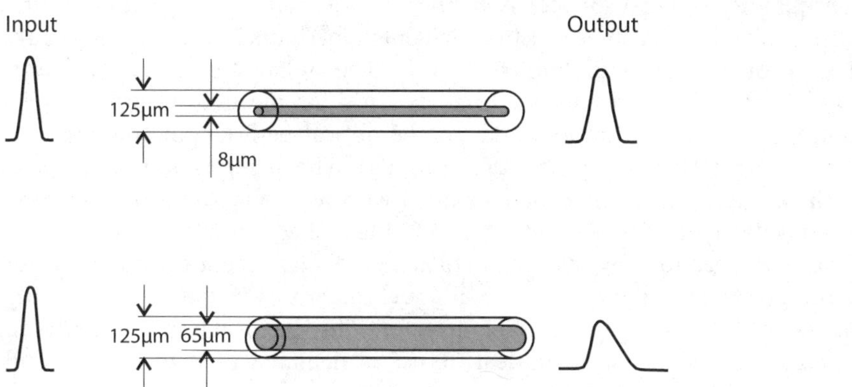

Figure 8.4. *Top*: single-mode optical fibre, *bottom*: multimode fibre

The refractive index n_0 belongs to the medium surrounding the fibre. For air, $n_0 = 1$. Incident light beams must stay inside a defined cone so as not to absorb the fibre wall. This cone is given by the numerical aperture. Exiting light at the remote end of the fibre will be enclosed by an equal cone.

The phenomenon of evanescence (Chap. 2, Sect. 2.1.2) is important with light guides since light propagates by internal total reflection. Every reflection at the interface between the media of different optical densities generates an evanescent wave (Fig. 8.3) which penetrates only a short distance but transfers energy to the medium. Hence, an interaction takes place which may be the source of a signal. To utilize the evanescent wave, a direct contact with the sample must be established by removing the cladding.

Optical fibres can be used as single fibres or in the form of fibre bundles. Fibre bundles often are used as carriers for chemical receptor layers in optodes. Single fibres become increasingly meaningful for miniature chemical sensors.

8.2
Fibre Sensors Without Chemical Receptors (Mediators)

The optical properties of analytes can be evaluated in many cases to obtain concentration-dependent chemical signals without a *mediator*. The term *mediator* is used here to characterize a chemical receptor which forms an optically active reaction product in contact with the sample. If the analyte itself is optically active, i.e. if it appears to be coloured in visible light, then the concentration of the coloured substance can be evaluated by means of *Beer's law* (also *Beer–Lambert law*) given by Eq. (8.2). The law is valid for monochromatic light only. Discrepancies occur if the spectral width is too broad to speak about monochromatic radiation:

$$A = \varepsilon \cdot c \cdot l \, . \tag{8.2}$$

The quantity A (absorbance) is defined as the common logarithm of the ration of light intensities I_0 (the incident light) and I (after having travelled through the sample): $A = \lg(I_0/I)$. The molar *absorptivity* (*absorption coefficient*) ε is a constant characteristic for the light-absorbing substance, c is its concentration, and l is the optical path length (the distance that the light travels through the material). The latter is normally given by the dimensions of the container filled with light-absorbing solution (the *cuvette*). Beer's law is utilized by a well-established analytical method, the *absorption spectroscopy* (*spectrophotometry*). A classical spectrophotometer is composed of a light source, a monochromator, the cuvette containing the sample solution, and a light detector. The classical arrangement can be emulated by means of optical fibres, as depicted in Fig. 8.5 (left and centre).

Quality requirements for optical sensor components are not lower than for classical spectrophotometers. It is possible to utilize classical photometers

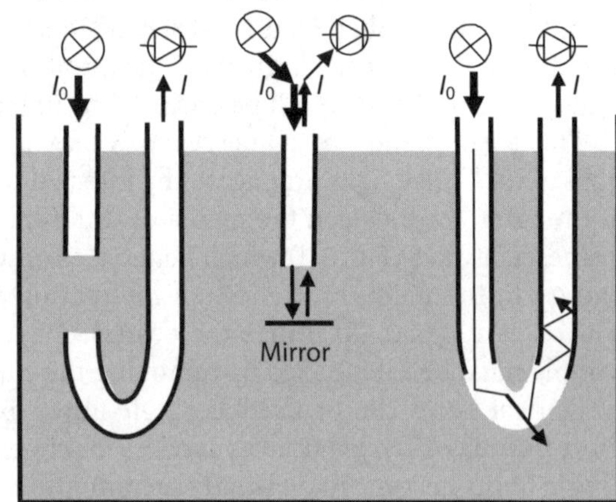

Figure 8.5. Light guides used as spectrophotometers (*left* and *centre*), or as microrefractometer (*right*). The intensity of incident light I_0 drops either by absorption in the sample or by loss due to incomplete internal total reflection

for use with optical fibre sensors. Instead of a cuvette, a metal block with optical guides inserted at both sides is introduced. The classical monochromator provides one light conductor with monochromatic light, whereas the second optical fibre delivers the output light to the light detector in the classical photometer. Lenses for focusing light beams were used to improve the efficiency of such arrangements. Also spectrofluorometers (Hirschfeld 1985) and different IR spectrometers (Bellon et al. 1989) have been converted for use with optical fibre sensors in a similar way. Adaptations of this kind are not very efficient, since important parameters, among them the numerical aperture, cannot be optimized sufficiently. Classical components like tungsten lamps, gas discharge lamps, prism or grating monochromators, and the photomultiplier are not consistent with the requirements of sensor technology. Like other chemical sensors, optodes and their peripheral equipment should be mobile, since the intention is to take the instrument and go to the sample. Hence, miniature dimensions are assumed. Meanwhile, the requirements are fulfilled by commercial instruments. Preferably the photomultiplier tube is substituted by miniature photodiode arrays. Such arrays can be used like any other photodetector, but in addition they offer the possibility of recording a complete spectrum in one go given sufficient resolution of the array. The number of integrated diodes is crucial for this problem. Arrays with 1024 elements are common. Since individual contacts for each diode in the array cannot be produced easily, *charge coupled devices* (CCDs) are becoming increasingly common. CCDs are used widely in digital cameras. Due to mass production, their prices are dropping. The pieces of information collected during illumination are somehow 'pushed out' of these arrays in a following step. Thus, a certain time is necessary to record a complete spectrum. The time requirement is commonly not larger than about 10 ms.

It is not necessary in every case to make use of a complete instrument with elaborate spectral dispersion. Often the set-up can be restricted to the demands of the actual measurement. A useful arrangement is the combination of a light-emitting diode (LED) with a photodiode possessing an absorption spectrum congruent with the emission spectrum of the LED. Such a small and cheap device is mobile and fulfils the demands of sensor use. An approach in this matter has been devised by Smardzewski (1988).

The refractive index of liquids can be measured efficiently by means of optical fibres. The refractive index depends on the composition of solutions. The mixing ratio of ethanol-water mixtures can be determined by means of classical refractometers. In chemical synthesis, the Abbé type refractometer is widely used to test the purity of organic liquids. A simple microrefractometer results if an optical fibre is bowed strongly up to the point where the internal total reflection becomes insufficient to conduct all the incident light to the detector (Fig. 8.5, right). The angle of total reflection depends on the ratio of refractive indices of the fibre material and the liquid surrounding it. At the place of insertion, of course, the covering layers of the fibre must be removed.

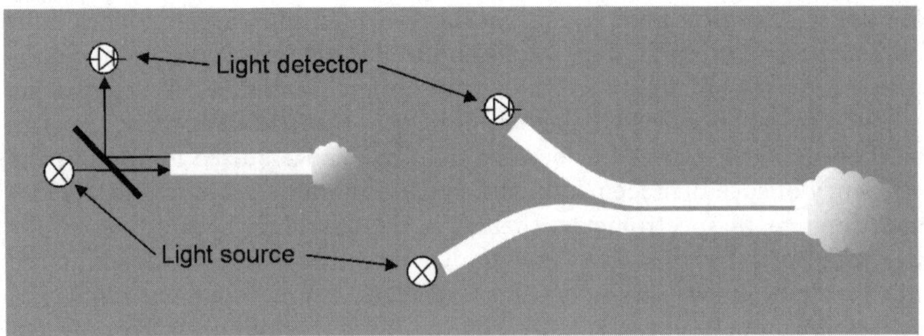

Figure 8.6. Sensors for investigation of samples with luminescence or diffuse reflection

By appropriate choice of materials, such microrefractometers can be adapted to the requirements of the specific analytical task. Arrangements of this kind played a certain role as detectors in liquid chromatography but have been displaced by more sensitive detectors.

If solution constituents appear as turbidity (i.e. if they reflect light in diffuse manner), or if they show luminescence, then their concentration can be measured by means of an arrangement shown schematically in Fig. 8.6. Samples with diffuse reflection scatter part of the intrinsic light. The intensity of this part can be evaluated by means of a light detector. The latter can be coupled with the optical fibre, as outlined in Fig. 8.6 (left), by means of a beam splitter. In this case, only one optical fibre is used to transmit light in both directions. A second, more common way is to use *bifurcated* optical fibre bundles, where two bundles are unified at the front end, which is exposed to the sample. One bundle comes from the light source, the other one goes to the light detector (Fig. 8.6, right). Sample components, which show luminescence (fluorescence, phosphorescence or chemoluminescence), emit light with a wavelength which differs from that of the incident light (the *exciting* light). Fluorescence sensors are characterized by an excellent detection limit, since they are relatively independent of external distortions.

Some examples of fibre sensors without chemical receptor are given in Table 8.1.

Table 8.1. Optical fibre sensors without chemical receptor

Analyte	Quantity measured
Cu^{2+}	Light absorption at 930 nm
Organic substances	Fluorescence
Hemoglobin in blood	Diffuse reflection at 600–750 nm

8.3
Optodes: Fibre Sensors with a Chemical Receptor

8.3.1
Overview

It has long been known that colour changes caused by chemical interactions may lead to conclusions about the concentration of samples. The well-known test strips make use of this fact.

Dyes which have been found to be useful indicator substances for test strips there can be combined with optical fibres. The results are devices now called *optodes*. Later, a further step of development started when experiences with ion-selective electrodes (ISEs) were applied to optical sensors. The term *optode* appeared in the early 1970s. It stands for a device which can be inserted into a solution like an electrode and from which some kind of a 'cord' leads to a measuring instrument. From the very beginning, optodes were connected with fibre optics. Later it became clear that a simple geometric characterization would be insufficient. Sensors through which a solution is travelling or where a light conductor crosses a solution can, of course, be considered optodes also, assuming they carry a chemical receptor layer to distinguish them from the simple fibre optic probes described in Chap. 8, Sect. 8.2. Hence, participation of a chemical reaction is the characteristic of optodes as defined here. Of course, there is no sharp boundary between both groups of fibre optical sensors. For example, consider the microrefractometer described in the preceding chapter. If its exposed surface is coated with a lipid layer, the interaction of this layer with organic sample molecules can be strong and selective, with results quite similar to those of a chemical reaction.

The preceding chemical reaction is the decisive source of sensor selectivity. In many cases, sensor designers benefited from achievements in the field of ISEs. On the other hand, there are good reasons to switch from electric to optical measurement technology, i.e. to apply optodes rather than electrodes. Commonly, the following advantages of optodes are cited (Wolfbeis 1990):

- Optodes do not require a 'reference electrode'. There is no electric circuit to close, hence it is not necessary to work with two probes.
- Optodes can be miniaturized easily.
- Optical fibres with high transparency allow signal transmission over long distances (up to 1000 m) without loss of quality. This characteristic is valuable in particular for long-distance measurements where the instrument cannot be placed near the sensor.
- The sensor signal is primarily optical in nature and consequently insufficient to block electrical interference from the environment. Such interference may provoke serious problems for electrochemical sensors.
- Materials for optical sensors, such as fused silica, are highly inert and may keep in contact with aggressive media for a long time.

On the other hand, optical sensors have some disadvantages:

- Ambient light may distort the measurement.
- Some of the reagents immobilized at the sensor surface, mainly dyes, are unstable and can be bleached by UV radiation or washed out by solvents.

The mediator present at the front end of the optode must be able to form an optically active compound in an amount which definitely depends on sample concentration. Useful optical effects are either spectral light absorption (formation of coloured products) or luminescence.

Many shapes for optodes have been proposed (Fig. 8.7). Most commonly, the reagent is immobilized at the surface of transparent microbeads made of glass or polymers. It can also be imbedded in a gel plug (Fig. 8.7a). The immobilized reagent stock is enclosed by a cap of dialysis membrane or similar semipermeable material. A reflecting layer at the bottom of the reagent reservoir can help to improve the efficiency (Fig. 8.7b). Less common are liquid pools at the end of the fibre bundle (Fig. 8.7c). Sometimes it is sufficient to impregnate the free end of the optical fibre by a reagent layer (Fig. 8.7d). In Fig. 8.7a–c, a double fibre bundle is used. One part acts as the sender, the other as the receiver. Figure 8.7d demonstrates that a single fibre may be used to deliver as well as to drain light. An alternative sensor shape is shown in Fig. 8.7e, where the immobilized reagent is applied at the surface of an optical fibre where the cladding has been removed. It is not easy to see why colour changes in this particular place should act on the light inside the fibre sufficiently. For an explanation, the phenomenon of evanescence must be considered (Chap. 8, Sect. 8.1). A further useful arrangement is the combination of two fibres oriented in parallel (Fig. 8.7f). One of the fibres acts as sender and is provided with a reagent layer that changes its colour as the result of a chemical reaction, or it is excited to luminescence. The receiving fibre is in close contact with the sender fibre. It takes over the light, possibly involving the evanescent field. At both fibres, the cladding has been removed along the contact region.

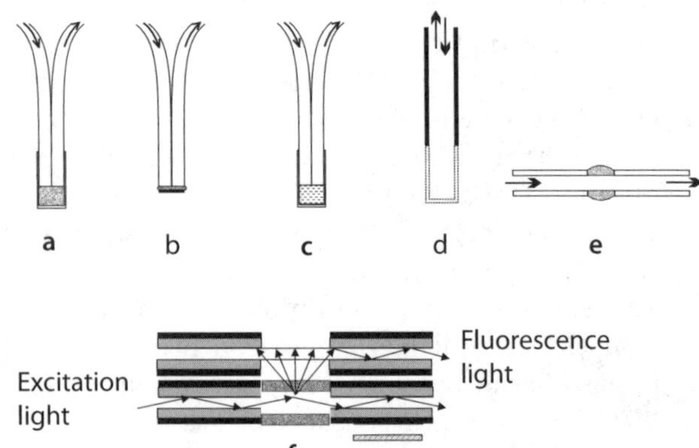

Figure 8.7. Common optode shapes

Different compositions of the chemical receptor layer have been proposed. The following ones predominate:

- A single reagent is used. It forms either a coloured or a luminescing substance in equilibrium with the analyte. Immobilized pH indicator substances are typical examples of this case.
- A combination of two or more reagents is used, one of them acting as chromophore. Common examples are ligands which release acid as a result of complex formation with the sample ion. A pH indicator substance, as the second reagent, changes its colour when responding to the acid released.

Immobilization techniques with optodes are less multifaceted than those with electrochemical sensors, since the selection of sensor materials is smaller. The most common immobilization procedures are the following:

- The reagent is fixed at the pretreated optical fibre surface by adsorption. At the surface, at first reactive tether groups form. They should bind as strongly as possible with the receptor molecules in order to avoid bleeding into the sample solution. Some examples are listed in Table 8.2.
- The reagent is lipophilized and entrapped subsequently in a hydrophobic layer at the sensor surface. Many common reagents can be made hydrophobic by attachment of a long alkyl chain. They are then protected from washout by the aqueous sample solution.

As with all sensors, it is a rule for optodes that the response should be *reversible*. The sensor should react on sample concentration changes as fast as possible, even if this process is repeated many times. The determining equilibrium between reagent and sample must be *mobile*. The most ordinary case for reaction of sample A with the immobilized reagent $\overline{R}$ can be written as

$$A + \overline{R} \rightleftarrows \overline{AR} \, . \tag{8.3}$$

The equilibrium constant K is given by

$$K = \frac{[\overline{AR}]}{[A] \cdot [\overline{R}]} \, . \tag{8.4}$$

Table 8.2. Immobilization procedures for receptor reagents at optodes

Carrier	Functional group	Surface modification	Attached receptor molecules
Cellulose	Carboxyethyl	Chloroacetic acid	Sulfonic acids
Glass; silicagel	Aminopropyl	Silanization	Carboxyl
Polyacrylamide	Carboxyethyl	Strong bases	Amines, proteins

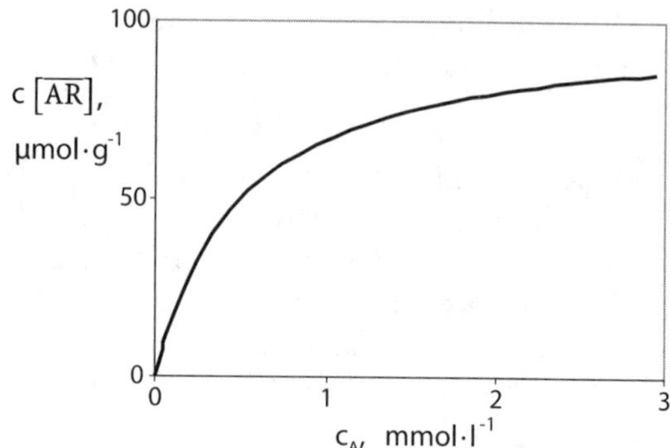

Figure 8.8. Surface concentration $\overline{[AR]}$ of coloured product as function of sample concentration c_A, calculated using Eq. (8.6) for $K = 2000\,l\,mol^{-1}$ and $c_R = 10^{-4}\,mol\,g^{-1}$

For the concentration dependence of the immobilized coloured product $\overline{AR}$ follows

$$\overline{[AR]} = K \cdot [A]\,[R]\ . \tag{8.5}$$

The equilibrium concentration of A is approximately equal to the solution concentration c_A. Assuming that $\overline{[R]}$ is equal to the difference of total reagent concentration c_R and $\overline{AR}$, Eq. (8.5) yields

$$\overline{[AR]} = \frac{K \cdot c_A \cdot c_R}{1 + K \cdot c_A}\ . \tag{8.6}$$

Under these conditions, the surface concentration of the coloured product $\overline{AR}$ depends on the sample concentration in a non-linear relationship (Fig. 8.8). For very low analyte concentration, i.e. for $c_A \ll 1/K$, the calibration graph is a straight line; for high concentrations it approaches a limiting value.

8.3.2
Optodes with Simple Receptor Layers

Optodes are different in respect to the composition of their receptor, independent of their shape. An increasing number of complex systems developed starting from the original primitive layers.

Actually, most common are pH-sensitive optodes. Immobilized pH indicator reagents act as receptors. Classical indicators as well as fluorescent indicator reagents are utilized.

In addition to pH-sensitive optodes, fibre sensors for numerous other analytes are available. Among them, sensors for physiologically active cations as well as for dissolved oxygen are worth mentioning in particular. If the principles of ISEs are to be transmitted to optodes, it is necessary to include a chromophoric group in the receptor layer. Commonly, this is an indicator reagent which has been made insoluble (lipophilic) and which responds secondarily to pH changes caused by the complex formation of the primary receptor

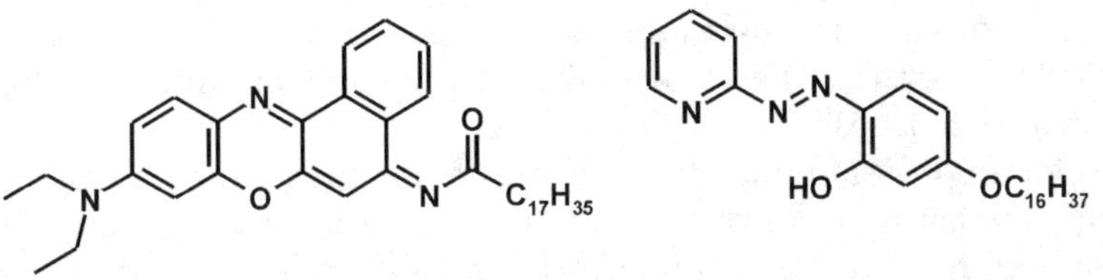

Figure 8.9. Lipophilized colour indicator substances (chromophores) for optodes. *Left*: pH indicator nil blue with C_{17} alkyl chain. *Right*: metal indicator pyridyl azo resorcinol (PAR) with C_{18} chain

with the analyte. The dye nile blue with an attached C_{17} alkyl chain has proved useful for calcium-sensitive neutral carriers as primary reagents (Fig. 8.9 left). Complexometric metal indicators like PAR also have been converted into chromophores by lipophilization (Fig. 8.9 right). Their selectivity is low, but they are useful for indicating the presence of certain groups of heavy metals.

The so-called *ionophores*, which are known under the name *neutral carriers* if used in ISEs, became meaningful also for optodes. Ionophores which carry a chromogenic group are called *chromoionophores*. However, this term is sometimes used for the combination of ion-exchanging ligand and lipohilic colour indicator.

Table 8.3 lists some examples of optodes with simple receptor layers. Among them are calixarenes, which are interesting for ISEs also, since they are highly selective ligands for cations (Chap. 7, Sect. 7.1.2). Above all, calixarenes are highly efficient receptors for certain organic analytes. Optical properties of such systems have been used successfully for chiral recognition of stereoisomers. As an example, when a special calixarene was combined with the

Table 8.3. Optodes with simple receptor layers

Analyte	Receptor/carrier	Measured quantity
pH (strongly acidic)	Congo red/cellulose acetate	Light absorption
pH (weakly acidic till neutral)	Fluoresceinamine/glass	Fluorescence
Al^{3+}	Morin/cellulose	Light absorption
Ca^{2+}; Na^+	Calixarene + lipophilized nil blue/PVC	Light absorption
K^+	Valinomycin + MEDPIN/ PVC	Light absorption
Cl^-	Fluoresceine/collodal silver	Fluorescence
SO_4^{2-}	Ba-chloranilate/glass	Fluorecence
NH_3	Porous PTFE/p-nitrophenol/gel	Light absorption
O_2	Ru-trisbipyridyl/silicone rubber	Phosphorescence quenching

(R)-enantiomer of 1-phenylethylamine, the optical absorption band maxima shifted to 538 and 650 nm, whereas with the (S)-enantiomer a shift to 515 nm resulted (Kubo et al. 1996).

Optodes for anion determination are less common than those for cations. In Table 8.3, the active agent of the chloride optode is silver fluoresceine, which itself does not fluoresce. By interaction with dissolved chloride, silver chloride is formed, which enhances fluorescence significantly. The controlling precipitation reaction is slow. Hence, the reversible response of the sensor cannot be expected. A similar mechanism operates also with the sulphate-selective optode listed in Table 8.3.

Not as many optodes exist for neutral molecules and for dissolved gases. A selective ammonia sensor is assembled by combining a porous PTFE membrane with successional reagent layers that react with ammonia under the formation of indophenol blue. Considerable efforts have been expended to develop sensors for dissolved oxygen. In most examples, the fluorescence quenching of oxygen has been utilized, which is pronounced with this molecule. This process also applies for the example mentioned in Table 8.3. In this case, a luminescing transition metal complex in silicon rubber is the active agent. Equation (8.7) expresses the relation between luminescence intensities I_0 and I (corresponding to the conditions without and with quenching, respectively) and oxygen concentration c. The process of fluorescence quenching is based on the deactivation of the emitting molecule as a result of collision with oxygen molecules. This process is fast and reversible:

$$\frac{I_0}{I} = 1 + K_q \cdot R_0 \cdot c \,. \tag{8.7}$$

The evanescent field is important for special shapes of optodes. The microrefractometer depicted in Fig. 8.5 is converted into an optode by coating its exposed area with a chemical receptor layer. This layer can be very thin, since the penetration depth of the evanescent wave is similar to the wavelength of light. Nevertheless, the light propagation in the optical conductor is affected by the change of refractive index when the reagent interacts with the sample. The changes are sufficient to allow concentration determination at the end of the light conductor.

Utilizing the evanescent field allows elegant designs of luminescence sensors, as demonstrated e.g. by the arrangement in Fig. 8.7f. In this arrangement, two optical fibres with zones freed from cladding are in contact with each other as well as with the sample solution. One of the fibres is coated with a reagent layer which is excited to luminescence depending on the light wavelength and on the analyte and its concentration. Luminescence is caused by the evanescent field. The second fibre acts as light receiver. It accepts the fluorescence radiation and conducts it to the fibre end.

The evanescent field plays an important role particularly with planar light conductors.

8.3.3
Optodes with Complex Receptor Layers

Ordered molecular layers have been used for optodes. Such layers can be generated by the Langmuir-Blodgett technique and related procedures. Every important receptor function (mobilizing ions, ion exchange, colour change) can be performed by its own individual reagent layer, which can be optimized individually. Transport processes can be controlled independently. An impressive example of such a high-tech optode is a sodium-selective sensor assembled of six ordered layers (Fig. 8.10). The layers have been generated successively by Langmuir-Blodgett technology on a silanized glass surface. Valinomycin attached with an octadecanol group acts as neutral carrier for potassium ions. This reagent was immobilized in a hydrophobic layer that had been covered by arachidic acid [n-eicosanoic acid, $H_3C-(CH_2)_{18}-COOH$]. The colour change is generated by immobilized rhodamine B.

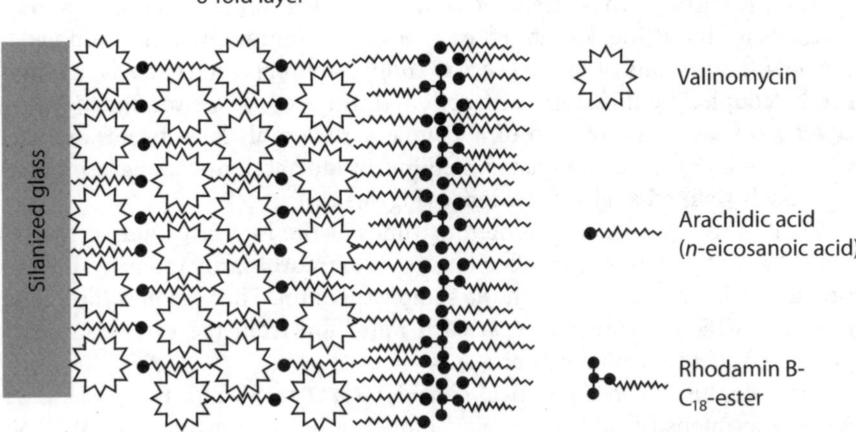

Figure 8.10. Receptor of potassium-selective optode composed of six stacked ordered layers (Schaffar et al. 1988)

8.4
Sensors with Planar Optical Transducers

8.4.1
Planar Waveguides

Optical chemical sensors can be assembled of thin planar layers instead of optical fibres. The evanescent field is always the basis of such sensors. It is an older idea to utilize the interface between a transparent plate and the medium studied in order to obtain optical or spectroscopic information. The well-known method *attenuated total reflectance spectroscopy* (ATR) is the result of such considerations. The theoretical background was developed first for macroscopic optical elements, i.e. for transparent disks with a thickness of some millimetres. When switching to chemical sensors, very thin transparent layers became interesting. On the other hand, a trend from fibre sensors towards thin planar layers can be noted. An important advantage of this development is the substantially larger useful interface of the planar layer.

The particular efficiency of planar optical sensors can be utilized only if the optical layers are very thin, i.e. in the range of a few microns. Such layers can be generated by the following procedures:

- Vapour deposition of transparent materials on a support.
- Spin-coating of polymer films
- Diffusion of components into the surface. For example, by introducing silver ions in this way into a glassy surface, a thin layer with a higher refractive index is formed. The latter makes up the core of the optical sensor.

The principle of measurement is visible in Fig. 8.11. There are different methods for inputting light into the active layer. Commonly, light is coupled to the device by means of prisms (*prism coupler sensor*). Alternatively, the light can be coupled by miniature diffraction gratings at the sensor ends (*grating coupler sensor, GCS*). With grating couplers, preferably laser beams are used which are coupled to the planar medium by diffraction at the grating with a precisely defined angle of incidence (Fig. 8.12).

Light is totally reflected numerous times at the interfaces due to the low thickness of the active layer. This means an extraordinarily intensive interaction with the interface, i.e. with the sample medium. The reason is the energy quantum which is transferred at each reflection from the extension of the evanescent field across the interface.

More sophisticated theoretical considerations result in a relationship between the intensity ratio of in- and outgoing light intensities and the concentration of sample constituents. For light-absorbing analytes, an analogue

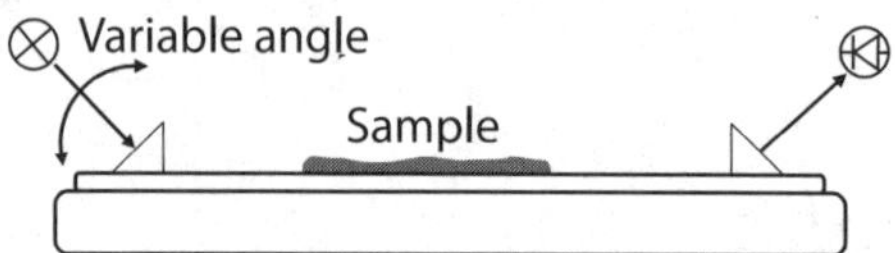

Figure 8.11. Arrangement for measurements with planary optical transducers (not to scale). The prisms for light input and output can be used with fixed as well as with variable angles

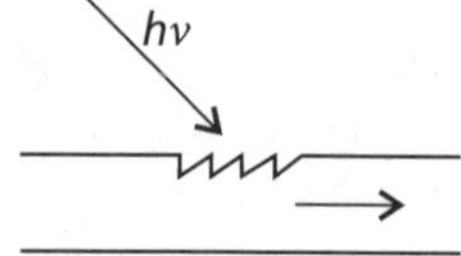

Figure 8.12. Light coupling via diffraction grating

Table 8.4. Chemical sensors with planar optical transducers

Analyte	Receptor	Principle of measurement
NH_3	Berthelot reaction in organic polymer	Light absorption, He-Ne laser
NH_3, toluene	Photopolymerization of organic films	Film-thickness variation (swelling) due to absorption
Dissolved chlorine	Phthalocyanine on SnO_2, indium-doped (ITO)	Change in layer colour, can be reset electrochemically
Na^+	Ionophore; anionic dye/ PVC membrane	"active" wave guide changes colour
Antigene	Immobilized antibody	Fluorescence

to Beer's law is observed. This relationship is strictly valid only for polarized light, but chemical sensors in planar optics can be handled commonly without special precautions, as with classical spectrophotometric instruments.

Some examples of sensors with planar transducers are listed in Table 8.4. With the ammonium sensor, blue colour is formed by the *Berthelot reaction* (formation of indophenol blue by a nitroprusside-catalysed reaction). The blue colour causes attenuation of the outgoing light intensity, which is related to analyte concentration by the aforementioned Beer-analogue relationship. In the second example, even the dimension change of active film caused by physisorption of gaseous samples (ammonia or toluene) is sufficient to obtain a measurable concentration-dependent signal. A sensor for dissolved chlorine makes use of the colour change of a phthalocyanine layer spread on a transparent, electrically conductive film of indium-doped tin dioxide. The colour change is irreversible, but the sensor can be regenerated by an electrochemical process which uses the transparent oxide layer as a redox electrode. The last example in Table 8.4 belongs to the group of optical biosensors. It has been included simply to demonstrate that fluorescence phenomena can be utilized in planar sensors as well. In this case, fluorescence results from antigen molecules with attached fluorescing markers. These molecules are adsorbed selectively by a layer of immobilized antibody molecules. Fluorescence radiation is excited by light coupled via mobile prisms. In this way, a spectroscopic operation is possible which is called *total internal reflection fluorescence* (TIRF).

8.4.2
Surface Plasmon Resonance and Resonant-Mirror Prism Couplers

Sensors which make use of the phenomenon called *surface plasmon resonance* (SPR) have become very popular in recent years. Recently, researchers have managed to transfer the principle to fibre optic sensors, but nevertheless

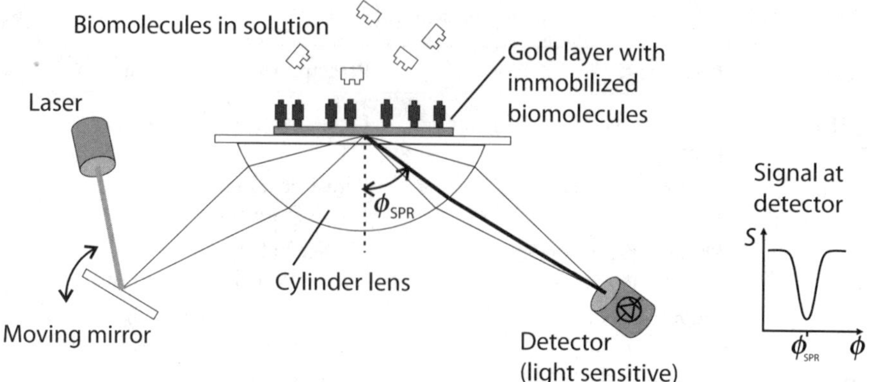

Figure 8.13. Set-up for measurement of surface plasmon resonance

SPR is commonly considered a branch of planar optical device technology. The arrangement according to the scheme shown in Fig. 8.13 is used nearly throughout an experiment. The upper surface of a 'prism' (indeed, a half cylinder made of glass is used generally) is coated with a very thin metallic film, which is covered by a chemical receptor layer. Most common are sputtered films of gold or silver. Receptors may be polymers or ordered monolayers of biomolecules. A light beam incident from the glass side is subject to total reflection at the interface and generates an evanescent field, which interacts with the metallic film. During this process, resonance with *surface plasmons* can occur. Surface plasmons are oscillations of the electron gas in the metallic film. A certain amount of energy of the incident light is absorbed. Depending on the wavelength and material properties of the metallic film, absorption maxima appear at specific angles of incidence. Such maxima can be recorded by means of photodiode arrays. Since the material properties of the metallic film are affected by the receptor layer, the optical signal is influenced by the sample concentration. After all, changes in the refractive index in the receptor layer are crucial. Hence, the SPR sensor more or less represents an extremely sensitive refractometer.

The sample can be in liquid or in gaseous phase. For liquid sample studies, the SPR arrangement throughout is designed in the form of a flow cell, where the sample solution streams along the receptor layer. Gaseous samples are also commonly allowed to stream through channels over the receptor surface.

SPR sensors are used preferably for biochemical analysis. The latter is discussed in more detail in Chap. 8, Sect. 8.5. In gas analysis, NO_2 sensors are of particular importance. Ordered phthalocyanine layers act as receptors. They are generated by the Langmuir–Blodgett technique on thin gold films. Among the exciting light sources were laser beams. SPR sensors for organic vapours with polymer receptor layers have been proposed, among them dextran gels.

Measurement technology with a *resonant-mirror* (RM) *prism coupler* is similar to the SPM method. Several thin transparent films are located on one plane of a prism made of a material with a high refractive index. One of them, with a low refractive index, is the coupler, another one, with a high refractive index, is used as waveguide. The optical films are covered by the receptor layer. If the angle of total reflection is achieved, light enters the waveguide and starts to interact with the receptor interface by manifold total reflection. Opposite to the side of incidence, at the prism an interference pattern can be recorded in a similar way as with SPR. It is generated by affecting the evanescent field of the light propagating in the waveguide.

8.5
Optical Biosensors

8.5.1
Fundamentals

Biosensors with optical transducers are distinguished from optical sensors for inorganic species by certain specific features. Differences exist in the importance of specific optical phenomena.

Light absorption and diffraction are less meaningful than with inorganic sensors. For biosensors, the most important optical effects are *luminescence phenomena,* in particular *fluorescence.* The reason is that many biochemically active fluoresce substances or they may be linked easily to fluorescing groups. Chemoluminescence and bioluminescence are important phenomena also. Commonly, oxygen is included. In some sensors, the signal is formed as a result of changes in the refractive index.

Immobilization of receptor and indicator substances is done by means of the same methods discussed in connection with electrochemical biosensors (Chap. 7, Sect. 7.4). With optical sensors, however, conditions to be considered are different from those with electrochemical sensors. As an example, some of the materials must be transparent.

8.5.2
Optical Enzyme Sensors

Optical enzyme sensors are designed preferably as extrinsic sensors, i.e. as optodes. There are, however, some examples of intrinsic sensors. As an example, an optical fibre has been described which was manufactured on the basis of a polystyrene fibre coated with adsorbed enzyme and indicator molecules. The colour changes brought about by the enzymatic reaction was detected using the evanescent field.

Examples of enzyme optodes are listed in Table 8.5. In examples based on light absorption or diffraction, the receptor is adsorbed at the surface of a finely

Table 8.5. Enzyme optodes

Analyte	Receptor	Principle of measurement
p-nitrophenyl-phosphate	Alkal.phosphatase/nylon membrane	Light absorption/diffraction
Glucose	Glucoseoxidase + bromothymol blue/light-diffracting membranes	Light absorption/diffraction
Glucose	Glucoseoxidase + Ru complexes/acrylamide polymer	Fluorescence quenching of O_2
Bilirubin	Bilirubinoxidase + Ru-dye/adsorbed on glass fibre	Fluorescence quenching of O_2
Cholesterol	Cholesterinoxidase + Ru phenthroline complex/graphite layer on Si	Fluorescence quenching of O_2
Lactate	Lactatedehydrogenase/nylon membrane + NADH	Fluorescence
Penicillin	Penicillinase + fluoresceine/polymer membrane	Fluorescence
Xanthine; hypoxanthine	Xanthin-oxidase + peroxidase/polyacrylamide gel	Chemi-luminescence
Diverse sugars	Peroxidase + luminol/polyacrylamide gel	Chemi-luminescence

woven nylon mesh or other light-diffracting media (Fig. 8.14). An intensively coloured reaction product can be generated by the enzymatic reaction, as is the case with the analyte *p*-nitrophenylphosphate. Alternatively, a pH indicator is used, for instance, with glucose sensors where pH changes result from the catalysed reaction.

Sensors utilizing luminescence phenomena are very common. The well investigated fluorescence quenching by molecular oxygen can be used to great

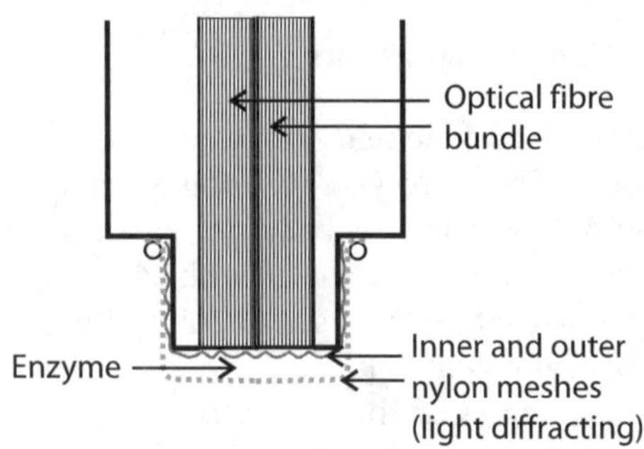

Figure 8.14. Enzyme sensor utilizing light absorption or diffuse reflection

effect. Preferably, ruthenium complexes are used. Their fluorescence increases if oxygen is consumed as a result of an enzymatic reaction. In this way, quite heterogeneous samples have been determined, like glucose, bilirubin and cholesterol (Table 8.5). The degree of quenching is related to sample concentration by Eq. (8.7).

The sensors for lactate and penicilline in Table 8.5 make use of fluorescence itself rather than fluorescence quenching. The lactate sensors stand for a group of optodes where the inherent fluorescence of NAD (nicotinamide-adenine-dinucleotide, see Chap. 7, Sect. 7.4.2) is evaluated. Reaction (8.8) can be influenced by establishing a specific pH so that it runs either from left to right or right to left. In this way, lactate as well pyruvate sensors can be manufactured.

$$\text{lactate} + \text{NAD}^+ \xrightleftharpoons{\text{lactate dehydrogenase}} \text{pyruvate} + \text{NADH} . \tag{8.8}$$

Chemoluminescence sensors often make use of the reaction of luminol with hydrogen peroxide formed by catalytic action of peroxidases (Fig. 8.15). Optical sensors for xanthine and hypoxanthine, as well as for glucose and other sugars, are based on this principle (Table 8.1). No exciting light source is necessary for such reactions. Another advantage is that the analyte is indicated directly at the sensor surface, without the need to diffuse into a membrane. Diffusion from bulk solution towards the sensor surface is the determining process. Continuous stirring is necessary to avoid depletion of the substrate hydrogen peroxide.

Sensors based on bioluminescence (biochemoluminescence) preferably utilize the action of luciferine, the luminophore of the firefly. In the enzyme-

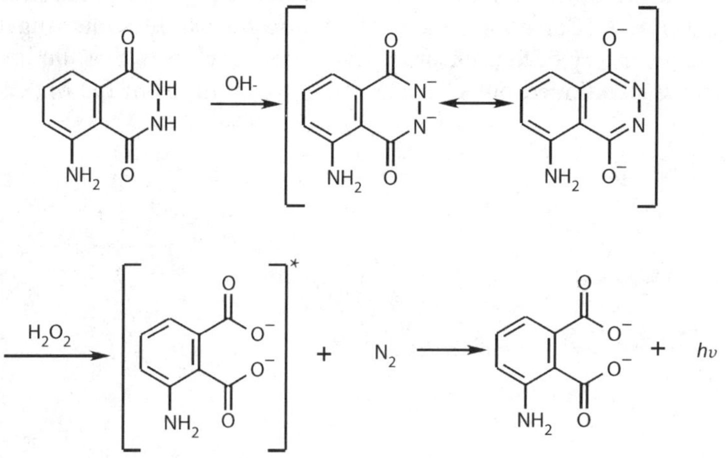

Figure 8.15. Chemoluminescence reaction of luminol

catalysed reaction with oxygen [Eq. (8.9)], light with a wavelength of 562 nm is emitted:

$$\text{luciferine} \xrightarrow{\quad O_2 + \text{luciferase} \quad} \text{oxyluciferine} + h\nu \,. \tag{8.9}$$

Combining reaction (8.9) with cofactors like ATP (adenosine triphosphate) results in extremely sensitive bioluminescence sensors. As an example, trace concentrations (down to femtomolar) of creatinkinase have been determined. This enzyme plays an important role in diagnostics of myocardial diseases. The reaction follows the scheme given in Eq. (8.10), where AMP (adenosine monophosphate) is a reaction product of ATP:

$$\text{AMP} + \text{creatinphosphate} \xrightarrow{\quad \text{creatin kinase} \quad} \text{ATP} + \text{creatine}$$

$$\text{ATP} + \text{luciferine} + O_2$$

$$\xrightarrow{\quad \text{luciferase} \quad} \text{AMP} + \text{diphosphate} + \text{oxyluciferine} + CO_2 + h\nu \,. \tag{8.10}$$

8.5.3
Optical Bioaffinity Sensors

In nearly all cases, bioaffinity sensors are identical to immunosensors, i.e. the complex formation of antibody and antigen is utilized. The glucose sensor based on concanavaline A (Fig. 8.16 and Table 8.6) is an exception to that rule. In other respects also, this sensor is exceptional. At the end of an optical fibre, a piece of dialysis membrane is attached where the protein concanavaline A is immobilized at the inner surface, mostly by crosslinking with glutaraldehyde. Concanavaline A (Con A) is capable of binding glucose and other sugar molecules. The geometry of the sensor is designed in such way that the immobilized layer is positioned out of reach of light coming from the optical

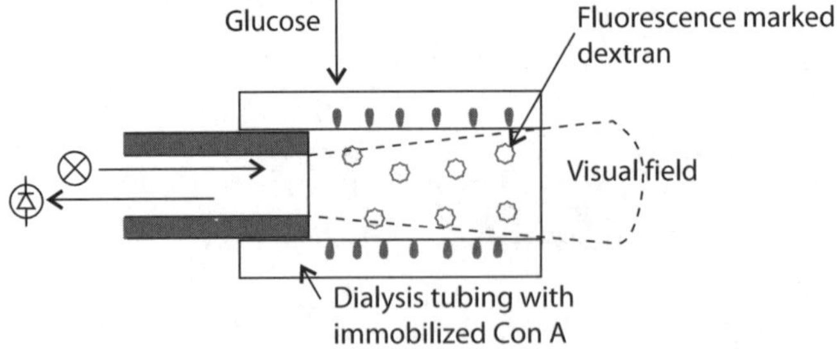

Figure 8.16. Bioaffinity sensor with concanavaline A and dextran marked by fluorescing group

Table 8.6. Optical bioaffinity sensors

Analyte	Receptor	Principle of measurement
Glucose	Concanavalin A + fluoresceine + dextran/ dialysis tubing	Fluorescence with indicator
Benzo[a]pyren- metabolite	Antibody/silica gel beads	Direct fluorescence fibre optics
Atrazin	Antibody marked with Eu complex	Competitive fluorescence, fibre optics
Morphin	Albumine-morphine conjugate + fluoresceine marked antibody/ quartz fibre	Competitive fluorescence, fibre optics
Theophyllin	Treated porous glass as filter/ theophylline marked with fluoresceine	Competitive fluorescence, flow-through cell
Antigens	Anti-IgG/quartz; antigen; fluoresceine modified antibody	Sandwich fluorescence
Antigens	Antigen marked with luminol + antibody marked with fluorophore	Chemiluminescence/ fluorescence

fibre. Dextran marked by fluoresceine is placed in the sample compartment. In the actual determination process, the analyte glucose and the marked dextran compete for the binding sites of concanavaline A. Glucose displaces some of the dextran molecules, which arrive at the inner part of the dialysis membrane, in this way reaching the field of view of the optical fibre. The blue exciting light from the fibre will generate the typical green fluorescence. The intensity of this emission can be evaluated as a quantitative measure of glucose concentration.

Optical immunosensors have attained a high degree of perfection. They are widely used due to the fact that classical immunoassays traditionally operate preferably on the basis of optical recognition techniques. The latter could be adapted for sensor use. Three procedures are the most commonly used

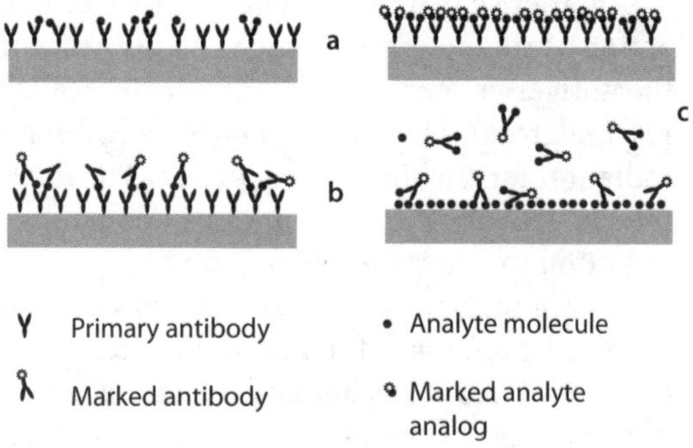

Figure 8.17. Measurement procedures with optical immunosensors. **a** Direct assay. **b** Sandwich assay. **c** Competing assay

(Fig. 8.17). They are specific for bioaffinity sensors. Analogues in other techniques of sensor signal evaluation are virtually non-existent:

- Direct determination (Fig. 8.17a): No marking of analyte or receptor. Analyte molecules bound to the surface are detected directly (e.g. by their intrinsic fluorescence).
- Sandwich indication (Fig. 8.17b): Binding of the analyte to the sensor surface modified by an antibody is followed by fixation of a second, marked antibody at the first layer of bound antibody molecules.
- Competitive determination (determination by displacement, Fig. 8.17c): The analyte competes with marked molecules for free sites at the sensor surface.

A choice of the manifold shapes of optical immunosensors is presented in Table 8.6. The list contains sensors with direct determination (a sensor for a metabolite of benzo[a]pyrene with intrinsic fluorescence), as well as some simple fluorescence sensors with competing modes of operation (for atrazine, morphine and theophylline). In the sensor for determination of the herbizide atrazine, the active antibody is modified by a fluorescing europium complex. For the morphine sensor, in a preceding operation a conjugate of the analyte with albumine is prepared and immobilized at the surface of a quartz fibre. This conjugate combines with the antibody marked by fluoresceine. Finally, the sample is added. It releases an equivalent amount of antibody with fluorescent marking. The decrease in fluorescence is evaluated by means of the evanescent field. The sensor for theophylline in the table is of planar shape. Its peculiarity is a filter matrix of porous glass positioned in front of the sensor. The matrix is covered by adsorbed protein A. Its function is to catch interfering antibodies which are specific to theophylline. The sandwich sensor also listed in Table 8.6 is based on the function of the antibody anti-IgG, which was immobilized at pretreated fibres or plates of quartz by crosslinking with glutaraldehyde. This layer interacts with the sample (an antigen) as well as with a fluorescent marked antibody. The fraction of antibody bound is finally measured in an evanescent field. The result depends on analyte concentration.

Another sensor mentioned in Table 8.6 operates on the basis of a combined action of chemoluminescence and fluorescence. In this less common example, the antigen is marked by luminol, whereas the antibody is marked by a fluorescing group. The latter is chosen such that the emitted light acts as exciting radiation for the fluorescing group. The analyte (the unmodified antigen) interacts with the components as a competitor, thereby decreasing the emitted radiation to a measurable degree.

Immunoprobes in surface plasmon resonance and in RM arrangements are used preferably for research. They are very useful for investigating the kinetics of immunochemical reactions and for obtaining information about the structure of the antigen-antibody complex. In this respect, it is doubtful whether such systems can be considered sensors at all. In addition to structure determination, SPR instruments are useful for concentration analysis as well. The concentration of human IgG, acting as an antigen in this special case, has been determined by SPR. It was adsorbed at a silver film on glass support. The sample, i.e. the antibody anti-IgG, binds to the immobilized antigen and affects the angle of total internal reflectance. Concentration is determined by evaluation of the measured effects.

Important insights into the structure of the avidine-biotin complex have been obtained by SPR. The technique was useful also in determining the absolute amount of protein molecules on metallic films. For that purpose, a gold film on glass plate was coated with a dextran layer exposed to a solution containing certain monoclonal antibodies in a flow-through cell. Optical quantities determined in this way were calibrated by means of radioactively marked antibodies.

A RM arrangement was used to recognize certain strains of bacteria which carry strands of protein A at their surface. A layer of immobilized human immunoglobuline G was used as receptor. The RM arrangement was positioned in a flow-through cell.

8.5.4
Optical DNA Sensors

As with electrochemical probes, indication of hybridization is the most important application of optical DNA sensors as well. When immobilized single strands form a double strand with the complementary counterpart by hybridization, the presence of a specific biologic species can be recognized.

The completed formation of a double strand traditionally is indicated optically, preferably by fluorescence measurements. When going from classical methods to chemical sensors, the most important requirement was to immobilize DNA strands at the surface of waveguides, regardless of whether they were planar or fibre shaped.

A useful section of the DNA molecule (commonly a synthetic oligonucleotide with 15 to 40 base pairs) is immobilized at the sensor surface. Formation of ordered layers is preferred due to their advantages for optical investigations. Ordered monolayers are thin enough to be studied by means of the evanescent field. A very good substrate for the formation of SAMs are gold films, which are present anyway in SPR. At best, the oligonucleotide to be immobilized is attached at one end with an SH group (via a molecular group acting as linker) and at the opposite end with a fluorescing group. An ordered layer of perpendicular molecules is formed which bind at the gold surface over their sulphur bridge, whereas the fluorescing group is exposed to the solution side similar to redox active amperometric hybridization sensors. A frequently used alternative way (Fig. 8.18) starts with a SAM formed of single-strand oligonucleotide molecules. The target molecule is marked ahead of time by combining it with a fluorescing complementary strand. The performed hy-

Figure 8.18. Indication of DNA hybridization by fluorescence measurement

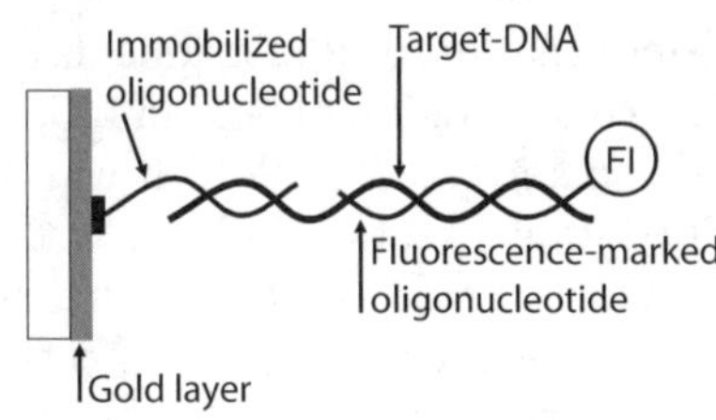

bridization with the immobilized 'oligo' is marked by a measurable increase of fluorescence intensity.

Oligonucleotides are immobilized at non-metallic surfaces most commonly by means of the avidine-biotin reaction (Chap. 7, Sect. 7.4.1).

Examples of optical DNA hybridization sensors can be found in Table 8.7. In the example in line 1 (Abel et al. 1996), a fibre surface was functionalized first by treatment with (3-aminopropyl)-triethoxysilane (APTES), or alternatively it was silanized by mercaptomethyldimethylethoxysilane (MDS). At the short-chained alkylsilane layer formed, biotin was attached, and this layer was subsequently treated with avidine or streptavidine. In this way, oligonucleotides could be fixed which had been biotinylated at their 5′ end. The reason is that each avidine molecule can bind several biotin molecules on opposite sides. The complementary strands were marked by fluorescing groups. Optical measurement makes use of the evanescent field. In the example at line 2 of the table (Piuno et al. 1994), an isonucleotide with 20 base pairs was attached at the surface of an optical fibre. The latter had been functionalized by treatment with APTES. As a *spacer*, a chain of 1,10-decanediol-bis-succinate (with the terminating group dimethoxytrityldeoxythymidine) was bound covalently. Spacers are useful to control the distance between DNA strands and to avoid uncovered surface sites. The oligonucleotide layer was capable of hybridization with the

Table 8.7. Optical DNA hybridization sensors

Receptor	Transducer	Principle of measurement
16fold oligonucleotide/avidin-biotin	Fibre optics	Fluorescence of target
20fold oligonucleotide with substituents/fluorescing intercalator	Fibre optics	Fluorescence of indicator
40fold ligonucleotide/ streptavidin-biotin/waveguide	Planar waveguide	Refractive index by resonant mirror (RM)
Oligonucleotide/avidin-biotin/metal	Planar waveguide/ metallic layers	Refractive index by surface plasmon resonance (SPR)
Oligonucleotide/linker/Ta$_2$O$_5$/glass	Planar grating coupler	Fluorescence of target

complementary ssDNA. The fluorescence dye ethidinium bromide (EB) acted as indicator for completed formation of dsDNA. The dye was intercalated selectively into the DNA double strand. The enhanced fluorescence intensity could be measured via the evanescent field of the fibre.

Line 3 of Table 8.7 (Watts et al. 1995) is related to a RM study. Biotinylated oligonucleotides with 20 base pairs, functioning as probe molecules, were bound at the surface of a waveguide via streptavidine. Hybridization initiates a variation of the virtual refractive index of the layer that can be measured by common RM methods. SPR also is very well suited for investigation of DNA hybridization.

In line 4 of Table 8.7 (Jordan et al. 1997), different versions of a procedure are covered where commercially available SPR chips were provided with biotinylated oligonucleotides of variable length. The latter act as probes for complementary strands in the sample solution. The kinetics of hybridization as well as further fundamental problems have been studied using this procedure.

In line 5 (Duveneck et al. 1997), a planar grating coupler of a very thin tantalum pentoxide film (thickness 100 nm) on glass substrate has been studied as a basis for nucleic acid sensors. The very high refractive index of the oxide assures a particularly strong evanescent field in the biologic receptor layer on the waveguide. The exciting radiation was coupled in via an etched grating. The emitted light from luminescence excited by the evanescent field was coupled out and focused via an optical lens system. The waveguide surface had been treated in advance by a special trimethoxysilane in order to bind the oligonucleotides acting as receptor molecules. In a process of solid-phase synthesis, the oligos were allowed to grow on the functionalized surface via a linker group of hexaethylene glycol. The target DNA strands were marked by fluorophene. After hybridization, as few as 100 attomoles were detected.

8.6
Sensor Systems with Integrated Optics

The technology of *integrated optics* (Chap. 2, Sect. 2.3.4) is a branch of microoptics, which emerged from the successful techniques of microelectronics together with micromechanics and other microtechnologies. The processes of integrated optics allow one to fabricate complete optical instruments like spectrometers, refractometers etc. in strongly miniaturized form and are extremely cost efficient, since fully automated mass production is common for this industry.

Sensor systems manufactured in integrated optics are often characterized by a high degree of complexity, hence they should be considered miniature instruments. On the other hand, naive users often are not aware of this complexity and apply the instruments like litmus strips or optodes.

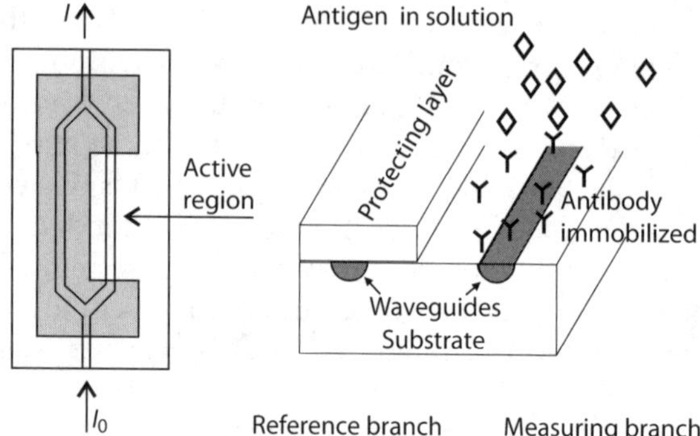

Figure 8.19. Mach-Zehnder interferometer in integrated optics used as immunosensor

The interferometers in integrated optics, mentioned in Chap. 2, Sect. 2.4.3, are considered to be sensor systems. They are the only complex systems currently in routine use. In such systems, intrinsic light of intensity I_0 is divided and guided in parallel in two optical paths. One of these branches is partially exposed to the sample medium, commonly via a receptor layer. The latter has a refractive index lower than that of the waveguide. The evanescent field penetrates the receptor layer. If the refractive index is changed by the interaction of the receptor with the sample, then the angle of total reflection and, consequently the light path length in the sample branch, is varied. In this way, the optical properties of the waveguide itself are modulated. For characterization, an 'effective refractive index' (n_{eff}) of the conductor is defined. The actual value of n_{eff} can vary also with the thickness of the receptor layer. This behaviour can be utilized also to obtain analytical information, e.g. if the receptor layer changes its thickness by swelling due to absorption of gas molecules. The thickness of the receptor layer must be lower than the depth of penetration of the evanescent field in this case. An increase or decrease in light path length in the sample branch of the interferometer results in interference phenomena when both light paths are unified. These phenomena can be evaluated by means of intensity modulation at the output of the instrument. Equation (8.11) characterizes the ratio of incident light intensity I_0 and of the intensity of the light coupled out:

$$\frac{I}{I_0} = \frac{1}{2}\left[1 + \cos\left(\Delta n_{eff} \cdot k_0 \cdot L\right)\right] . \tag{8.11}$$

In the above equation, Δ_{eff} denotes the difference in the effective refractive indices of reference and sample branches, k_0 denotes a constant (the so-called *wave vector*), and L is the length of the receptor region (the measuring window).

Immunosensors play a particularly important role among the interferometric sensors in liquid phase. The working principle is depicted in Fig. 8.19, where a sensor on a glass substrate was chosen as an example. The waveguides have been generated by diffusion of Ag^+ ions into the glass surface.

8.7
References

Abel AP, Weller MG, Duveneck GL, Ehrat M, Widmer HM (1996) Anal Chem 68:2905–2912
Bellon V, Boisdé G (1989) Proc SPIE, Int Soc Opt Eng 1055:350–385
Duveneck GL, Pawlak M, Neuschäfer D, Bär E, Budach W, Pieles U, Ehrat M (1997) Sens Actuat B 38-39:88–95
Hirschfeld T (1985) J Instrum Soc Am 85:305–317
Jordan CE, Frutos AG, Thiel AJ, Corn RM (1997) Anal Chem 69:4939–4947
Kubo Y, Maeda S, Tokita S, Kubo M (1996) Nature 382:522
Piunno PAE, Krull UJ, Hudson RHE, Damha MJ, Cohen H (1994) Anal Chim Acta 288:205–214
Schaffar BP, Wolfbeis OS, Leitner A (1988) Analyst 113:693–697
Smardzewski RR (1988) Talanta 35:95–101
Watts HJ, Yeung D, Parkes H (1995) Anal Chem 67:4283–4289
Wolfbeis OS (1990) Chemical Sensors – Survey and Trends, Fresenius J Anal Chem 337:522–527

9 Chemical Sensors as Detectors and Indicators

As a matter of course, sensors will yield measurable, concentration-dependent signals. It is not self-evident, however, that concentration values necessarily must be calculated from the sensor signal. In this chapter, analytical methods are discussed where the concentration of interest is not determined primarily by evaluating the sensor signal, but by means of other phenomena. Chemical sensors have an auxiliary function only in this case. As an example, in a titration process a sensor functions to indicate a certain value of the 'degree of reaction', i.e. the extent to which a reaction has been completed. In a titration, the amount of reagent consumed up to this point is the source of analytical information.

Specifications for sensors used as indicators generally are less strict than for sensors used for direct concentration determination. Long-term stability, selectivity and linearity play a less important role. In the case of detectors for separation techniques (chromatography and electrophoresis), even non-selective behaviour is preferred.

Some of the sensors presented in the following sections are utilized nearly totally for detection or indication. Direct concentration determination is not usual with such sensors.

9.1
Indicators for Titration Processes

Titration is one of the classical methods of analytical chemistry. It is widely used since the first half of 19th century. It has not lost its significance till our time. Figure 9.1 depicts the working scheme of a titration process for the particular case of volumetric titration. The *titrant*, a reagent solution of exactly known concentration (a *standard solution*) is added stepwise under continuous stirring to the sample solution, until *equivalence* has established, i.e. until reagent and sample have reacted to a degree of approximately 100 percent. In an acid-base titration, this point is reached if the sample (which was either acidic or alkaline before) reacts neutral, i.e. if it is *neutralized*. The indicator is announcing this state. The amount of titrant consumed up to this point can be determined from the scale of the burette. The consumed volume is used to calculate the result. Titrations have been performed long before sensors had been

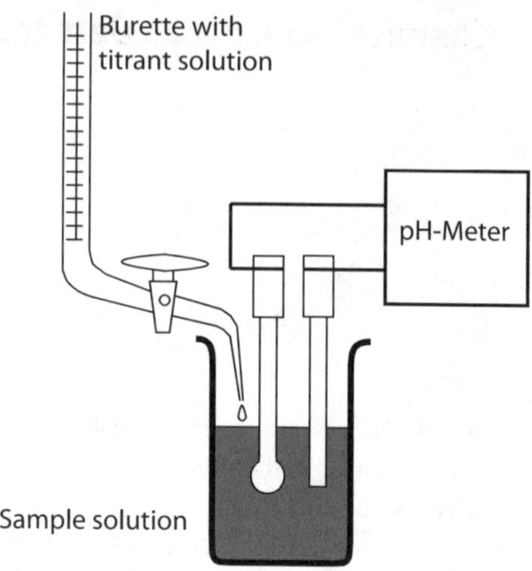

Figure 9.1. Volumetric titration with potentiometric indication

available. Traditionally, *indicator substances* have been used, i.e. dissolved dyes which change their colour markedly in the vicinity of the equivalence point. Beginning with the 1930s, instrumental indication by means of electrochemical probes also became increasingly accepted.

In the example of Fig. 9.1, a potentiometric set-up of indicator and reference electrodes is used to indicate the equivalence point. For acid-base titrations, the pH-sensitive glass electrode makes up a useful indicator probe.

Potentiometric techniques are prevalent for instrumental titration processes. Often the typical sigmoidal (S shaped) potentiometric titration curve is used as a synonym for the term 'titration curve' as a whole, although curves with linear branches are found also. Titration curves are recorded by plotting the results of potentiometric measurement (the voltage differences between reference and indicator electrodes) vs. the amount of titrant used. A typical example for titration of a strong acid by means of a strong base is shown in Fig. 9.2. The equivalence point (the point where the amount of titrant added would just correspond to 100% reaction degree) in this case is reached at

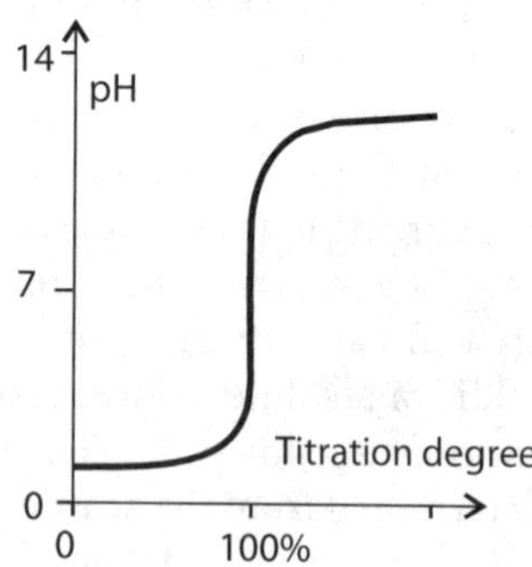

Figure 9.2. Titration curve of acid-base titration

pH = 7. The function of the indicator circuit is to announce the operator if this point is reached. Subsequently, the amount of titrant is read from the burette scale and the result calculated.

Nearly all chemical sensors useful for liquid samples can be utilized to indicate titrations. Besides the preferred potentiometric, other electrochemical probes are also used, mainly amperometric and conductometric sensors. The so-called biamperometric titration works with simple wire pairs. Photometric and thermometric indication techniques are less common than electrochemical methods. Miniaturization does not play an important role for titration probes. Classical arrangements predominate to this day. Commercial titration instruments are only slowly starting to make use of the achievements of modern sensor technology. As an example, optodes have achieved a certain popularity in recent years for titration applications.

In volumetric titrations, the titrant is added in the form of a solution. An important alternative is *coulometric titration*. In this method, the titrant is generated in situ by electrolysis of the solution containing the analyte. The amount of charged used until the equivalence point has been reached is evaluated to calculate the amount of sample. Calculation is based on Faraday's law [Chap. 2, Eq. (2.36)]. Coulometric titrations are easily performed on a miniature scale. If the generator electrodes are extremely small, then mechanical stirring is not necessary. Diffusion is sufficient to transport reactants. Such microtitrator systems can be combined with an electrochemical indicator arrangement in order to establish a sensor-actuator system. Such a set-up is an example of miniaturized total analytical systems (μ-TAS) which are presented in Chap. 10. Such systems are much more complex than simple chemical sensors, although in practical use this is not always transparent.

9.2
Flow-Through Detectors for Continuous Analysers and for Separation Techniques

Flow-through detectors are chemical sensors, since their function is to 'report' about the concentration of certain analytes in a flowing stream. As with titration, the content of a sample solution is not determined by evaluation of the magnitude of a sensor signal alone. In fact, the preceding operations in a flowing stream are crucial. As an example, the flowing stream will cross a separation column, where the dissolved substances are separated into 'packages' which stream consecutively through the detector. At the detector, an image of these packages is expected. Hence, the sensor should reflect the concentration change of each sample plug as a function of time. The geometric properties of these images are evaluated to obtain the content of the sample solution. Preferably, either the signal height or its area is determined. A strict concentration proportionality of the flow-through detector is not mandatory, but the signals should be reproducible. Selectivity also is not an important feature.

On the contrary, detectors for separation processes should generate signals with a magnitude equal for a given concentration value of each substance, independent of the kind of substance.

The 'images' of sample concentration profiles should be as true as possible. Thus the following specifications should be fulfilled as strictly as possible:

- The *dead volume* of the flow-through detector should be as small as possible. Every detector requires a minimum sample volume to form a signal. As an example, the channel of a photometric detector, which is crossed by a light beam, must be filled completely with sample solution. The electrode surface of an electrochemical detector must be covered completely with sample solution. On the other hand, the *resolution* of the detector (the ability to distinguish between two closely adjacent signals) is determined predominantly by its dead volume. This is understood if one imagines what happens if a very steep concentration gradient or a very narrow peak is incorporated completely inside the detector's dead volume. Such extremely narrow peaks are found with separation techniques like chromatography and electrophoresis. Hence, miniaturization of detectors is always the goal.
- The *response time* of flow-through detectors should be as short as possible. Sluggish response will result in low resolution. A quantitative characterization of response time is commonly given by the time elapsed until 99% of the final signal has been achieved (t_{99}, see Chap. 1, Sect. 2.3.2).

9.2.1
Continuous Analysers

In continuous analysers, classical wet chemical operations are performed in non-classical manner, i.e. in a flowing stream. The sample flows through the instrument, where reagents are added or separation processes are conducted, until finally the content is determined by evaluation of detector signals. Two types of devices are in use, *segmented* and *non-segmented analysers*.

In automatic analysers of the air-segmented type (*segmented flow analysis, SFA*), although not each single sample portion has its own reaction vessel, it does have an individual solution plug, the *segment* (Fig. 9.3, top). This segment is provided with necessary reagents, and mechanical as well as thermal operations are performed at this spot. In a later stage of development, methods without segmentation appear. Measured increments of sample solution are *injected* into a continuously flowing stream. This is the working principle of the most important non-segmented technique, the *flow injection analysis (FIA)*. The schema is depicted in Fig. 9.3 (bottom). The detector plays an important role in segmented as well as in non-segmented flow analysis. The detector yields a characteristic signal shape for each individual sample increment. The sample concentration is determined, in the best case, by evaluation of signal height.

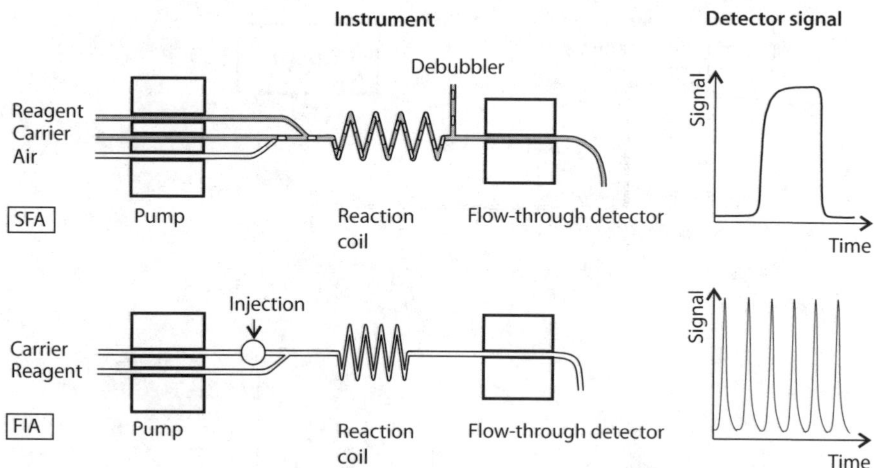

Figure 9.3. Working schemes of continuous automatic analysers of segmented flow analyser (SFA, *top*) and flow-injection analysis (FIA, *bottom*) types

The signal source of automatic analysers, as a rule, is the chemical reaction. The instruments are designed to minimize the high expenditure of human labour which is commonly necessary for performing classical chemical operations manually. As an example, a spectrophotometric determination is automated by means of the instruments depicted in Fig. 9.3. Manually, such an analysis starts with careful pipetting and mixing sample as well as reagent solutions. Subsequently, the reaction is allowed to proceed for a well-controlled time interval, normally between 15 and 60 min. Time is controlled using a stopwatch. Finally, the reaction mixture, where a concentration-proportional colouration has appeared, is poured into the cuvette of the spectrophotometer, where subsequently the absorbance is determined as the basis for calculating the concentration. In an automatic instrument, all these operations occur in a flowing stream without human intervention. Further operations can be included, in particular those which require high manual effort. An example is the separation of volatile components by distillation to determine its content independently of interfering substances.

FIA is an example of modern analysers rationalizing wet chemical analysis. The basic scheme is simple (Fig. 9.4, top). For the case of a simple photometric analysis, the sample is injected into a *carrier stream*. A photometric reagent dissolved in this stream will react chemically starting from the border of the analyte plug during the formation of a coloured product. According to the sample concentration in the plug, at an optical detector, a signal is formed which is characterized by a specific geometric shape, but with a magnitude reflecting the sample concentration (Fig. 9.4, bottom). The instrument can be operated even without a mechanical pump, although normally a *peristaltic*

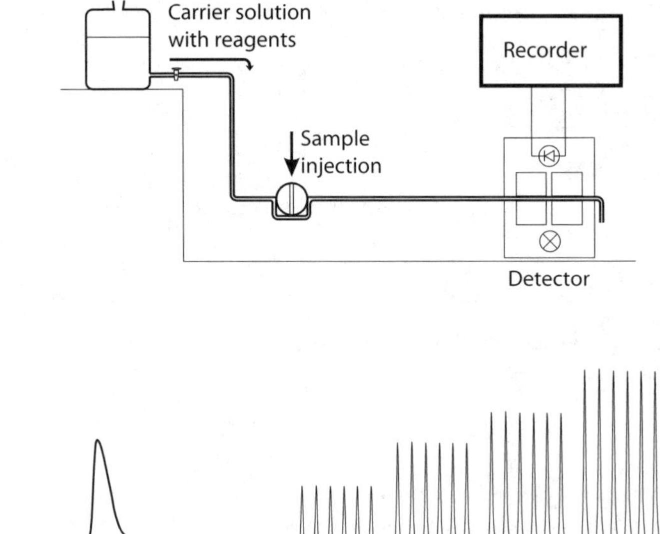

Figure 9.4. FIA. *Top*: minimal experimental arrangement; *bottom*: signal shape at detector (*left*), multiple signals of repeated injections with increasing concentration

pump is used which has the advantage that all the solutions come into contact only with inert materials. The sample is introduced by means of an injection valve, which must ensure that well-measured, reproducible volume increments are injected into the carrier stream. Commonly, the valve contains a rotating disc with an inner channel filled with sample solution. In the proper position, the carrier solution sweeps out the volume inside the channel and transports it through the system like a plug.

Photometric detectors predominate in FIA, since photometric determinations are adapted most commonly to the method. Two simple detector arrangements are shown in Fig. 9.5. The carrier stream with imbedded sample plugs streams through a non-transparent block, which is provided at opposite sides with a light source and a light detector. In the set-up at left, a light-emitting diode (LED) is used as light source, whereas a facing photodiode acts as light detector. A red lighting LED with a red-sensitive photodiode is a good combination, e.g. for more or less greenish products of the photometric reaction. Also, diodes with green or yellow emission combined with photodiodes of corresponding sensitivity work well. A more versatile arrangement is formed when a traditional spectrophotometer is converted into a FIA instrument by means of optical fibre bundles (Fig. 9.5, right). In commercial instruments, the light beam commonly does not cross the solution stream in a perpendicular fashion, but in parallel with it in order to achieve a longer light path and con-

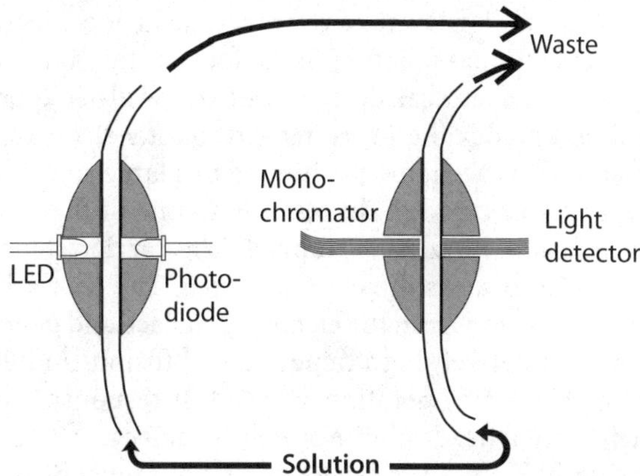

Figure 9.5. Simple photometric detectors for FIA

sequently a better sensitivity. Detectors of this kind are in use also for liquid chromatographic analyses (Fig. 9.7).

Electrochemical detectors commonly work amperometrically. Less common are potentiometric detectors. In both cases, the intention is to minimize the dead volume. The widely used devices shown in Fig. 9.6 are optimized to fulfil this requirement. In thin-layer cells (right-hand side of picture), the solution flows in a very narrow gap (a few microns) alongside the surface of the working electrode. The dead volume corresponds to the channel volume currently in contact with the electrode surface. The working electrode area is rather large, but the layer in contact with it is very thin. As a result, the amperometric mode of operation might switch to coulometric mode, where all electroactive substances in the dead volume are completely converted into their corresponding reaction products. At the output of the detector, no trace of the introduced material remains. The sensitivity of coulometric detectors is outstanding. With potentiometric indication, the geometry of the cell and its streaming behaviour are less important. A specific problem of potentiometric

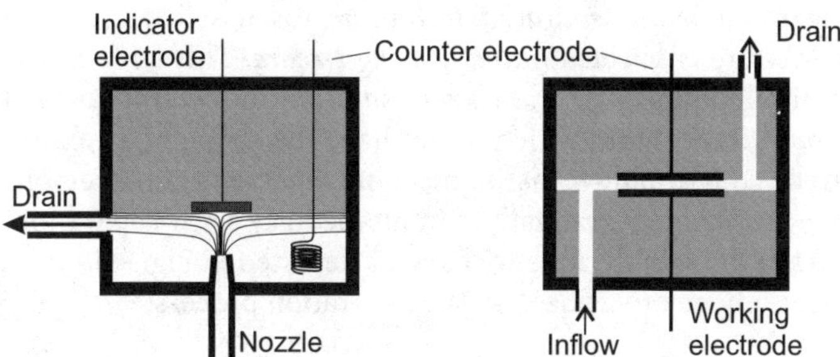

Figure 9.6. Electrochemical flow-through detectors. *Left*: wall-jet arrangement; *right*: thin-layer cell

detectors is it requires a reference electrode. Electrodes of the second kind are not easy to manufacture in the form of thin layers.

Electrochemical detectors of the wall-jet type (Fig. 9.6, left) are simple, robust and flexible. Working and counter electrodes (if necessary, the reference electrode as well) are positioned in a large solution reservoir. The disc-shaped working electrode is in plane with one of the walls of this reservoir. In front of it, a fine nozzle is arranged. The carrier solution is ejected as a sharp jet directed towards the electrode area. The jet washes away quickly residues of older solution from the electrode surface and generates a powerful convection. Thus, a relatively high value of the diffusion-limited current is achieved with an amperometric operation, as is a fast response if operated potentiometrically. The virtual dead volume of wall-jet detectors is extraordinarily small. It cannot be derived from the cell geometry, but must be determined experimentally by calibration.

Amperometric *microelectrodes*, discussed in Chap. 2, Sect. 2.2, also make up highly efficient flow-through detectors. The diffusion-limited current of such electrodes (i.e. their concentration-proportional signal) is largely independent of streaming phenomena in the detector. This is a consequence of the fact that the determining concentration gradients are of very low dimensions that are beyond the reach of external streaming processes.

Obviously, chemical sensors to be used as flow-through detectors assume shapes which are quite different from those common for direct concentration determination in quiescent solution. An even more specific design is found with detectors for separation methods.

9.2.2
Separation Methods

In analytical separation techniques, the components of a sample mixture are separated into individual *portions* which are apart from each other. At best, each component has its own volume increment inside the volume of a separation column. With all the *instrumental separation techniques* discussed here, a small amount of sample mixture is *injected into a flowing stream*. This stream (the *mobile phase* in chromatography or the *carrier solution* in electrophoresis) flows through a separation column which might have the shape of a column or of a capillary. In the course of flow, distinguishable separate sample regions (plugs or portions) are formed gradually. Simultaneously with separation, a certain broadening of the sample zones occurs, as depicted in Fig. 9.7.

The following phenomena are utilized in the separation process:

- Interaction of sample components with the stationary phase in chromatography, i.e. the equilibria of adsorption and extraction (partition) as well as related phenomena. A simplified explanation is that components with a stronger interaction stay at the surface of the stationary phase longer

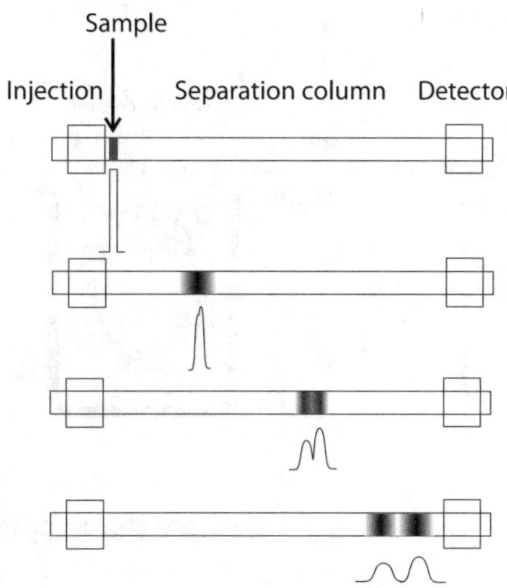

Figure 9.7. Partition of sample mixtures along separation column

than components with a weaker interaction. Hence the former tend to 'lag behind' the latter.

- Different mobility of ions in an electric field.

With modern separation techniques, analytical information is extracted from the signal of the detector positioned at the end of the separation route. Detectors for separation techniques must fulfil stricter specifications than detectors for continuous automatic analysers. In particular, the dead volume must be very small due to the small sample volumes typical for modern separation instruments. Some of the detector devices are useful exclusively for separation methods. For example, usual gas chromatographic detectors are not found in any other field of analytical chemistry. Figure 9.8 shows the differences among the most common separation techniques.

Chromatography

The most important chromatographic techniques are gas chromatography and high-performance liquid chromatography (HPLC).

The repertory of available flow-through detectors for chromatography is surprisingly small considering the enormous importance of the method. For gas chromatography, two standard types of detector are of overwhelming predominance, namely the flame ionization detector (FID) and the electron capture detector (ECD). In both detectors, components of the streaming gas are ionized, either by flame plasma or by a beta radiation source. An electric field is used to draw off the charged particles. Depending on their amount, an electric current of variable amplitude arises. The latter is evaluated to

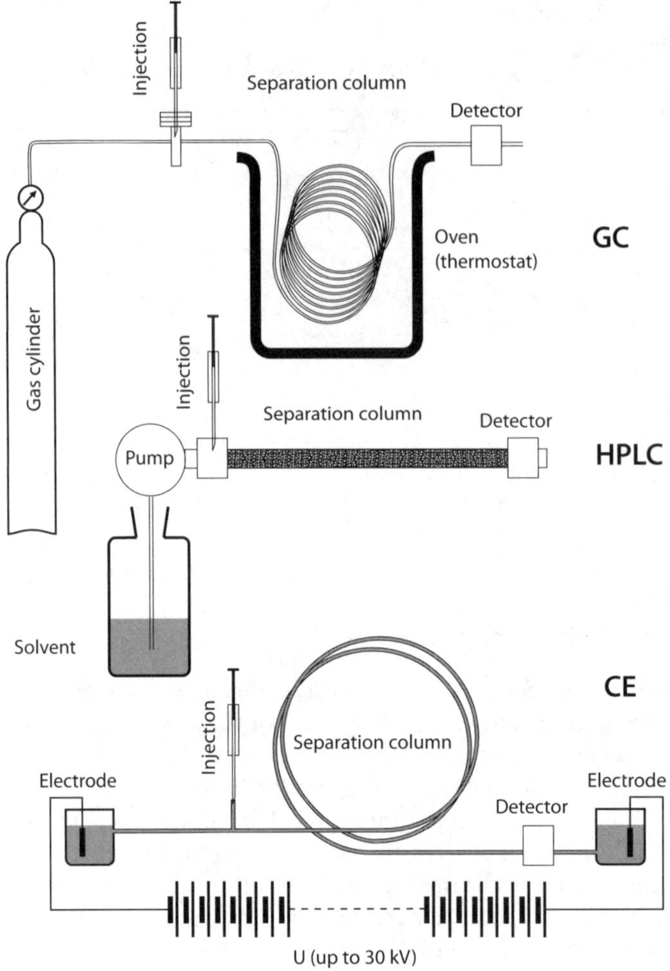

Figure 9.8. Instrumental set-up for separation methods gas chromatography (GC, *top*); high-performance liquid chromatography (HPLC, *centre*) and capillary electrophoresis (CE, *bottom*)

obtain a concentration-dependent signal which can be recorded continuously (Fig. 9.9).

In HPLC, photometric detectors predominate, mostly utilizing UV radiation and often shaped as a 'Z cell' (Fig. 9.10). UV light is preferred since the majority of analytes studied by HPLC are organic substances which absorb light in the UV region. Hence, only one detector is sufficient for all relevant analytical problems. The Z shape is useful as it displays a relatively long light path whereas the dead volume is kept relatively small.

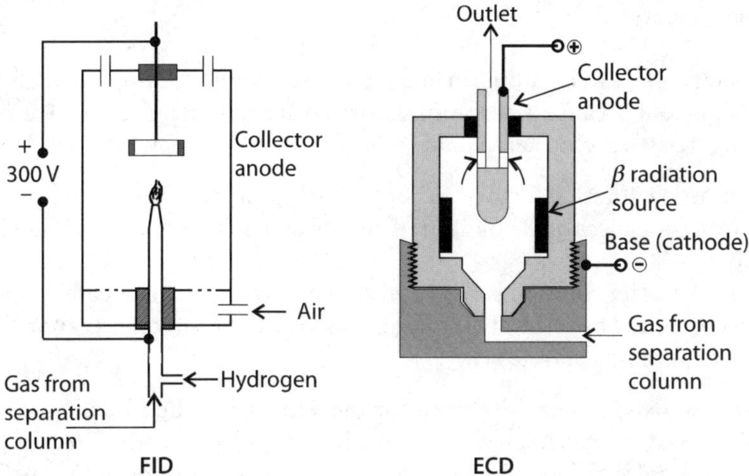

Figure 9.9. Detectors for gas chromatography. *Left*: flame ionization detector (FID); *right*: electron capture detector (ECD)

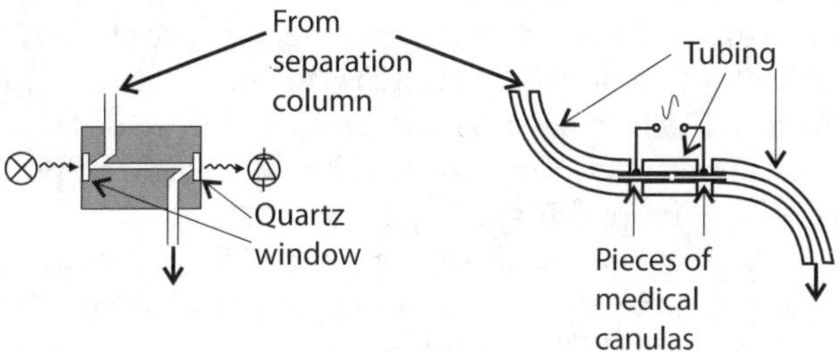

Figure 9.10. Detectors for HPLC. *Left*: photometric UV detector in Z-cell shape; *right*: conductance detector (example), where electrodes are formed by sections of syringe needles mounted by pieces of polymer tubing

The so-called *ion chromatography* (IC) is a variant of HPLC with specific importance for inorganic analysis. Electrolyte solutions containing diverse ions (salt solutions, acids and bases) can be analysed by separation of their ion content at a separation column filled with an ion exchanger. A conductance detector (similar to the example given in Fig. 9.10) is useful in most cases to obtain a reasonable signal.

Electrochemical detectors similar to that useful for continuous automatic analysers (Fig. 9.6) have been used for HPLC as well. Actually, they do not play an important role. This might change depending on advances in the relatively new field of capillary electrophoresis.

Capillary Electrophoresis

Capillary electrophoresis is gaining in importance rapidly as its sample demand is extremely low and because it can be miniaturized and automated easily. There are two characteristics which are vital for the success of the method:

- The separation column is 'empty' in contrast to chromatographic columns. Thus, one of the causes of the undesired peak broadening of the signals does not apply.
- Transport of carrier solution is accomplished not by mechanical pumps but by *electroosmosis*. The resulting sample zones are much narrower than with mechanical pumping (Fig. 9.11).

In electroosmosis, the driving force for movement of a liquid column is not brought about by mechanical pressure. Instead, electrostatic forces are exerted on ions in a 'diffuse double layer' close to the column wall. Such a layer is established if electrolyte solutions have contact with a solid polar material like glass. Ions in the diffuse double layer are hydrated, i.e. they are surrounded by water molecules. Following an electric field acting along the axis of the capillary, the ions in the diffuse part of the double layer start to move, taking along with them the whole liquid column which is virtually 'pumped' in this way towards the detector. *Electroosmosis* and *migration* are two different phenomena caused by the same thing, the electric field. Movement of both processes can be unidirectional or opposing.

Sample zones in capillary electrophoresis are very narrow. Consequently, only a minimum amount of substance is required for successful analysis. This advantage can be utilized only with high-resolution flow-through detectors. There are, however, limits to the miniaturization of traditional detection techniques. Photometric detectors, for example, can be built on a smaller scale only to a certain degree. This is one of the reasons for the renewed popularity of electrochemical detectors in recent years. Much work has been done on developing new types of miniature flow-through cells (Matysik 2003). Amperometric detectors, in principle, have a low substance demand, since only the electroactive particles inside the Nernst diffusion layer are necessary to get a signal. Useful signals have been obtained with sample amounts in the range of only picomoles; sometimes even femtomoles have been sufficient. A detector designed especially for capillary electrophoresis is shown schematically in Fig. 9.12. The working electrode, a thin glassy carbon fibre, protrudes into

Figure 9.11. Streaming profiles of liquids either pumped mechanically (*right*) or moved by means of electroosmosis (*left*). Parabolic deformation of sample zone is largely avoided with electroosmotic conveyance

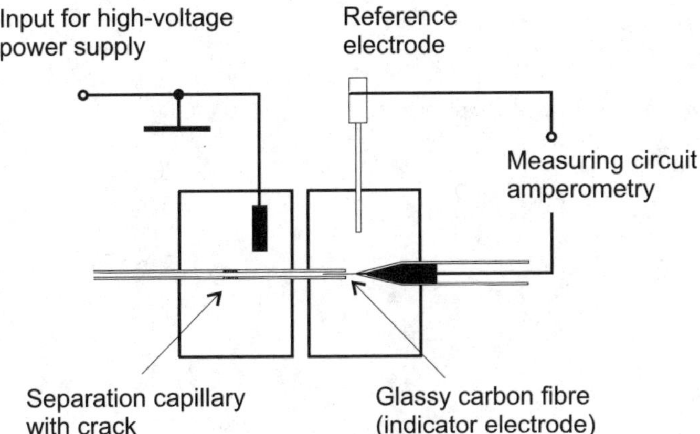

Figure 9.12. Amperometric detector for capillary electrophoresis

the capillary end, where the solution escapes. Electrical contact with the high-voltage circuit is made not via this capillary end, but by a thin crack in the capillary not far from the end. Due to this trick, the amperometric working electrode is positioned outside the high voltage circuit and its highly sensitive measuring circuit remains free from distortion (Wallingford and Olefirowicz 1989).

9.3
References

Matysik FM (2003) Anal Bioanal Chem 375:33–35
Wallingford RA, Olefirowicz TM (1989) Anal Chem 61:292A

10 Sensor Arrays and Micro Total Analysis Systems

10.1
Two Trends and Their Causes

A priori, sensors should be small and be available in large volume. The origins of their development can be traced, in part, to the idea of creating technical sense organs. On the other hand, they became part of the arsenal of analytical chemistry. Considering all these characteristics from a holistic point of view, it becomes comprehensible that sensors may help to achieve objectives which would not be within reach using classical scientific tools. With the advent of chemical sensors, new ideas arose of 'non-classical' analytical chemistry. Many of these new ideas will be of utmost importance in future.

The following examples are characteristic of the trends which appeared when chemical sensors had become available generally and when they had found their place in the minds of technicians and scientists:

- A combination of several different sensors on a common base should allow one to analyse a variety of analytes simultaneously. A combination of many of the same type of sensors should allow one to analyse the spatial distribution of analytes on a surface.
- Arrays of several sensors should offer better functionality than is common with traditional arrangements. Objectives could be better interference resistance, mutual control of different sensor types, or better adaptation to changing environmental variables.
- The efficiency of individual sensors and sensor arrays is impressive. On the other hand, it must be stated that not all analytical problems can be solved in this way. In many cases, the effort for *sample preparation* was larger than for the actual analytical measurement. Typical sample preparation operations like powdering of solids or dissolving substances commonly are done conventionally even if sensors are applied. In such cases, miniaturization of sensors does not help much. A possible answer to such considerations could be the development of miniature analytical instruments or even of miniature automatic analysers. Chemical sensors can be expected to be a substantial part of such devices.

The trends listed above brought about three important results, i.e. *smart sensors* (utilization of artificial intelligence in sensors systems), *sensor arrays*, and finally *micro total analysis systems* (μ-TASs). The latter became popular under the name '*lab-on-a-chip*'.

10.2
Smart Sensors and Sensor Arrays

10.2.1
Intelligence in Sensors

Why Intelligence?

Right from the start, the development of chemical sensors was connected with the development of modern microelectronics. In some cases, both use an equal technological basis, i.e. the silicon chip. Integrated electronic circuits on this basis have reached an impressive technological standard, and they are available for reasonable prices. Obviously, it should be useful to provide chemical receptors not only with electronic transducers, but additionally with integrated circuits for signal processing. Moreover, even deficiencies in sensor response could be corrected by electronic means rather than improvement of chemical receptors (due to the thesis 'chemistry is expensive, mathematics is cheap'). Indeed, today even very-large-scale integrated circuits (VLSI circuits) with hundreds of thousands of transistors on one chip are not a cost problem. On the other hand, research efforts to improve the efficiency of sensor chemistry would incur enormous costs to achieve similar results. The modern systems considered in this chapter make use of the achievements of intelligent signal processing as well as techniques of artificial intelligence like *fuzzy logic* and *neuronal networks*.

Self-Test, Self-Diagnosis and Self-Calibration

The term 'intelligence' in sensor systems can stand for the following operations:

- Communication of the sensor system considered with other sensor systems or with actuators.
- Adaptation of the system to varying environmental parameters, e.g. temperature.
- Automatic calibration of the sensor system or automatic baseline correction.
- Self-diagnosis of system in case of errors.

The sensor system's *ability to communicate* is dramatically improved by the insertion of electronic signal processing units. Electronic systems are important particularly if the measuring result is not read and evaluated by a person but if it is done by automatic processing. A simple case is to activate an acoustic or optical signal if the sample concentration determined by the sensor exceeds a certain threshold value. The other extreme would be to control extensive operations completely by sensor systems, e.g. industrial chemical processes or military operations. Electronics has a versatile intermediator function in such applications. From an engineering point of view, this is a routine job not worthy of special mention.

Adaptation of analytical systems to varying values of environmental parameters is a classical problem which was important long before the advent of chemical sensors. A typical example is compensation for the temperature dependence of pH measurements by means of a glass electrode. Rising temperature first causes a steeper slope of the calibration graph $E = f(\text{pH})$. Furthermore, the calibration graph will shift with temperature in parallel to the axes of the coordinate plane. Traditionally, one tries to minimize the temperature dependence by specific choice of internal buffer solution of the glass electrode and their internal contact. In this way, it is possible to position the intersection of isotherms into the electric zero point of the instrument, as in Fig. 10.1 (left). If this partial compensation is successful, only the correction of temperature-dependent change of the slope remains to be corrected. This can be accomplished easily by changing the amplification factor of the instrument in the opposite direction. A thermistor in the feedback circuit of an amplifier (Fig. 10.1 right) can fulfil this task. More serious problems arise if the intersection cannot be shifted into the zero point. In this case, a temperature-dependent shift of the graphs as well as the changing of their slope must be compensated for. The traditional way of doing this is to calibrate the instrument first by means of two buffer solutions of known pH. To compensate for both temperature effects, the amplification factor must be changed, and furthermore an auxiliary voltage of balanced magnitude must be added. With traditional instruments, repeated

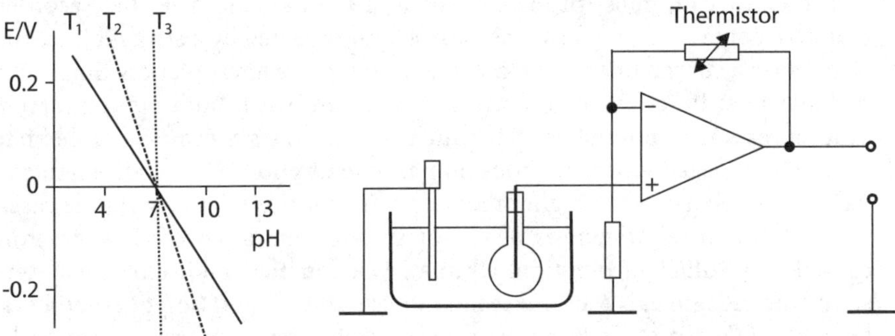

Figure 10.1. *Left*: temperature dependence of slope of calibration graph in potentiometric measurements. *Right*: automatic slope correction by means of thermistor in feedback branch of amplifier

alternating variation of two different variable resistors is necessary following a certain algorithm. For automation of these operations, simple analogue electronic compensation circuits would be insufficient. A microprocessor in the instrument would solve the problem. Nowadays, microprocessors (or even microcontrollers, i.e. processors provided along with ADCs and DACs) are available. They are small and cheap. Hence it is not a problem to combine them with a pH-sensitive chemical sensor.

The concept of microprocessor use in temperature compensation can be expanded to establish methods for *auto calibration*. To this end, the magnitude of ambient temperature influences should be known precisely. Normally, this is not the case, or it is connected with an individual sensor arrangement. Furthermore, to implement a true self-calibration, it would be necessary that the system itself might generate well-defined concentration changes. Such requirements could be fulfilled only for a few special cases, e.g. for a system designed to determine sulphide or hydrogen sulphide in the environment (Jeroschewski et al. 1994). In this system, well-defined increments of the analyte are generated by coulometry, i.e. by reduction of elementary sulphur imbedded in the cathode. According to Faraday's law [Chap. 2, Eq. (2.36)], defined charge amount values will generate defined amounts of substance. In this way, real concentration changes are generated.

Automatic baseline correction is much easier to achieve than automatic self-calibration. Analytical signals in many methods sit on a baseline in the form of peaks. Normally, the peak area contains the useful analytical information. The baseline varies only to a certain extent. The signal is extracted commonly by subtracting the baseline and subsequently integrating the peak. Traditionally, the baseline is determined by means of two successive measurements, one with and another without the sample. Subtraction of the baseline determined in this way may result in error, since the conditions in both measurements can be different. Often it is better to perform only one measurement (with sample) and to extract the most probable course of the baseline from the recorded graph. In some cases, the baseline can be approximated by a straight line, and sometimes a curved line can be drawn, e.g. by spline interpolation. Since only mathematical functions of the measurement are used, both approximation procedures can be automated. Nevertheless, it is always a technical problem to analyse the measurement without human intervention. The requirements in technical intelligence are higher than expected for this relatively simple case.

Self-test and *self-diagnosis* of sensor systems aim at correction of errors caused by insufficient function. Ultimately, a complete failure of the system should be recognized. A useful technique is parallel operation of several systems of equal structure in an array (see below). Much more difficult is the recognition of a gradual malfunction of the system. For the widely used amperometric sensors, a typical malfunction is gradually lowering sensitivity as a result of progressive deactivation of the electrode surface when substances are poisoning the surface by formation of passive layers. Gold electrodes are poisoned already by traces of sulphur-containing substances. Such failures can be detected only if the sensor function is compared continuously or periodically with the results of an independent analytical method. Methods for comparison must be truly independent, i.e. it would not be sufficient to compare an electrochemical sensor with another electrochemical sensor, or optical sensors with another optical method.

True self-test and self-diagnosis remain rare. Instead, the approach commonly taken is to maintain the activity of the system by periodic regeneration procedures as long as possible. On the other hand, disposable sensors are a useful alternative. For amperometric sensors and detectors, complex electrochemical pulse sequences which renew the electrode surface are applied widely.

10.2.2
Sensor Arrays

Why Sensor Arrays?

There are good reasons to combine sensors to form an array:

- *Redundancy*
  Increased reliability is provided with several sensors operating in parallel. If one of the sensors in the array fails, the others can take over its tasks. The failure can be recognized easily by comparison of the signals. Reliability is of particular importance for medical applications.
- *Multidimensional information*
  – Two-dimensional imaging information (mapping) of substance distribution on an area is obtained by combining sensors of the same type in an array.
  – *Additional variables* (temperature, light wavelength etc.) can be acquired by sensors of different types in an array.
- *Multicomponent analysis*

Requirements for specifications of sensors in an array often are stricter than those for single sensors. Calibrating arrays is more difficult than calibrating single sensors, hence *reproducibility* and *long-term stability* are highly important parameters. Sensors in arrays must be as small as possible; consequently not all existing types are useful for arrays. The *manufacture* of arrays is laborious; they cannot be assembled in small or cottage enterprises. Modern, fully automated production is virtually always necessary. *Data processing* of the results of array studies is demanding and often based on highly sophisticated statistical methods or methods of pattern recognition. Consequently, *instrumentation* is also more ambitious. Classical analogue instruments, strip chart recorders and similar devices are no longer useful. The state of the art is to connect the sensor systems to multichannel-input-output cards in a PC and evaluate the results by graphical programming software packages like LabView.

The so-called *electronic noses* and *electronic tongues* are special types of chemical sensor arrays. Their working principles are quite different from those commonly used in analytical chemistry as well as in sensors discussed so far. These arrays are presented in a separate section.

Multidimensional and Multicomponent Analysis

Arrays for a two-dimensional imaging presentation (mapping) can be realized in an elegant manner by means of *light-addressable potentiometric sensors* (LAPS). Figure 10.2 presents the scheme of such an array. On a thin silicon chip (ca. 0.5 mm), two insulating layers are applied. The lower one, a silicon dioxide layer, is covered by a somewhat thicker one of silicon nitride. The latter acts as protecting layer against the aqueous solution in contact with the chip. A potentiostat stabilizes the potential difference between the reference electrode in solution and the silicon base. Specific spots of the chip can be illuminated from below (through the silicon base), but alternatively also from the solution side. The spots are illuminated with modulated light by means of light-emitting diodes (LEDs). In this way, certain spots of the chip are activated (*addressed*) and excited to supply alternating current at the output of the chip, which can be measured as shown in the figure. The term 'potentiometric sensor' is somewhat misleading, since potentials are not measured directly. There is, however, a special operating mode where the potential dependence of the light-induced alternating current values are evaluated. If each LED is

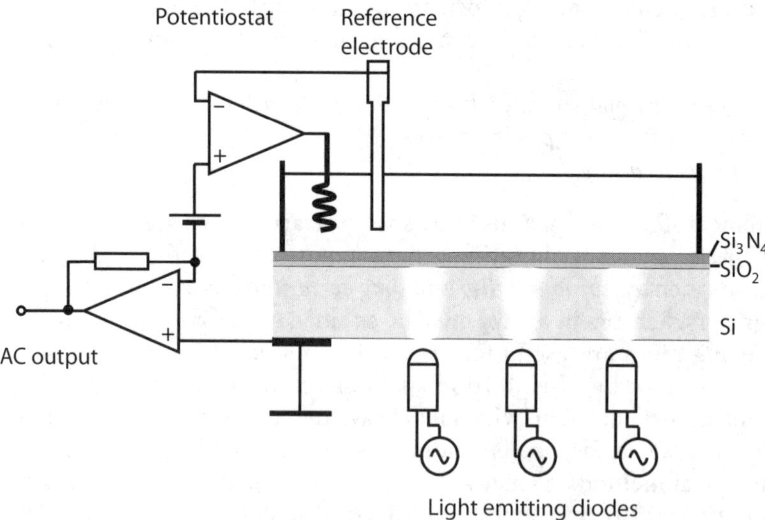

Figure 10.2. Structure and measuring set-up of LAPS sensor arrays

activated individually, the result is a pair of values containing two pieces of information, i.e. the signal magnitude and the position on the chip. The latter is given since it is known which of the LEDs was active. The results of the measurement series can be combined to yield a two-dimensional 'map' which shows a concentration distribution in an area.

Voltammetric arrays consisting of a multitude of individual metal areas on a common substrate are another example of sensor arrays useful for multidimensional analysis (Hoogvliet et al. 1991). Each individual electrode is polarized to assume its individual potential value (Fig. 10.3). The potentiostats can be programmed in such way that the difference between each electrode and its next neighbour is a constant increment (e.g. $\Delta E = 5\,\mathrm{mV}$). Each potentiostat at its output yields a single value which corresponds to one point in a voltammogram. Hence a complete voltammogram is present at one moment, without potential scanning. Such arrays allow one to consider *time* as an additional independent parameter. If the array is positioned in a flow-through system, a three-dimensional image is obtained, reflecting the voltammograms of substances subsequently flowing through the apparatus. Figure 10.4 depicts the results of such measurements in a simplified manner.

Obviously, by means of sensor arrays several components can be determined simultaneously. This way of doing analytical chemistry is useful if the analytical probe is small, but on the other hand only a few components have to be studied in a restricted concentration range. In medical diagnostics, arrays of potentiometric sensors have been around for a long time. Such arrays, which have not been really subject to miniaturization, are used to determine the medically important ions Ca^{2+}, K^+, Na^+ and Mg^{2+} simultaneously with low time demand. The quick availability of results is important particularly during an operation if physiologic specifications are kept constant within a certain range.

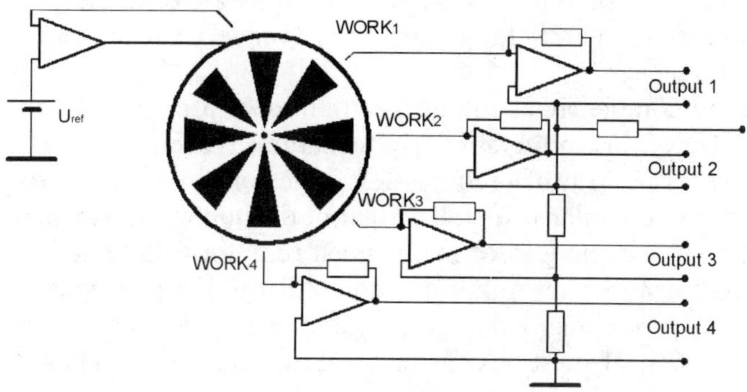

Figure 10.3. Voltammetry with a multielectrode array. Each electrode is controlled by its own potentiostat

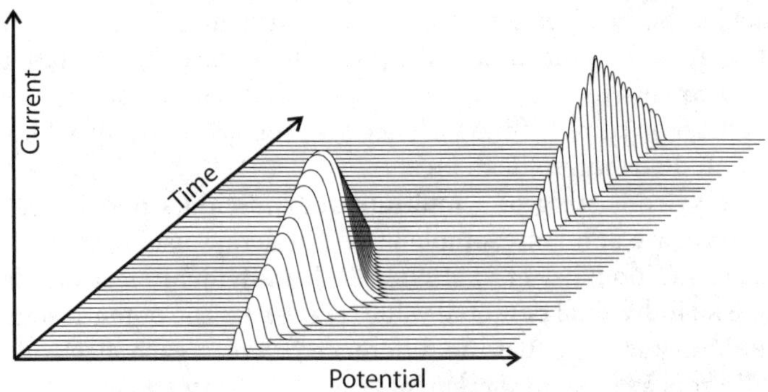

Figure 10.4. Signal of voltammetric electrode array in a flow-through system when several electroactive substances are flowing through (simplified)

Votammetric sensor arrays for simultaneous determination of heavy metals are known in numerous variants. As in the example presented in Fig. 10.3, electrodes as a part of an array can be polarized individually at defined potentials. This is useful for stripping determination of metal traces (Chap. 2, Sect. 2.2). It can be advantageous to use electrodes made of different materials. In a stripping determination, in a first step, a well defined potential is imposed at the working electrode which is left constant for a certain amount of time. Under stirring, the metal cations are deposited, each at its corresponding electrode. In the subsequent anodic dissolution (stripping) process, analytical signals are obtained. By means of such arrays, the elements zinc, cadmium and lead are determined simultaneously at electrodes of gold, silver, copper and platinum.

Electronic Noses and Electronic Tongues

'Electronic sense organs' represent a quite uncommon way to analyse the environment. 'Nose' means an array for gas analysis, 'tongue' an arrangement for operation in liquid phase.

Arrays of this kind completely abandon the traditional principles of analytical chemistry. They work quite similarly to natural sense organs. The standard arrangement is an array of a more or less large number of single sensors of a common type. A similar principle is that of the human nose, where the mucous membrane is composed of many smell receptor cells. The idea is to characterize multicomponent mixtures. In contrast to the principles of analytical chemistry, it is not intended to identify and to quantify single components reliably. The classical questions of analytical chemistry are still posed, i.e. we are interested in obtaining knowledge about the quality and quantity of substances in the studied mixture. We cannot expect, however, the familiar answers of analytical chemistry. In principle, sensor arrays yield some kind of 'collective' or summarizing information, like that of natural sense organs. The latter provide us with impressions like 'musty', 'acerbity' or 'fruity flavour'. We obtain something like a 'chemical portrait' or a 'signature' of a mixture of substances. Nevertheless, the result is a chemical analysis of a certain part of the environment.

The sensors that are combined in an array should be arranged in such way that they can be read individually. A large number of signals is obtained in this way. It is impossible to evaluate this mass of information by traditional means. There is a need for modern calculation methods which can handle large amounts of data as well as multidimensional systems.

From a mathematical point of view, all the sensors discussed here (including those in the arrays) are so-called *first-order sensors*. This means that each single measurement results in a *data vector* for each sample. This vector sorts the signals of individual sensors into groups. Such a property of a sensor array can be illustrated by comparison with a spectroscopic measurement. A single measurement at a defined wavelength in spectroscopy corresponds to a measurement with a single sensor in an array. The resulting spectrum (composed of measurements with several different wavelengths) contains information about the quality and quantity of each component in the mixture. With sensor arrays, a useful display of the results must be found, i.e. the *'chemical portrait'* of the mixture must be represented visually.

Evaluating the results of sensor arrays means solving a quantitative calibration problem. Prior to calibration, commonly a qualitative characterization of the sample matrix is attempted, generally by means of *pattern recognition*. For quantitative multicomponent analysis, multivariate regression models are used.

Most common are the following calibration techniques:

- Parametric analysis
  - linear regression
  - non-linear regression
  - *discriminant function analysis* (DFA)
- Non-parametric analysis
  - pattern recognition
          including primarily *principal component analysis* (PCA)
  - cluster analysis
  - fuzzy logic
  - neuronal networks

Fuzzy logic and *neuronal networks* are methods of *artificial intelligence*. With neuronal networks, an extensive analogy to the function of natural sense organs is obtained.

The outstanding singularity of electronic sense organs is demonstrated by some unexpected properties. For example, the requirements for sensors combined to give an array are relatively modest. Fulfilling the following characteristics is sufficient:

- 'Nose' or 'tongue' arrays consist of a number of *partially selective* individual sensors. Here, selectivity means *preferred response to a single sample substance*, so that all the remaining substances can be considered formally as *interfering substances*.
- Each individual sensor must have a certain power to distinguish between components of the mixture, i.e. it should distinguish between *sample* and *interferent*. The sensor should respond to both, but with different sensitivity.
- The signals of individual sensors should differ somewhat, i.e. they should be characterized by a certain *individuality*.

The above requirements are easily met. Roughly speaking, certain shortcomings are used to one's advantage. Partial selectivity, usually considered a weakness, is made the source of information. However, information can be extracted only by application of extensive calculation procedures.

Examples of electronic noses are listed in Table 10.1. It is amazing to see how the simplest receptors yield considerable results. Simple layers of conductive resins change their resistivity when exposed to volatile flavours to such an extent that types of beer can be distinguished. The gas phase on top of the beverage is analysed by means of the corresponding array. Generally, electronic noses are suitable for characterizing beverages. A less demanding problem is distinguishing types of whiskey (e.g. to distinguish real scotch whisky from articially flavoured surrogate). In this case, even today's sensor arrays can substitute well-educated (and well-paid) tasters.

A special type of electronic nose is the *freshness sensor*, which is gradually finding its place in butchery and the meat industry. In the near future, lunch parcels etc. could be equipped with such sensors.

Table 10.1. Examples of electronic noses

Sensor type	Number of sensors in array	Analytical method	Application example
Conducting polymer layers	12	Conductometry	Flavours in beer
Metal oxide ceramics	12	Conductometry	Flavours in coffee
Metal electrodes	16	Amperometry	Toxic vapours in cereal storing houses
MOSFETs	20	Capacitive photocurrent	Vapours of alcohol, ammonia etc.
Lipid layers	6	Piezo crystal	Discrimination of alcoholic beverages

Electronic tongues, i.e. sensor arrays designed to characterize liquids, are not as widely used today as electronic noses. However, there are applications which should be taken seriously, as is clear from Table 10.2.

Not all the examples given in Table 10.2 follow the standard configuration given above. For example, the voltammetric electronic tongue by Ivarsson et al. (2001) is assembled with different sensor types, not identical ones, as would be the standard practice. Three sensors made of different noble metals are included. This is a rather low number of individual sensors in an array. Nevertheless, several sorts of tea could be distinguished from one another rather clearly. A further interesting example is listed in the table, i.e. characterization of mineral waters by means of an array composed of chalkogenide glass electrodes (Di Natale et al. 1999). Waters originating from different sources could be recognized and clearly distinguished from one another. Even an intentional contamination yielded unequivocal results.

In the actual state of the art, the primary goal is qualitative characterization. As the examples in the table demonstrate, the task common to all methods is to distinguish variants of complex substance mixtures. Consequently, in the international literature, results are often presented by means of pattern recognition techniques. Often, principal component analysis (PCA) is used. The core of this procedure is *eigenvector analysis* of data. The eigenvector represents the direction of maximal variance of a data population. The data are scaled by subtracting the average values and dividing by the standard deviation of the sensor signal. Thus, each sensor is given equal weight in evaluation. The results of eigenvector analysis can be displayed in a new coordinate system as a function of eigenvectors. An example is given in Fig. 10.5 for a three-dimensional analysis. Often characteristic data agglomerates arise which allow clear assignment. Outliers among data can be recognized easily.

Fuzzy logic is a mathematical concept designed to study systems or processes which are characterized more qualitatively. Human language, for example, is

Table 10.2. Examples of electronic tongues

Sensor type	Number of sensors in array	Analytical method	Application example
Ion-selective glass electrode (chalkogenide glasses)	4–6	Potentiometry	Beverages; mineral waters
Metal electrodes	3	Voltammetry	Tea flavours; heavy metals
Noble metal layers in flow stream arrangement	12	SPR	Absorbable organic traces in ultrapure chemicals

Figure 10.5. Presentation of results of PCA containing data of a sensor array

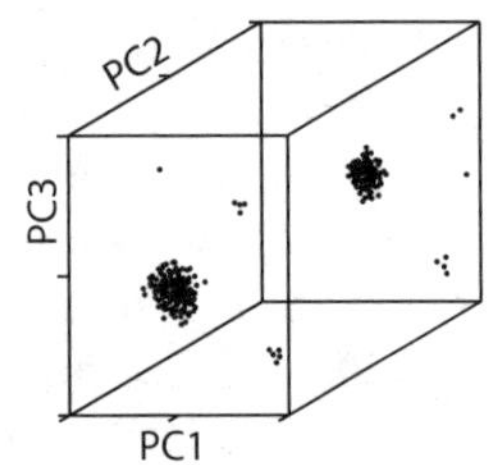

a system which is capable of transmitting and interpreting sufficient information by means of low-precision procedures. Fuzzy logic emulates such diffuse descriptions (like, e.g. 'hot' or 'cold' to characterize temperature) on the computer. In sensor science, results of arrays are subdivided into *fuzzy sets*. In this way, multicomponent sample mixtures can also be distinguished from each other.

Neuronal networks (Rojas 1993) simulate brain functions. In sensor science, they are used to construct non-parametric, non-linear models of the results of sensor arrays. Neuronal networks are made homogeneously of elements having the same basic structure, the so-called *neurons*. Often three-layer networks of the *feed-forward* type are built, where neurons are arranged in *layers* (Fig. 10.6, left). The number of *input neurons* in such networks corresponds to the number of received sensor signals. The numbers of *hidden neurons* and of *output neurons*, respectively, depend on conditions. The network is 'trained' by standard samples. In this way, the number of hidden neurons can be optimized. Neuronal networks are suited pimarily to obtain qualitative information, but less to a lesser extent for quantitative analysis. Graphical representation in the form of 'radar plots' (Fig. 10.6, right) has proven useful.

A specific type of sensor array is *DNA chips*, mentioned in Chap. 7, Sect. 7.4. This chapter describes how a biologic species or even a specific individual can be identified by the process of *hybridization*, when two single strands of DNA

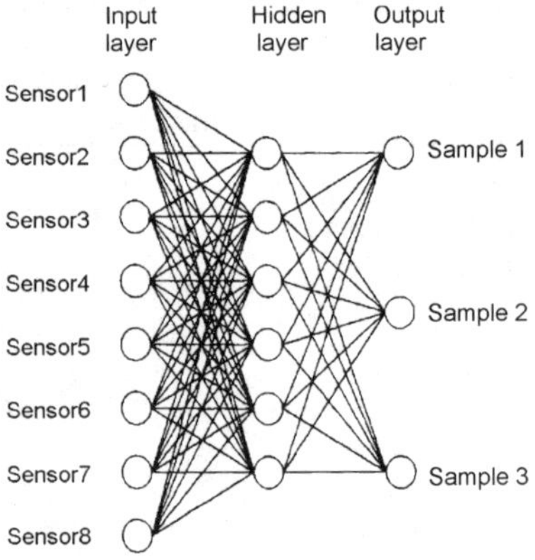

Figure 10.6. Scheme of three-layer neuronal feed-forward network for evaluating results of sensor arrays (*left*). Graphical representation of signal pattern as 'radar plot' (*right*)

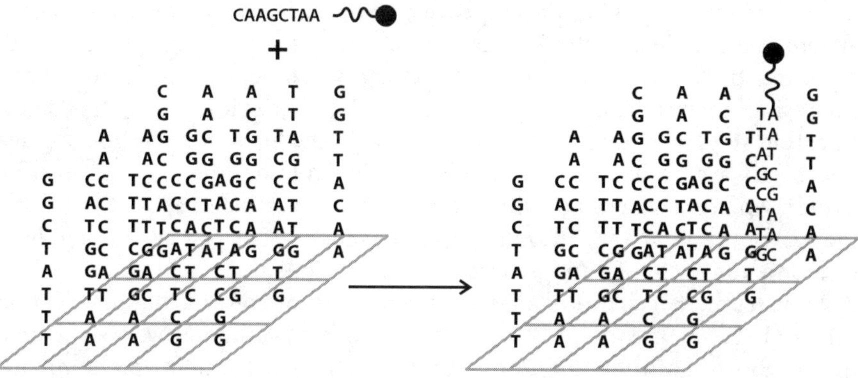

Figure 10.7. Scheme of sequence analysis by means of DNA chip

combine to give double-stranded DNA. Probes for this process are made by immobilizing oligonucleotides (characteristic sections of the single-stranded DNA molecule) at a solid surface. If the complementary strand (as a part of the organism to be identified) is present in the sample solution, the duplex of single strand and its corresponding complementary strand is formed, i.e. *hybridization* occurs.

To decode a certain genome region of an individual, it is necessary to analyse a large number of DNA sections simultaneously. For this purpose the so-called DNA chips have been developed. Commonly, such chips are glass plates divided into numerous single segments. Each segment contains a certain oligonucleotide sequence immobilized by common techniques. A single segment of 25×25 mm can contain up to 50000 different oligonucleotide sequences. Molecules of the DNA to be studied are cut by enzyme action into fragments, to which are attached indicator groups. Indicator groups are mostly fluorophores, but redox indicators are also useful. A fluorescing spot (or, alternatively, an island with redox activity) appears at places where double strands have formed. By scanning the chip surface, one can determine the kind of marked molecules (by evaluating the position on the chip) and the quantities in which they occur. In a single experiment, up to 50000 gene sequences can be tested in this way. The processes are illustrated in Fig. 10.7.

10.3
Micro Total Chemical Analysis Systems (μ-TASs)

10.3.1
History

Analytical chemistry has made great progress due to the introduction of chemical sensors. Many traditional tasks can be solved cheaper, and many new and unconventional opportunities, especially in the field of on-site analysis, have arisen. Nevertheless, nobody assumes that in future chemical sensors will replace entirely all the traditional analytical instruments. This is unlikely to occur since many peripheral operations must be done prior to the actual determination step. Even if the sample is available in the form of a solution, dosage operations are necessary. A precisely metered volume, the aliquot, must be withdrawn from stock, and this aliquot must be transferred to the instrument. Also, standard separation operations must be performed, among them filtration, ion exchange and precipitation operations.

In instrumental chemical analysis, a general trend towards miniaturization can be observed. In around 1950, many analytical instruments were so big that they occupied entire laboratories. Today, tabletop instruments predominate. This development results mainly from the enormous technical progress in electronics, which made available highly complex, but cheap, integrated circuits. Consequently, one may observe a strong trend towards miniaturization of all measuring instruments, independent of their intended use. There are, however, specific reasons to minimize the dimensions of analytical instruments. There is a growing demand for mobile instruments which can be taken to the sample, e.g. to the environment. Furthermore, very small instruments are expected to work faster than big ones, since small-scale chemical reactions generally happen faster.

With growing experience in the field of chemical sensors, and with the awareness that smaller scale does not necessarily mean a deterioration in performance, it seemed worthwhile to consider miniaturizing entire laboratories. An interesting aspect also was automation, which is much easier to carry out with smaller instruments. A higher degree of automation improves the immunity of the instrument against interference and can be operated by less qualified personnel.

Macroscopic analytical instruments cannot be scaled down linearly if miniaturization is intended. Physical parameters have different impacts on the device when dimensions are changed. For example, gravitation forces have a diminishing effect with miniaturization, whereas miniaturization magnifies the effects of adhesion and viscosity. Diffusion as well as heat dissipation gain in importance with smaller dimensions. These effects can be put to use in different ways. Heat dissipation of very thin wires or leads is much better than with macroscopic conductors. Hence, microcircuits tolerate much higher electric power per cross-sectional area. In chemical systems, often diffusion alone is sufficient to transport reacting species. This is utilized, e.g. in microtitrations. The theoretical basis of technical *scaling down* was worked out originally for microelectronic circuits, but universal relationships are available today (Pagel 2001).

Precursors

Total analytical systems existed before the term TAS was suggested. Many techniques that are based on flowing solution could be considered to be total analytical systems, including *flow injection analysis* (FIA), *electrophoresis* and *chromatography* (see also Chap. 9). In FIA, as an example, traditional chemical processes are automated by transferring them to a flowing stream. Traditionally, the reagent is added to the sample, then one must wait until a characteristic colour has developed, and finally the mixture is poured into a cuvette and its intensity measured by a photometer. Instead, in FIA, a sample aliquot is injected into a flowing reagent stream. The streaming liquid transports the sample which reacts with its dissolved reagents. Subsequently, the stream flows through a detector which generates a concentration-dependent signal. In this way, the throughput of a photometric analysis is improved markedly. Similar efficiency is achieved with chromatographic methods, which can be considered to be flow-through methods as well. The step from TAS to μ-TAS started with existing flow systems. Three objectives were envisaged with miniaturization, namely shorter analysis time, minimizing sample demand, and device mobility (to allow utilization outside the laboratory).

Even true μ-TAS existed before the term was defined. Such systems did not receive much attention from analytical chemists. They originate from a different source, namely instant picture photography, a technique invented and promoted since the 1940s by the Polaroid company. In this technique, complex chemical processes like photographic development and fixation, removal of silver negative etc. are not performed step by step, with individual reagent solutions in a darkroom, but only by controlled diffusion in the photographic film inside a camera. Indeed, instant picture films are special forms of a miniature laboratory. This highly complex technology was later transferred to systems designed for analytical chemistry. *Multilayer film elements* for chemical analyses, at first glance, seem to be ordinary test strips. However, they are complex film systems on a transparent base like the cellulose acetate well known from photography. On this basis, several layers of well-defined thickness are positioned. They consist either of dry gels or of semipermeable membranes of selective permeability. The layers contain all the necessary reagents. Analysis starts with application of a drop of sample solution. The liquid spreads quickly in the top layer, the *spreading and reflecting layer*. Spreading results in the formation of a sample solution spot of defined thickness. Next, constituents of sample solution diffuse into the adjacent reagent layer where they start to react. Interfering substances are separated by means of filtering layers. Finally, a coloured reaction product is formed in an amount proportional to the sample concentration. Light absorption of this product can be measured by means of a reflection photometer as a measure of sample concentration. Determination of urea in blood is presented in Fig. 10.8 as an example. In reagent layer 1, ammonium carbonate is formed by enzymatically catalysed hydrolysis. Since

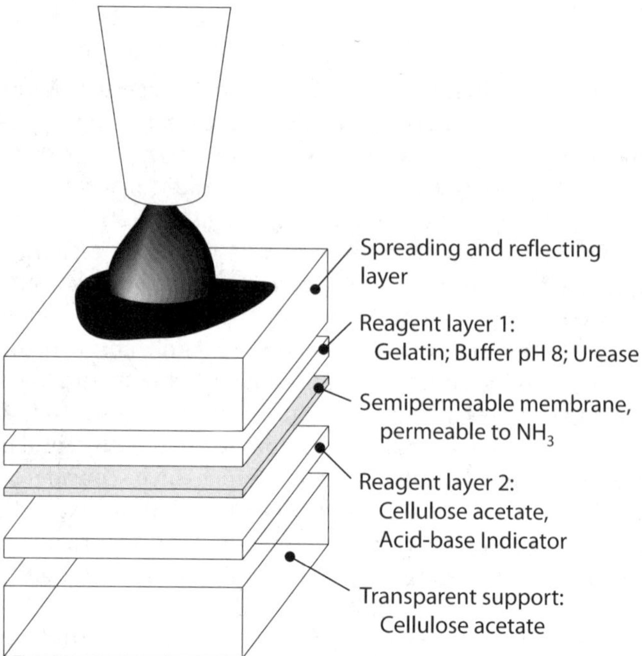

Figure 10.8. Multilayer film elements for determination of urea in blood as an example of an old *μ*-TAS (*left*). Evaluation by means of a reflection photometer (*right*)

pH = 8 is imposed by buffering agents in the layer, ammonia is present in equilibrium. Ammonia permeates a gas-permeable membrane and cause a colour change of the pH indicator substance immobilized in reagent layer 2. The absorption of this layer is a linear function of urea content in sample solution. With photometric evaluation, light crosses the coloured layer two-fold, first with incident reflection and second after diffuse reflection in the spreading and reflecting layer on top of the strip. All processes run spontaneously. The transport of substances is brought about exclusively by diffusion. It is desirable to keep constant the temperature as well as relative humidity during measurement. The small incubator shown in Fig. 10.8 (right) is designed to maintain these conditions.

μ-TAS and Lab-on-a-Chip

Micro total chemical analysis system (*μ*-TAS) and *lab-on-a-chip* are recent terms. Commonly, they are used for analytical systems with flowing streams. The development of such systems started with attempts to scale down traditional flow-through instruments, including, in particular, chromatographs. When techniques derived from microelectronics became available, the term *lab-on-a-chip* appeared. At first, the term stood for systems based on a single silicon chip in *monolithic* arrangement. Later, also different materials were accepted as 'chips'. Nowadays, the term is used for any miniature analytical system on planar support.

Development of the systems started in the 1970s. The advent of micromechanics as a derivative of microelectronics allowed manufacturing of micropumps and microvalves. Nearly at the same time, SAW sensors and ISFETs appeared, i.e. successful examples of combined chemical and electronic functional units. On this basis, a coulometric microtitrator was proposed which combined a coulometric generator compartment with an integrated ISFET as indicator unit. Different titrants were generated coulometrically (van der Schoot and Bergfeld 1985). In 1979, a miniature gas chromatograph was mentioned for the first time (Terry et al. 1988), which was said to be useful for analysis of mixtures within a few minutes. An injection valve and separation column were integrated on common a silicon chip. A second chip, coupled with the first one, contained a heat-conductivity detector in silicon technology. The announcement of such an instrument caused a sensation at the time. However, the device was never presented to a larger public, and it never became commercially available. Possibly, this resulted from military interests. For a certain time, the interest in miniaturized laboratories diminished.

In 1990s, the need for μ-TASs and their importance was discussed for the first time. The term μ-TAS was established by Manz et al. in 1990, simultaneously with the presentation of a flow-through concept on a silicon chip with integrated sample pretreatment, separation and detection. The idea of utilizing electroosmosis rather than mechanical pumps for liquid transport also originated in the 1990s. Electroosmosis had proven highly efficient in capillary electrophoresis, a technique which appeared in the 1980s. In 1990, a miniature liquid chromatograph with open capillary on a silicon chip was presented. It made use of a high-pressure pump and conductivity detector (Manz et al. 1990).

The development of novel miniature modules for FIA with a fibre optic detector (Manz et al. 1991) had a positive effect on further development. In 1992, a complete system for capillary electrophoresis on a planar glass carrier was established successfully (Harrison et al. 1992). Thus, electroosmosis proved to be of value for solution transport as well as for injection operations.

The boom in microanalysers started around 1994. Laborious sample treatment procedures and other peripheral operations were performed in reactor columns connected upstream or downstream from the central part. Highly efficient instruments, like a miniature mass spectrometer, became available. Amperometric detectors became the standard device for flow systems since they could be miniaturized easily. They are found primarily in liquid chromatographic μ-TASs as well as in such systems which work on the principle of capillary electrophoresis.

The term *micro total chemical analysis systems* (μ-TASs) is used now nearly exclusively for systems with a basic structure described in Chap. 9, Fig. 9.8. A *liquid column* moves through the device. It is driven either by a micropump or by electroosmosis. This moving column, also called the *carrier stream*, accepts a *small sample volume increment*, which is *injected* at the appropriate position. In the adjacent line, either a separation takes place (this is characteristic for chromatography and electrophoresis) or the sample reacts with ingredients of the carrier stream, thus forming a product which can be indicated. In both cases, at the end of the separation or reaction line, a detector is located which yields a concentration-dependent (better concentration-proportional) signal.

10.3.2
Technological Aspects

Essentially, the technology of μ-TASs is aimed at manufacturing the channels required for flowing liquids and to combine them with other necessary units at a planar base. A special challenge is the development and manufacture of miniature detectors which should generate concentration-dependent signals in tiny volume elements.

Most of the techniques applied originally emerged as a by-product of procedures in microelectronics. The terms micromechanics, microfluidics, micromechanical systems, microsystems technology and micromachining refer to silicon technology which starts from a wafer of monocrystalline pure silicon. Complex structures are fabricated by microlithography and by etching procedures. Even three-dimensional structures are produced in this way, as e.g. a microcantilever designed to measure mechanical acceleration. To fabricate channels by means of photolithography, a high *aspect ratio* is necessary, i.e. the 'side walls' of an etched depression must be equal to, or even higher than, its base plane.

In addition to silicon technology, alternative processes have appeared in recent years. Moulding by casting resins proved useful for making microchannels, as has the so-called *LIGA technology*. This is an acronym from the German words for lithography, electroplating and moulding (*L*ithographie, *G*alvanik, *A*bformung). For this technology, primarily electrochemical metal deposition (electroplating) is used to generate structures.

Figure 10.9 depicts the essential steps in making microchannel structures, exemplary for microlithography. First, specified areas are covered by a *mask* to form a *structure*. Starting with this structure, material is removed from the substrate. The intention is to manufacture channels with a rectangular cross section. This is managed by means of two consecutive etching steps. In the first step, a metallic layer is etched down to the substrate. In the next step, the substrate itself is attacked, resulting in a somewhat rounded channel bottom. After removing the mask, a cover plate is attached by gluing or melting. In this operation, clogging or contamination of the fine channels must be avoided.

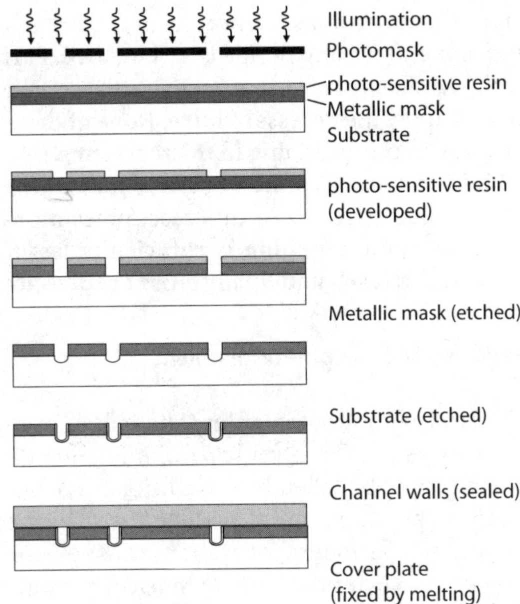

Figure 10.9. Microlithographic process for manufacturing microchannels

Mostly, the side walls of the channels must be sealed by an inert material. This is of particular importance if the channel walls must be made of silicate materials as a precondition for electroosmosis. At the surface of pure silicon, by heating in air a thin layer of SiO_2 is generated.

10.3.3
Characteristic Operations and Processes in Micro Total Analysers

Electroosmosis in Microchannels

In most cases, micro total analysers make use of flowing liquid streams with inlaid sample 'plugs'. Such plugs are regions of *solution inside a solution* where both liquids are miscible without restrictions. The plugs form distinguishable regions only for a limited amount of time. They will diminish gradually if left undisturbed, caused by natural convection and by diffusion. Movement of the liquid stream along the capillary promotes the undesired broadening of substance zones, since at the capillary wall adhesion forces bring about adhering solution layers resulting in a parabolic streaming profile, as discussed in Chap. 9. Considerable progress in streaming techniques has been achieved by the introduction of *electroosmosis* as a substitute for mechanical pumps for solution promotion (Chap. 9.2). In contrast to mechanically pumped solution, zone broadening does not occur with electroosmotic pumping. As a result, sample demand is reduced drastically. With miniature systems, this advantage is most prominent, since the ratio of wall area to volume increases with decreasing dimensions. Hence, in miniature systems the negative effect of the parabolic streaming profile is accentuated. Furthermore, the mechanical resistance against pumping is extreme with very narrow channels. Micropumps can hardly apply the necessary force. None of these problems apply with respect to electroosmotic pumping in microsystems.

Ions to be determined possibly may move opposite to the flow direction of the liquid stream. In this case, precautions must be taken to ensure that electroosmotic pumping is sufficiently fast to ensure that all ion packages formed can reach and finally cross the detector.

Sampling and Sample Pretreatment

Classical methods of sampling and sample pretreatment are not well adapted to microsystems. The advantages of miniature dimensions are lost if e.g. a blood sample must be taken by a traditional syringe or if manual extraction or filtration operations must precede the final measuring process. On the other hand, such peripheral operations must be done also with miniature systems. Obviously, such preceding or subsequent operations attracated increased interest with the advent of analytical microsystems. Some examples of successful integration of sample preparation are discussed below.

A novel sampling method in living organisms is microdialysis (Fig. 10.10). A semipermeable tube, commonly made of a cellulose derivate, is introduced into the body. Small molecules can permeate the tube wall and are accepted by an acceptor solution which transports the material towards the analyser. A sophisticated technical solution offered by the BAS company (Fig. 10.10) is based on a concentric arrangement of two tubings, which also allows infusion of agents into the organism.

Microdialysis can evolve its efficiency only in combination with microanalysers. Low demand in the sample volume allows one to utilize very thin dialysis tubes. Very thin tubes allow for a much faster operation than those of ordinary dimensions since the increased ratio of surface to volume has a positive effect.

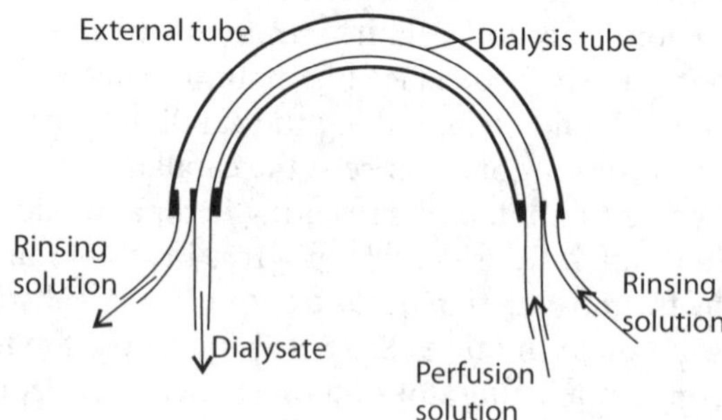

Figure 10.10. Microdialysis as a method of sampling in organs of a living organism

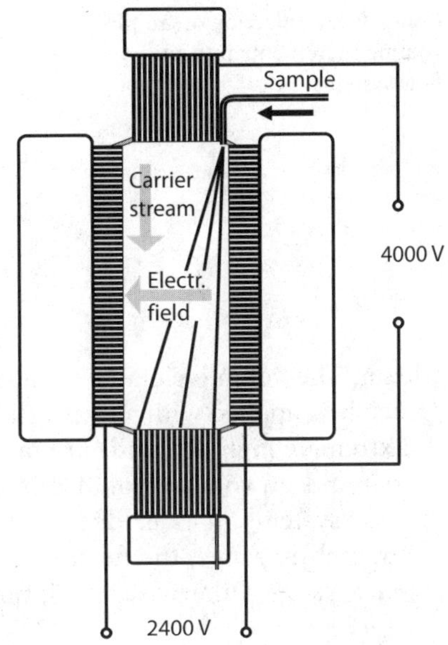

Figure 10.11. Free-flowing electrophoresis as a method to remove interfering ions from a flowing stream

The rediscovery of dialysis was not the only result of the search for universal pretreatment operations. Free-flow electrophoresis (FFE) has proven useful in the removal of interfering ions. The sample stream is conducted through a flat cell. An electric field is then imposed perpendicular to the stream direction. This field more or less deflects the ions, depending on their mobility. In this way, interfering ions are hindered from flowing through the detector. Certain electrolyte fractions can be separated selectively by variation of the flow rate and field strength. A useful design was based on a silicon chip of $10 \times 15\,\mathrm{mm}$ with comb-shaped longer edges (Raymond et al. 1994). A voltage of ca. 50 V is applied between the opposite sides. Part of the solution can flow out through the openings of the comb-shaped leads. In this way, interfering ions can be removed (Fig. 10.11).

Sample Injection and Detection

Miniaturized flow systems contain extremely narrow capillaries or channels. The internal diameter is in the range of ca. $10\,\mu\mathrm{m}$. It is not easy to inject a well-metered volume increment into conduits of such dimensions, regardless of how the solution is pumped, i.e. mechanically with micro-HPLC or electroosmotically with micro capillary electrophoresis. There remains only one way to perform such an injection, namely by applying crossed channels (Fig. 10.12). As shown there, the carrier stream flows from point 1 to point 2. If the liquid column 2 is moved for an instant by application of a driving force between points 3 and 4, then a well-defined sample volume is shifted into the carrier

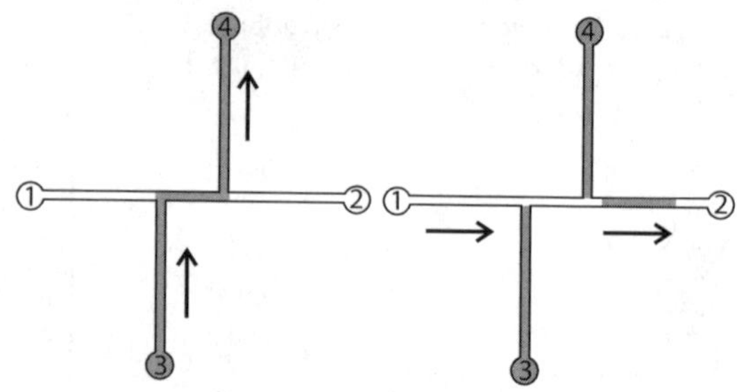

Figure 10.12. Injection of sample volume increments into micro flow systems

stream. The volume element shifted can be determined by calibration. This principle is applied with all existing modern micro flow systems.

Extremely high demands are made on *detectors in micro flow systems*. An oversized dead volume would frustrate all efforts at improving the efficiency of such systems. It is evident that with an overall total system volume in the microlitre range, the dead volume of the detector should not exceed the picolitre range. Otherwise, resolution would be worse than with macroscopic systems.

Among photometric detectors for micro flow systems, only those with laser-induced fluorescence (LIF) were successful. Absorption detectors would have to send their light beam in a longitudinal direction through an extremely narrow capillary to achieve a light path of sufficient length with appropriate dead volume. Technical problems could not be overcome. Lasers, when used as exciting light source, can be focused with extreme precision to a point established as the flow-through cuvette. Photomultiplier (PM) tubes are the only light detectors which fulfil these requirements, although their dimensions are not very well suited for microsystems.

Electrochemical detectors have been attracting increased attention for microsystems. The best chances for miniaturization are offered by *amperometric detectors*. They predominate in real instruments. The advantages of microelectrodes have been discussed elsewhere (Chap. 2, Sect. 2.2). Most advantageous is the fact that their function is not disturbed by changing the flow rate. Construction details follow principles given in Chap. 9. Primarily wall-jet as well as thin-layer detectors with decreased dimensions are used.

Microsystem mass spectrometers have been developed (Feustel et al. 1994), but they are not a standard device so far. It is expected that the combination of capillary electrophoresis and mass spectrometers will be of tremendous importance in the near future.

10.3.4
Examples of μ-TAS

The only significant examples of micro total analysers so far are devices for capillary electrophoresis. Actually, they are the only commercially available

μ-TASs. FIA has not reached this stage of development, although there are several technical proposals in the literature.

Capillary Electrophoresis

Originally, it monolithic silicon technology was expected to become widespread in analytical chemistry. This expectation has not been fulfilled. Planar glass substrates might be better substrates.

The archetype for all later constructions has been a micro-CE chip on glass support. One of the first examples (Effenhauser et al. 1993) of such a chip consisted of a glass base of 70 × 80 mm, containing etched channels 12 μm deep and 50 μm wide. The arrangement of the chips with overlapping channels (Fig. 10.13) allowed for the injection of three different sample volume sizes (80, 180 and 240 pl), depending on how the sample channel (connection between points 1 and 4) was inserted into the carrier stream. The sample channel is filled completely with sample solution prior to injection. Switching between different dosating volumes is controlled by high-voltage positions. In this example, dosation is performed exclusively by electroosmosis. The separation capillary is the longest line (50 mm) visible in the figure. A set-up for laser-induced fluorescence has been used as a detector. The laser set-up could be placed at an optional position on the chip. Similar devices have been proposed, frequently containing amperometric rather than photometric detectors.

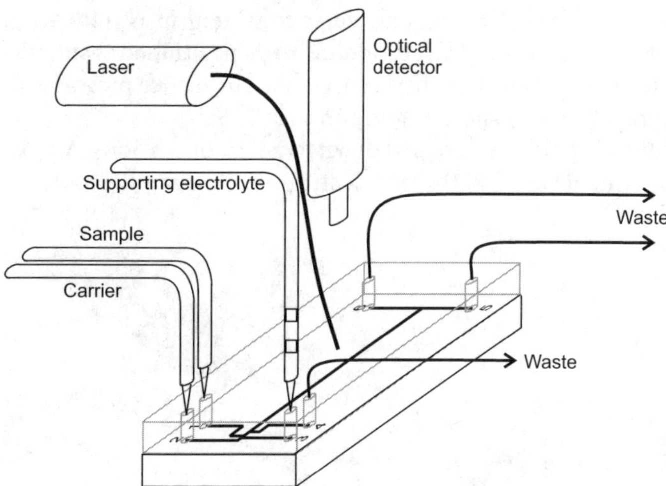

Figure 10.13. Design of microcapillary electrophoresis chip with electroosmotic sample injection. Detection by laser-induced fluorescence

Titration Devices

Traditional titration operations are costly and laborious. Nevertheless, they are indispensable due to their inherent precision. There has always been great interest in miniaturizing and automating titration equipment. A well-known method of substituting the precision mechanical parts necessary for exact dosation of titrant solution is electrochemical (coulometric) generation and dosation of titrants. However, coulometric titrators also require skilled handling. Commercially available autotitrators still largely follow the volumetric principle. Their requirements in laboratory space and their time demand are not very different from those of classical manual set-ups.

An interesting attempt at a microtitrator has been mentioned already, namely the combination of a generator electrolysis unit (actor), where an acid or base can be generated coulometrically, with a pH-sensitive ISFET acting as sensor (van der Schoot and Bergveld 1985). Inside a thin diffusion zone, the acid or base content of the surrounding solution is neutralized. The titration curve recorded by evaluation of the ISFET response is used to calculate the content. The process is a true titration, although the classical procedure is not simulated completely. The titrator is fast enough to be inserted into a flow system (Olthuis and Bergveld 1995). The result is a kind of 'higher order' microtitrator. On the one hand, the combination of generator electrode and ISFET sensor (called a *sensor-actor system* by the authors) meets the requirements for being considered a μ-TAS. On the other hand, this system is included in a higher-order TAS. In the version shown in Fig. 10.14, two ISFET arrangements are present. One of them acts as a reference system in combination with a pseudo-reference electrode. The second ISFET is positioned inside the titration zone and is used to record the titration curve. The device promises to become the prototype of a new generation of complex TASs.

Volumetric titration has also been scaled down to microdimensions. A most interesting example (Guenat et al. 2001) uses electroosmotic pumping to trans-

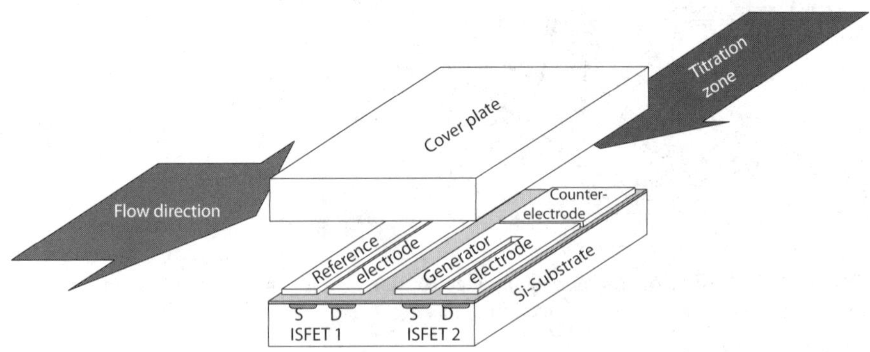

Figure 10.14. Coulometric microtitrator in flow-through system

port solutions to a micro mixing device, where the titrimetric reaction takes place. Sample as well as titrant solutions are transported in this way. In the micro mixer, an indicator electrode is positioned which allows one to record a common titration curve. This so-called 'nano-titrator' allows flow rates of 2 to 65 nl/s. Some micromoles of iron(II) have been titrated cerimetrically.

10.4
References

D'Amico A, Di Natale C, Paolesse R (2000) Sens Actuat B 68:324–330

Di Natale C, Mazzone E, Mantini A, Bearzotti A, D'Amico A, Legin A, Rudnitskaya A, Vlasov Y (1999) Alta Freq 10:1–3

Effenhauser CS, Manz A, Widmer HM (1993) Anal Chem 65:2637–2642

Feustel A, Müller J, Relling V (1994) A microsystem mass spectrometer. In: Van den Berg A, Bergveld P (eds.) Micro Total Analysis Systems. Kluwer, Amsterdam, pp 299–304

Guenat OT, Ghiglione D, Morf WE, de Rooij NF (2001) Partial electroosmotic pumping in complex capillary systems. Part 2: Fabrication and application of a micro total analysis system suited for continuous volumetric nanotitrations. Sens Actuat B 72:273–282

Harrison DJ, Manz A, Fan ZH, Ludi H, Widmer HM (1992) Anal Chem 64:1926–1932

Hoogvliet JC, Reijn JM, van Bennekom WP (1991) Anal Chem 63:2418–2423

Ivarsson P, Holmin S, Höjer NE, Krantz-Rülcker C, Winquist F (2001) Sens Actuat B 76:449–454

Jeroschewski P, Schmuhl A, Haushahn P (1994) Elektrode zur Herstellung von Schwefelwasserstoff/Sulfid-Standardlösungen. DE-Patent Nr. 4244438

Manz A, Graber N, Widmer HM (1990) Sens Actuat B1:244–248

Manz A, Miyahara Y, Miura J, Watanabe Y, Miyagi H, Sato K (1990) Sens Actuat B1:249–255

Manz A, Fettinger JC, Verpoorte A, Haemmerli S, Widmer HM (1991) Micro Syst Technol 1991:49–54

Olthuis W, Bergveld P (1995) Proc Electrochem Soc 1995:24–32

Pagel L (2001) Mikrosysteme: Physikalische Effekte bei der Verkleinerung technischer Systeme. Schlembach, Weil der Stadt

Raymond DE, Manz A, Widmer HM (1994) Anal Chem 66:2858–2865

Rojas R (1993) Theorie der neuronalen Netze. Springer, Berlin Heidelberg New York

Terry SC, Jerman JH, Angell JB (1988) IEEE Trans Electron Devices ED-26:1880

Van der Schoot B, Bergveld P (1985) Sens Actuat 8:11–22

Printing: Krips bv, Meppel
Binding: Stürtz, Würzburg